ALSO BY COLIN TUDGE

The Engineer in the Garden
Last Animals at the Zoo
Global Ecology
Food Crops for the Future
The Food Connection
Future Food
The Famine Business

THE TIME BEFORE HISTORY

5 MILLION YEARS OF HUMAN IMPACT

COLIN TUDGE

A TOUCHSTONE BOOK
Published by Simon & Schuster

TOUCHSTONE
Rockefeller Center
1230 Avenue of the Americas
New York, NY 10020

First Touchstone Edition 1997

TOUCHSTONE and colophon are registered trademarks
of Simon & Schuster Inc.

Designed by Irving Perkins Associates.

Manufactured in the United States of America

10 9 8 7 6 5 4 3 2 1
10 9 8 7 6 5 4 3   (Pbk)

Library of Congress Cataloging-in-Publication Data

Tudge, Colin.
  The time before history: 5 million years of human impact / Colin Tudge.
    p.   cm.
  1. Human evolution.   2. Evolution (Biology)   3. Mammals—Evolution.
  I. Title.
  GN281.T84   1996
  573.2—dc20   95-42026
                                                            CIP

ISBN 0-684-80726-2
     0-684-83052-3 (Pbk)

Illustrations on pp. 49–53 from M. A. J. Williams et al., *Quaternary Environments,* reproduced by permission of Hodder Headline Plc.

To my children: Mandy, Amy, and Robin

ACKNOWLEDGMENTS

The ideas in this book have largely been inspired by the long conversations I have enjoyed with biologists and other scientists over many years, men and women too numerous to mention; my thanks to all of them. But for this book I am particularly aware of my debt to the following: Professor Bernard Wood and Dr. Alan Turner of Liverpool University; Dr. Georgina Mace of the Institute of Zoology, London; Dr. Elisabeth Vrba and Dr. Andrew Hill of Yale University; Dr. Michael Archer of the University of New South Wales; Professor Bob Brain, formerly director of the Transvaal Museum, Pretoria; Dr. Chris Stringer of the Natural History Museum, London; Professor Bill Ruddiman of the University of Virginia; Dr. Tim Partridge of the Transvaal Museum, Pretoria; and Dr. David Webb of the University of Florida.

Also to Dr. Roger Lewin for his continuing advice and encouragement; Neil Belton of Jonathan Cape, who is among the most patient and supportive of editors; and Gillian Casey Sowell of Simon & Schuster. But of course, none of the above can be held responsible for any errors or other eccentricities of this book.

CONTENTS

CHAPTER 1

PROLOGUE: A PROPER HISTORY OF HUMANKIND

How odd it is to suggest, as historians conventionally do, that the Assyrians, the Egyptians, the Greeks, and the Romans were "ancient." They had advanced technologies for architecture, engineering, and agriculture; they made war, paid taxes, studied the stars, developed the arts and miscellaneous philosophies, and allowed themselves to be organized by priests, generals, and bureaucrats. If any of us were whisked into their midst we would be struck at first by their quaintness and their foreignness and by the alarming proximity of death, but we would soon feel more or less at home. At least, we would soon be irked by the same day-to-day necessities of housekeeping and social intercourse that beset us now—and stunned, when we found the leisure to look around, by the brilliance of their technologies and the excellence of their arts and crafts.

The truth is, of course, that those people were not "ancient" at all. The epithet was applied by eighteenth-century historians who thought that the world itself was new and that the people they called ancient had lived near its beginning. But the Assyrians, Egyptians, and the rest had inherited several thousand years of civilization, exactly as we have done—several thousand years of cities and of organized agriculture. Yet those cities, and the paraphernalia of farming, did not grow out of nothing. They in turn were preceded by tens of thousands of years of recognizable culture and invention. For all that time there had been trade of a kind, and formal social hierarchies, arts that included extraordinary draftsmanship, as seen in the caves of southern Europe and elsewhere, and a steadily increasing armory of tools and techniques from the toggle and the needle to the spear and the adze; all of which are still in use, and taken for granted, but each of which was a revelation that tightened the human grip on the environment as decisively as any present-day development in electronics. Tiny technical advances— hardly more than tricks—have a huge impact on biological success, as can be seen in any group of animals; and the particular trick of human beings is not to forget, but to accumulate, over time, all the tricks of other people and of previous generations.

A great deal happened, too, before the preagricultural cavepainters, potters, tailors, and traders of Europe and Africa; there were several million years in which the human stage was set, during which, increment by increment, our ancestors crossed the divide between apedom and humanity. We have a huge amount in common with our fellow creatures. Charles Darwin emphasized that the differences between "us" and "them" should be seen as matters of degree. Yet we have tricks—biological tricks—that make us different, and put us qualitatively into a new league, one of which is our supreme manual dexterity, and another of which is the ability to pool our thoughts, ultimately through the medium of speech. These tricks do not make us morally superior to other creatures, I believe, but they do give us enormous power over them and over our environment at large. In short, the people who preceded the people whom the historians called "ancient" had themselves inherited an ancient culture. Conventional history, in short, starts almost at the end.

The vauntingly presumptuous purpose of this book is to put that right. I want to tell the true history of human beings, beginning at the beginning, which, in truth, was around 5 million years ago. I also want to put that history into perspective, by showing what went before that and why our own history began in the first place. The past 2,000 years of conventional history are effectively the present. The past 40,000 years, when our own species truly got into its stride in recognizably modern form, are only yesterday. The 5 million years that are the main focus of this book are, well, the day before yesterday.

I want to look at our whole history, our 5-million-year history and the time that led up to it, for two reasons. First, because it is fun—it is a cracking good yarn; or, to put the matter more portentously, because a knowledge of our own history is a significant part of culture and, at present, that knowledge is absurdly curtailed.

But second, I want to look at our history in its entirety because it is necessary. The Spanish-American philosopher George Santayana commented that "those who do not learn from history are obliged to repeat its mistakes," and that surely is the case. How can we understand what is happening to us now without reference to the past? How can we even begin to guess how any human enterprise might turn out—making war, building dams, farming land that once was forest—except by seeing what happened before? But the 2,000 years or so that we traditionally treat as "history" simply does not tell us enough. It deals only with the *fait accompli;* the activities of a peculiar creature that had already taken over the world. It does not begin to show us just how peculiar we really are—how we broke away from other animals, and what an enormous effect we had upon them even before conventional history began. Still less can a study of the past 2,000 years reveal the biggest lesson of all: that the planet itself is labile; that we cannot take its benignity for granted; that it can be a very hostile place indeed; and—most strikingly—that we ourselves, feeble though we may seem, have the power to influence its course, to switch the entire Earth from one state to another.

THE FIRST PURPOSE: SCIENCE FOR AESTHETICS

This book deals mainly with ideas of modern science, and some may find it odd that science should be pressed into the service of aesthetics. Some even feel that science is an antiaesthetic and even a blasphemous pursuit—echoing William Wordsworth's somewhat chilling averral that "we murder to dissect." The comment is obviously not intended to be purely literal, although it is literally the case that the particular science of biology has progressed in large part by cutting up and seeing what is inside. Rather is his comment taken to imply that through our efforts to reveal mechanisms we snuff out the spirit or perhaps simply the mystery of the object. The suggestion is that simply by finding out we lose our sense of wonder and hence our respect for the thing in hand.

Yet this idea seems to me to be positively perverse. After all, no one doubts—do they?—that when, say, pianist Alfred Brendel listens to the music of Schubert he hears more than the rest of us do, or doubts that his aesthetic response is of a finer texture than that of the rest of us. Of course, Brendel is a special individual, perhaps endowed with a more subtle armament of emotions. But the thing that most obviously distinguishes the professional musician from the rest of us is knowledge. Most of us may hear a faint sadness creep into the music, or a lift of mood. Musicians could, if they chose, describe precisely what lies behind each effect—describe and admire: the shift of key, the half-allusion to some earlier melody. Of course, the knowledge generally remains subliminal. Musicians do not, in the normal course of listening, list each modulation as it passes. But because they know, they can hear with precision while the rest of us are merely vaguely aware, and because of that extra acuity they catch an infinity of nuances that pass the rest of us by. But no one says of musicians that their deep knowledge "murders" what they listen to. No one seriously doubts that the enhanced perception is a bonus, that the rest of us are deaf by comparison because our ears are not informed.

To me the parallel is precise. Living things are thrilling. If I did not feel this excitement, I would never have elected to study biology. Virtually all biologists whom I know feel the same. Knowledge of the subject is a way of getting closer to the thing you love, of maintaining contact, and in the end of refining the emotional, which means the aesthetic, response. When you know just a little of how the Universe works (and even the greatest scientists know only a little), you can feel much more of what is going on.

Specifically, this book is about change—about the rise and fall of landscapes and the evolution of living lineages over time. I suggest that once you become aware of the idea of evolution, once you begin to feel that things do change through time, then your perception of everything around you is enhanced. Another dimension is added to your view of the world, and that is the fourth dimension: time. You begin to perceive that an animal or a plant and the lineage to which it belongs, and the planet itself, are like a flame; not so much a thing as a performance, always becoming something else; and that each of us and our species as a whole are part of the overall unfolding.

This point, strange though some may feel it to be, is plain to see within the history of art. There can be no doubt, for example, that the Romantic painting of the late eighteenth through the nineteenth centuries was strongly influenced by the emerging science of the time. The perception of entire generations of landscape painters was transformed by their knowledge of the new geology. Look, for example, at James Ward's fabulous painting *Gordale Scar* in London's Tate Gallery. "Scar" is Yorkshirese for a face of rock, and Gordale is a gorge—battered cliffs of carboniferous limestone, almost 300 million years old. Ward has painted them from a low perspective, stretching hugely upward to a thunderous sky. At the foot of the scar is a rough white bull of some archaic breed and it feels—and the observer also feels through its tension and its rolling eye—the menace of the place. But why is this landscape so menacing? Because Ward has painted a wall of rock that is ancient but far from permanent—a cliff that is poised to crash. The gorge has been carved by a waterfall that one day will level it to the ground. The notion that landscape is transient was emerging from the revolutionary thinkers of the day who first perceived that the tops of mountains were once at the bottom of the sea and that, in the fullness of time, the greatest mountains are beaten to dust. Ward is not an impressionist who sees with Claude Monet's "innocent eye." His eyes are informed. He sees the history behind the rocks and feels the threatening future.

I now spend much of my time in Yorkshire, not far from Gordale Scar. Like Ward, who was here almost two centuries ago, I find that as my knowledge grows, so my appreciation deepens. There is a cave in the limestone hills close by that collapsed just a few months before I am writing, perhaps for the hundredth time, perhaps for the thousandth. That same cave has yielded the remains of lions, hyenas, rhinoceroses, and reindeer. Like all of northern Europe, these hills have experienced at least ten glaciations in the past million years, and those creatures passed through here to escape the encroaching ice, or followed in its wake as it retreated—as it did, for the last or at least most recent time, only about eight thousand years ago. The bleak hills are beautiful; but their excitement is enhanced when you see them in your mind's eye beneath the mile-high mountain of ice that carved them in their present form; or see the Arctic creatures that were never far from the glacier's edge; or the animals that now live only in tropical Africa that flourished here in the intervals of warmth; or the forest that was here for a few thousand years and was wiped away by the first farmers. And what applies here applies everywhere on Earth. It is thrilling, this thought: that anywhere you stand might at different times have been at the bottom of the sea, or the top of a mountain; might have been desert or tropical forest or buried three miles under ice; might, like my own native region of London, have been the home of sea lilies or straight-tusked elephants and sabertooths, and our own beleaguered and mystic-minded ancestors. It's a thrilling thought, too, that many of the landscapes and ecosystems of the past were like nothing that exists today. After all, there have at times been tropical forests at high latitudes—hot, vaporous, but with short winter days and long winter nights, or the permanent sunshine of near-polar summer—and these have no parallel in the modern world. Because it is thrilling, it is a pity not to be aware and to miss what is there to be felt if not directly to be seen. It is the science of paleontology that feeds the aesthetic experience.

To the notion that science is blasphemy—that it is demeaning to God to probe the mechanisms of the Universe—I can only suggest that the opposite applies. The point was made by Isaac Newton, widely acknowledged as the founder of modern science. Like most of his seventeenth-century scientific contemporaries, he was deeply religious. He felt that to study the works of God was to gain insight into the mind of God; and he felt it to be an obligation to use his own God-given intellect to probe his Creator's mind more deeply. For Newton, indeed, the pursuit of science was itself an act of reverence.

We live in a more secular age. The fashion of the past few decades (though not, in fact, when I first started studying science at school in Britain in the 1950s) is that science provides the means to produce "high" technologies that in turn generate wealth. To me, and to most of the best scientists whom I know, such material benefits are almost incidental. I suggest, rather—a secularized restatement of Newton—that the true purpose of science is not to change the Universe but to appreciate it. To revert to the language of religion: it is not blasphemous to describe how the Universe works; rather is it blasphemous to suggest that the Universe is in any way diminished by our feeble essays in description. God is not a conjurer whose tricks seem tawdry when exposed. Most true scientists agree that the more subtle their descriptions become, the more mysterious and wonderful the Universe appears.

So that is the first purpose of this book: to add the dimension of time to everyday perception and hence, I hope, to enhance the sense of wonder.

THE SECOND PURPOSE: THE LESSONS OF HISTORY

The common lack of true historical perspective has some bizarre consequences, although it demonstrates what an ill wind it must be that blows nobody any good. Thus, in 1968, Erich von Daniken published his extraordinary *Chariots of the Gods?*, which became a best-seller in several countries. I have the fourth reprint in English paperback, dated 1993. Von Daniken suggests, among other things, that the Pyramids might have been built by aliens because the Egyptians could not possibly have managed the technology themselves. Yet, to revert to my opening statement, the Egyptians were modern people. The Pyramids are extraordinary, but we might as well suggest that nineteenth-century Parisians could not have built the Eiffel Tower, or the Americans of the 1920s the Hoover Dam.

Again, to expand upon an earlier point, the pharaonic Egyptians had reached the peak that they did only after several thousand years of civilization. This point is made by Plato. In the dialogue *Critias*, the eponymous hero narrates a meeting that an acquaintance of his had had with an Egyptian sage, who simply points out that the Greeks have no history and therefore effectively have no culture. After all, says the sage, the Greeks were newcomers to southern Europe, while the Egyptians had been in situ for millennia, and had correspondingly ancient memories and traditions. His tone was positively disdainful: to him, the Greeks were parvenus. Traditionally, the *Critias* has been thought of simply as a piece of Platonic storytelling; but as Mary Settegast now discusses in *Plato Prehistorian* (New York: Lindisfarne Press, 1990), modern archaeology increasingly suggests that the pronouncements of that ancient sage were literally correct. Egyptians in pharaonic times were fully justified both in their feeling of modernity and in feeling themselves the inheritors of truly ancient civilization. The Greeks, by contrast, were newcomers. Indeed, the Egyptian sense of history—already embracing, as it did at that time, about six thousand years—was more realistically developed than that of today.

Intriguingly, the earliest memories of Plato's Egyptian sage would take us back roughly to the end of the last ice age. And this brings me properly to my main historical point—that humans are not merely cultural beings, minds on legs, but are a biological species, as prone as any to the deep rhythms of the Earth itself. History and archaeology must become one; archaeology and paleontology must become one. We can understand ourselves and what is happening to us, and our own impact upon the rest of the world, only when we look at ourselves on the grand scale of time. We count the rhythms of our own lives through the passing of days and seasons. As members of families, we note the passing generations. Historians traditionally deal in centuries. But for the most part we remain blissfully unaware of the deep rhythms that lie beneath—rhythms that must be measured in millennia, or in millions of years, or in tens of millions of years. We are like small boats that bob up and down on the surface chop and are oblivious to the deeper and longer surges of the swell and tide. In school we learn of ice ages in one set of books and of "history" in another and fail to see how the two are connected; we fail to perceive, therefore, that beneath the surface tremors of our lives there are much deeper and more powerful forces at work that in the end affect us and all our fellow creatures at least as profoundly as the events of day-to-day.

But why does this matter, this lack of perception, apart from the fact that we miss out on an exciting set of ideas? Because, in fact, the analogy between the human species and the small boat that bobs on the surface and blissfully drifts on the tide is not quite accurate. Small boats do not influence the sea on which they ride. We, tiny creatures though we are when taken individually, have had and are having a huge effect upon our fellow species and, increasingly, upon the fabric of the planet itself. Of course, we are vaguely aware of this. Everyone knows, after all, that we are wiping out our fellow creatures—that we might indeed eliminate half of those that are left within the next few decades. Everyone has heard of the greenhouse effect, by which we may heat the planet as profoundly as it was cooled in the ice ages; or of the destruction of the ozone layer which, had it not been spotted and analyzed, could (and I do not exaggerate) have wiped out almost all living things of the present day.

But we tend to feel that this present-day destructiveness is just an anomaly. We tend to feel, too, that in the fullness of time the world will settle down again into some form that may be slightly different from the present but nonetheless agreeable. The Gaia hypothesis, first mooted by the British physicist James Lovelock in the 1970s, has tended to reinforce this notion (whether it was intended to or not): that somehow or other the Earth will revert to some "normal" state whatever we do to it.

Yet if we look at our own history in its entirety, and briefly at the 100 million years or so of our prehistory, we see how dangerous this complacency is, the sheer precariousness of our existence, and of all other creatures. It is not enough just to state the fact. We need to feel that it is so. Only through knowing what has happened in the past, and therefore what can happen, will we finally see the need to change our ways. Sense in the end derives from sensibility.

The lessons that come roaring out of our history and our prehistory, but are missed if we look only at the past few centuries, are threefold. First, we can see that apparently small events can have huge consequences—literally global consequences—that can, thereby, change the flora and fauna of the whole world. Some people at least are bothered now by the pending greenhouse effect. If it occurs as fulsomely as is conceivable, then nothing can ever be the same again. But what is the cause of it? Just a rise in concentration of a gas in the atmosphere that in any case is present in barely more than a trace. That is all it takes to change the climate of the world. When we look at our own history and prehistory, we see this very effect in action, and the sheer magnitude of what results. Without such effects, indeed—albeit working in reverse: a loss in greenhouse gases rather than a gain—our own species probably would not have come into existence.

Second, we can see, when we take the long view, that matters of huge consequence can take many thousands or even millions of years to unfold. If an influential species is wiped out, then it may take that long for others to adjust fully to its absence, perhaps to evolve to fill the gap and perhaps not, depending on how things turn out. If a landscape is flooded or a sea is drained, or if two places are joined that hitherto were separate or divided that once were joined, then again, the readjustment may take millions of years. Add that point to the first and we begin to see how momentous it can be, and how long-lasting, to do the kinds of things that we do now as a matter of course: build highways across continents, remove forests, divert rivers.

Third, when we look at the world over time, we see that it can never be the same way twice. The physical environment depends upon the content of the atmosphere, and the heights and positions of mountains, and where the continents happen to be as they migrate around the world, and how the world is situated relative to the Sun, and no combination of circumstance can ever be recaptured. Thus it is that the lessons of history cannot be precise, but must be expressed in general terms. But the general lesson is terrifying, which is that our present-day world, to which we and all other animals and plants are adjusted, does not have to be like this at all. It can be tropical from pole to pole, or iced from the poles almost to the equator. It need not contain oxygen in its atmosphere—or not enough, at least, for any air-loving creature to breathe. It might if things were ever so slightly different contain so much greenhouse gas in its atmosphere that the world would cook, like Venus. Earth as it stands is a miraculous jewel of a place, an oasis in an astonishingly hostile Universe. But there are other ways for it to be, and at times in the past it has shown how different it can be. The Egyptians, according to *Critias*'s sage, could remember back to the last ice age. Even they—modern people—had some hint of how different the world can be.

Add those thoughts, those memories of how the world can be, to the fact that our own influence is now becoming so obvious and we begin to see what a dangerous game we are playing. More specifically, we begin to see how inadequate it is to think only in short periods of time;

and to be satisfied with politics and policies of economics that deal only with the here and now. Harry Truman, as president of the United States and hence the most powerful of all world leaders, opined that "a week is a long time in politics," and so it is. Few politicians in democratic societies think beyond the four or five years of their term of office. To economists, thirty years is "the long term." As British economist John Maynard Keynes pointed out, the "long run is a misleading guide to current affairs. In the long run we are all dead." Thus do world leaders make a virtue of their immediacy. Thus do they deal only in desperately trivial twinklings of time. But no politician nowadays can hope to succeed for long without at least mentioning the word "environment." And as we can see from our own long history—and will see in the course of this book—it is impossible to contemplate the environment unless we think as a matter of course in very long periods of time indeed. In fact, I will suggest, as the central message of this book, that we cannot claim to take the environment seriously until we acknowledge that a million years is a proper unit of political time. That is the general lesson of history.

But it is difficult to think in very long periods, as our own lives are so brief. Let me then suggest a device: the translation of time into distance.

A SENSE OF TIME

Begin by thinking of one year as one millimeter. A millimeter is not too small to see; it is the size of a perfectly respectable grain of sand, or an insect that is big enough to give you a nasty bite. Ten millimeters—a centimeter—then becomes a decade. A centimeter into the future takes us comfortably into the twenty-first century.

Ten centimeters—what was called 3½ inches when I was a lad; about the length of a man's middle finger—is a century. A finger's length ago Queen Victoria's reign was coming to an end, motor cars were up and running, Orville and Wilbur Wright were preparing to begin the age of air travel, and the world, though far from innocent, had not yet experienced the full horrors of world war. Twenty or so centimeters ago—the length of a man's hand—the French rose up against their aristocracy and the Americans finally told the British that enough was enough. Fifty centimeters ago—fingertip to elbow—Columbus first landed on islands off the eastern coast of America.

A meter is roughly a yard: another human-sized measure, initially conceived as the distance from the tip of a man's nose to the tip of his fingers when his arm is held outwards at shoulder height. It corresponds to a thousand years. It was roughly one meter ago that William the Conqueror invaded Britain and transformed its politics and the Maoris colonized New Zealand and transformed its ecology. A meter and a half, the height of a Jersey cow or indeed of Queen Victoria, takes us back to Mohammed and to the first human colonization of Madagascar and Hawaii. Two meters, the height of an average basketball player, takes us back to Christ, Herod, John the Baptist, and the Roman invasion of Britain. Thirteen meters, roughly the length of the kind of yacht that competes for the America's Cup, takes us back to the first human invasion of North America: from Siberia into Alaska, through the land bridge that now lies deep beneath the Bering Strait.

Go back 30 meters, a standard length for municipal swimming pools, and we find that the Earth is still populated by more than one kind of human being. The Neanderthals, *Homo neanderthalensis,* still shared a large chunk of Eurasia with modern humans, *Homo sapiens. Homo sapiens* began, probably in Africa, 100 meters back: the length of the straight of an Olympic running track, gloriously bestridden in the time it takes to draw breath by Carl Lewis and Linford Christie and their ilk.

A million years is a kilometer: the length of London's Park Lane. A kilometer ago human beings were of the species known as *Homo erectus,* including types such as Peking Man; and they shared the planet with the last of our extinct hominid relatives known as the robust australopithecines, alias paranthropines. *H. erectus* first appeared about 1.8 kilometers ago. Five kilometers or so ago, a pleasant Sunday afternoon stroll, takes us back to the first members of the Hominidae. In fact we are back at the start of the Pliocene.

The length of a marathon—40 kilometers or so—takes us back to the very first appearance of creatures that might reasonably be called apes. The last of the dinosaurs disappeared about 65 kilometers ago. I must ask you to envisage your own version of 65 kilometers—40 miles. For me, a South Londoner, traveling east, it is roughly the distance to Canterbury: a Chaucerian pilgrimage, now a quick dash down the M2.

Two hundred kilometers ago takes us back both to the start of the dinosaurs and, roughly, to the start of the mammals, for they both began their glorious reigns at about the same time. Traveling from New York City, this would take us halfway to Washington, D.C. Five hundred and fifty kilometers—550 million years—takes us roughly from New York City to Richmond, Virginia, the beginning of the eon known as the Phanerozoic. The first period in the Phanerozoic was the Cambrian, in which almost all of the main groups (phyla) of modern animals first appeared.

Two thousand kilometers and a bit more, New York City to Little Rock, Arkansas, takes us to the first eukaryotes: that is, the first of the creatures such as animals, plants, and fungi, whose cells had recognizable nuclei (as opposed to the prokaryotes, the bacteria, whose cells have no distinct nucleus). Somewhat more than 3,000 kilometers takes us to Austin, Texas, and to the beginning of life. Forty-five hundred kilometers takes us from New York to Los Angeles and to the beginning of the Earth itself.

This is the first surprise. When I began this exercise, with a year represented as a palpable entity, a unit quite easy to envisage, I thought that the beginning of the Earth would be on the other side of the world. But New York to Los Angeles is the kind of distance that St. Paul used to take, literally, in his stride.

The Universe itself began, so modern cosmologists suggest, with the Big Bang, about 15 billion years ago.* That translates into 15,000 kilometers: New York to Japan. With the aid of the jumbo jet, some people almost commute from New York to Japan. The distance is not so frightening. Thus can we envisage and encompass the vastest reaches of time; in fact, all the time that there has ever been, for before the Universe began, the concept of time is meaningless. In practice, however, the time scales relevant to this book do not extend beyond tens of millions of years; the width of a big city. That does not seem particularly daunting. But the conventional time scales of historians—dealing at best in centuries, the length of a man's middle finger—really are dangerously small.

But before I finally embark on this 5-million-year extended history, I want to explain the method briefly. That is, I want to tell the story in a way that some people feel is not respectable, not "scientific." But I think it is, and would like quickly to say why.

---

*Or this at least was the conventional view when I started writing this book. Observations made late in 1994 now suggest that the Universe might be only 8 billion years old, although this idea is embarrassing since other calculations suggest that certain stars must be older than this.

SCIENCE AS MYTH; MYTH AS SCIENCE

In 1979 a young historian of science at Yale, Misia Landau, was re-searching the history of paleoanthropology; that is, she was exploring the way scientists over the past century or so have described our own evolution. She realized that their descriptions took the form not of cold, "objective" science, as the scientists themselves liked to believe, but of myth. In their descriptions of our ancestry they perceived and projected the genus *Homo* as a hero of the kind who features in the folktales of every culture: a hero who begins life humbly, is faced with a series of hurdles, overcomes those hurdles, and in doing so is honed and improved until he emerges as *Homo sapiens.* Finally, in many ver-sions, *Homo sapiens* commits the sin described by the Greeks as "hubris" and so is destroyed. Hubris is rather a nice idea. The Greeks after all believed deep down that the affairs of the world and all the creatures therein were in the hands of the gods, and that if human beings ever succeeded in this world it was only with the gods' approval, or as Chris-tians would say, "by God's grace." Human beings should never make the mistake of believing that their destiny is in their own hands or that any success they achieve is gained simply through their own efforts. Such self-belief is presumptuous in the extreme. It is not merely con-ceited; it usurps the prerogative of the gods, for they alone dispense human fate. Heroes, after their earthly success, are particularly prone to hubris. But the gods do not tolerate such presumption. Hubris is punished. Death is the usual penalty; either that, or worse.

As Roger Lewin excellently describes in *Bones of Contention,* modern scholars of human history have responded in three main ways to Misia Landau's revelations. Some simply said, "Nonsense. Science is 'objec-tive,' and we sift the facts objectively, and tell it like it is." Others were shocked. They had been brought up to believe that science ought to be objective, but when Landau revealed the literary narrative that runs through so many of their "scientific" descriptions, they realized how deeply they had deceived themselves. Scientists find it just as difficult as the rest of us not to impose their own prejudices and stereotypes on what they see. Whether you are a scientist or not, it is all too easy to fit whatever you see into a story that is already inside your own head.

But a third group soon recovered from the shock of Landau and ef-fectively said, "So be it"; and so do I. For this seems to me to be pre-cisely in line with the modern philosophy of science. For science, try though scientists might, cannot trade in irrefutable, "objective" truths, as it is traditionally believed to do. It does deal with ideas of a particular kind, ideas that can be tested and in theory refuted. But all of its truths, up to and including Newton's laws of mechanics, must be seen to be provisional. Built in to any idea of science is the realization that it could be wrong; logically, the possibility must always remain that the theories of the time are flawed, and that the real "truth" has yet to be envisaged. Science progresses by the proposal of ideas—hypotheses— of the kind that can be tested. More broadly, it also proceeds through grand statements of the kind that can be called heuristic. Heuristic statements (whether in science or any other field) are of the kind that promote understanding, even though they may not literally be true themselves. In practice, a statement may be heuristic and may properly belong in science even though it is not directly testable. This would be the case if the statement, though untestable itself, suggested hypotheses that were testable. Such statements provide what the great twentieth-century philosopher Sir Karl Popper has called an "agenda" for science.

In short, it is perfectly reasonable for paleoanthropologists (scholars of fossil humans) to be quite unabashed by Landau's thesis—provided they also acknowledge the full significance of what she is saying. They should be unabashed because it is perfectly reasonable to behave precisely as the old-style (and new-style!) paleoanthropologists have done, and describe *Homo* as a hero—because the description of early humans qua tragic heroes is (or can be) heuristic. That is, such a description gets the show on the road. It is a coherent idea that pulls the facts together and gives an overall sense of what went on. Testable hypotheses are generated by, and nest within, this overall picture, and as they are tested, they modify the picture. That is the proper dialectic between the real world and our mind picture of it. In practice, such dialectic provides the only route to progress.

There is one final advantage in telling the history of human beings in the manner of myth. It is that the human brain—unlike a computer—is supremely adapted to grasp narratives. We love stories with a beginning, a middle, and an end. "In the beginning was the word," says St. John in the first line of his gospel; but whatever the theological import of this, as a statement on human psychology it is wrong. In the beginning was the story. Stories we remember. So it seems to me that a well-related myth is doubly heuristic: it may itself be a testable hypothesis, or at least contain or give rise to such hypotheses; but in addition it is easy to remember, because human brains are adapted to remember stories. There is only one danger, and it really is dangerous, but it is easily avoided once we know what it is. The danger is to imagine that the well-told tale, the myth, which should at best be seen as a working model, is in fact the truth itself. Many scientists, in other contexts, have made this mistake: they have erected hypotheses just to get their investigations on the road, and then mistaken the hypothesis for the truth they were trying to discover. In short, I see no harm in presenting human evolution as myth, and human beings as heroes, because this is a colorful and memorable way to convey ideas; but I will try to bear in mind, and ask you to as well, that the myth is not intended as the final truth. It merely provides a mental framework on which, in time and with luck, the truth may be hung.

To tell the story properly, however—to come to grips with all the forces and the characters involved—it is necessary to bring several apparently different threads together. It seems worthwhile to provide a guide.

AN INTRODUCTION TO THE REST OF THE BOOK

The first and crucial point is that we human beings have in the end been shaped by the world itself. Unless this planet had behaved in particular ways at a particular time in the history of our own lineage, we would not have come into being. Furthermore, the kind of changes that we are now threatening to make—such as global warming—are the kinds that have been seen in the past. It is necessary, then, to set the stage by describing the modus operandi of the Earth itself, the subject of the following chapter.

Then we should ask why our lineage—our ancestors—should have responded to earthly forces the way they did; and why all other animals have responded as they have done. In short, we should explore the underlying mechanisms of evolution. That is the theme of chapter 3.

History is not lived in isolation; indeed, it is largely the story of relationships. Nations interact with other nations; species with other species. The other creatures that have occupied our planet this past 50 million years or so helped to shape our evolution—we evolved in part to cope with them—and we, in our turn, have influenced them. Again, we cannot understand what we are doing to our fellow creatures now without knowing what has happened to them in the past; without that background knowledge, we cannot see what is different now. So in chapter 4, I introduce our fellow creatures.

With the scene set and the other characters in place, the rest of the book can focus on our own kind. Chapter 5 shows how the forces dis-

cussed in chapters 2 and 3 acted to produce humans. Chapter 6 asks
why this was such a big event, for "What's So Special About Us?" Then
in chapters 7 and 8, I discuss the effects that we have had upon the
world—not in the past few decades, which is obvious enough, but over
the past few tens of thousands of years, which is the time in which our
influence ceased to be that merely of a large mammal and we pushed
the world into a new ecological phase.

Finally, in chapter 9, I want to see how the lessons of history—our
proper, extended history—might apply to the future. In short, I consider the next million years: what effect we will have on our fellow creatures during that time, what might happen to us; and indeed what we
need to do if our species, or any other creature, is to last nearly as long.

small events can have huge consequences
impt things/events can take a long time

CHAPTER 2

HOW THE WORLD WORKS

People the world over congregate each year to give thanks for the harvest—and they keep their fingers crossed for the next one. The rituals are largely for appeasement. Every society has cause to know that the beneficence of the Earth cannot be taken for granted. Some of us in temperate islands to the north expect good times and complain when rain delays the tennis, but the folklores of at least five hundred cultures from the American Southwest to Australia recall horrendous floods: Noah's was only one of many that derive from Mesopotamia, where the Tigris and the Euphrates run in parallel into the Persian Gulf. Other people—or sometimes the same ones—endure five-year floods almost as routine. From California to Azerbaijan, people live in the knowledge that one day their cities will succumb to earthquake. Some places are far more vulnerable than others, but nowhere is safe. Two-thirds of the city of Lisbon was shaken to pieces one day in 1755. Volcanoes continue to harry us, too, and always will. In 1980, Mount St. Helens in Washington State threw out about a cubic kilometer of ejecta. The eruption of Vesuvius in A.D. 79 was about the same size, and completely obliterated the southern Italian city of Pompeii. The eruption of Santorini on the island of Thera north of Crete in 1470 B.C. produced ten cubic kilometers. The ash and acid rain from Santorini wiped out the Cretan civilization of the Minoans, which then dominated the Mediterranean, after which the way was clear for the mainland Mycenaeans and hence the rise of Greece. All that remains of Thera now is a sickle of rock around the caldera—the hole where once the volcano stood.

Volcanoes affect the climate, too, for the debris they throw into the upper atmosphere can block out the Sun, in the same way that nuclear war might lead to "nuclear winter." Ben Franklin first suggested that this might be the case after the eruption of Laki in Iceland in 1783 caused a bluish haze in Europe that spread as far as Siberia and North Africa, and led to the freezing winter of 1783–84, both in Europe and North America. The bluish fog is now known to be sulfur dioxide, SO_2, dissolved in atmospheric water vapor to form sulfuric acid. It fell in the months after Laki as 100 million tons of acid rain; the relicts of which can still be clearly seen in cores of ice taken from Greenland. Laki was about the same size as Santorini. So, too, was the eruption of Krakatoa in Indonesia in 1883; and the climatic changes that resulted from that volcano were the first to be studied by scientists worldwide.

But Santorini, Laki, and Krakatoa were all of them dwarfed by the eruption of Tambora on the island of Sumbawa east of Java. Between April and July of 1815, Tambora produced a series of eruptions which, taken all together, were ten times bigger than any of those three. The worldwide climatic effects of Tambora were not formally studied, but they were obvious enough. The eruptions may have caused the famously awful weather at the Battle of Waterloo. In the following year, 1816, there was effectively no summer: the people of New England spoke of it as "Eighteen Hundred and Froze to Death." In Italy, Lord Byron, Claire Clairmont, Percy Bysshe Shelley and his wife Mary made the best of their holiday by making up stories while the cold rain fell; and Mary's efforts resulted in *Frankenstein*. Crops failed across the whole northern hemisphere in 1816. In Europe, still recovering from the Napoleonic Wars, the weather may have encouraged the spread of the louse that carries typhus, and led to the European epidemic of 1816 to 1819. The crop failures and all the resulting misery in turn inflamed political unrest all over Europe. (In Britain in 1819, troops charged upon a crowd of 60,000 people who had met in Manchester at St. Peter's Fields to demand reforms. Four hundred people were injured and 11 killed in what became known as the Peterloo massacre.) Yet no society has ever known or can ever experience the full range of enormity of which the Earth is capable and to which it is subject. To see it all you would have to live an awfully long time. Two and a half billion years ago the composition of the Earth's entire atmosphere changed as profoundly as such change is possible, as free oxygen gas was introduced for the first time from the new organisms that practiced photosynthesis. More astonishingly yet, throughout the lifetime of the planet the continents that seem so irredeemably solid have moved—not a little, but right across the surface of the globe, splitting and rejoining as they went, floating like froth on the hot and slowly swirling rock of the mantle beneath. The earthquakes and the volcanoes that have done us so much harm are simply a side effect. Of course the earth shudders; of course the turbulent rock beneath bursts through now and again, when entire land masses are on the move.

Five times at least within the past billion years there have been changes of climate so dramatic that the fauna of the entire world has been transformed. The latest such mass extinction was 65 million years ago and wiped out the dinosaurs; and yet it was not the biggest. Since about 50 million years ago the entire Earth has become inexorably cooler, not steadily so but in a series of bursts, with occasional reversals to widespread tropics. There are still plenty of hot days at hot places, but the world has now become so cold that the temperature hovers on the point of freezing; and at least ten times in the past million years, the balance has tipped, ever so slightly, but enough to take us into an ice age. The ice sheets on Antarctica, North America, and Eurasia have themselves been as big as modest continents and as deep as mountains. So much water was locked within them that the sea level fell world-wide—sufficiently to join continents and countries that previously were apart: Siberia to Alaska via the vast but transient causeway of Beringia, which was as big as Poland; the islands of Indonesia to mainland Asia; New Guinea to Australia; and enough to drain the Mediterranean Sea and to deplete Africa's Lake Victoria, not once but several times. Falling sea level also combined with drifting continents to create the Isthmus of Panama just 3 million years ago, in the Pliocene, and so bring North and South America together for the first time. Why does this matter? Why is it pertinent to discuss the climate and the move-ment of continents in a book about the history and the evolution of human beings and of our fellow creatures? The answer is twofold. First, climate is not a backdrop, like stage scenery. We realize now it is a cru-cial force in the evolution of all living things. In general, it seems that the entire evolutionary machine moves toward an uneasy equilibrium unless the system is jolted every now and again. While conditions re-main the same, the interacting creatures seem to come to an agree-ment; in times of stability natural selection tends to encourage them to stay as they are. Conditions can be altered in several ways and for sev-eral reasons, but the most significant, frequent, and universal causes are those of climate. Indeed, Elisabeth Vrba of Yale University suggests that global changes of climate every now and then cause ripples of evo-lutionary change worldwide—a global shift in the fauna. She calls this "the turnover pulse hypothesis." Specifically, it seems that our own fam-ily, the Hominidae, came into existence after an episode of cooling in the late Miocene, while our own group within that family, the genus *Homo,* evolved as it did in response to another burst of global cooling 2.5 million years ago. Other creatures changed at those times, too. Cattle and gazelles, for example, came into existence in the late Pliocene, at about the same time as our own genus.

It is clear, too, that ever since animals and plants appeared on Earth they have changed their location; many indeed seem to have raced around the world several times. How and where they moved affected their own evolution and the ecology of all around them. Whether they chose to move was determined, more often than not, by climate: hostile ice behind, alluring sun and greenery ahead. The opportunity to move or the proscription on movement was provided or imposed by the shifting continents, and the rise and fall of sea level: whether South America was still joined to Africa, or to North America; whether Africa had yet collided with Eurasia, and whether the Mediterranean was full or dry; whether Madagascar had split from Africa. It matters, in short, what the climate and the continents have done. These are the forces of history.

There is a more practical reason, too, why this history matters. For although the underlying forces are so tremendous, although they seem to operate on a scale that is well beyond human compass, nonetheless it is becoming clear that we can influence the climate. By changing the climate we influence the ecology and evolution of the whole world effectively for the rest of time. Indeed, there is nothing that it is more pertinent to understand than the way that climate behaves, the extremes of which it is capable, and the ways in which it can be influenced. "Influence," of course, emphatically does not mean "control."

Until just a few decades ago, most of the phenomena that are now known to dominate the Earth's history in general and its climate in particular were not even known to exist. Such ideas as there were—for example, the notion that the continents do in fact move—were widely disbelieved or even ridiculed. Now, thanks to an extraordinary coalition of scientists—geophysicists, astronomers, chemists, paleontologists and climatologists, all crucially assisted by computers of remarkable power—we seem to be approaching a grand synthesis. The forces are beginning to be understood. More than that, we are beginning to understand their interactions: how the height of the land affects the chemistry of the atmosphere; how the tilt of the planet relative to the Sun affects the influx of energy—not how much, but where and when that energy falls; how the position of the continents affects the flow of ocean currents, and hence the distribution of energy around the Earth; and how indeed each combination of factors may give rise to feedback loops that sometimes suppress and sometimes enhance each other.

This intricate picture has not been easy to put together. It has required a succession of insights over three centuries, each one demanding a huge leap of imagination, and Earth scientists, though little recognized, have been among the most imaginative of all. But because the insights did require such imagination, each has met with incredulity. No scientists have been more comprehensively scorned than those who have sought to explain how the world behaves. Time after time, however, the most extraordinary ideas have turned out to be right, and the conservative notions have proved inadequate. Up until now, though, the Earth has always outstripped its interpreters. Always—so far—it has proved more complicated than anyone expected.

So now we can see at least in outline why the world has cooled this past 50 million years, and the explanation, I suggest, represents one of the great scientific insights of the late twentieth century. We can see, too, in general terms, why the cooling has not been smooth, why there have been jerks and reversals, and why there could be jerks and reversals in the future, leaving little or no time for planning or taking stock. We can begin to see how it is that small changes—like the apparently trivial rise in atmospheric carbon dioxide—can have huge sequelae. And finally we can see why those sequelae, in principle, are literally unpredictable, although we can make guesses, and those guesses are aided by knowledge of what has happened in the past.

There is no obvious place to begin the description, for everything depends upon and interacts with everything else—landscape, atmosphere, ocean currents, the whole Earth with the rest of the cosmos. The simplest course is to pick up on the main themes and then show how they have now been brought together. The full flavor of the ideas is brought out best if we approach them historically.

THE START OF THE MODERN IDEAS

The key ideas of classical geology and paleontology (the study of fossils) were laid down in the eighteenth and early nineteenth centuries,*

---

*For a detailed treatment of who thought what in this key period, see Anthony Hallam's excellent *Great Geological Controversies,* 2nd ed. (Oxford, Oxford University Press, 1990).

and among them was the crucial notion that the world has not always been the way it is now. For example, there are marine fossils at the tops of mountains, showing that those mountaintops were once at the bottom of the sea. Thus, God did not put all the present-day mountains in position and leave it at that. Mountains have risen, and they have collapsed. Rocks, too, change: mud hardens and debris is crushed into solid rock; and rock crumbles to sand and silt. There are rocks that are produced by volcanic action, known as igneous; and those that are deposited from debris over time, called sedimentary. Some of these sedimentary rocks form from deposits of rivers. Others are formed from the skeletons of marine creatures. Skeletons of the planktonic protozoa known as foraminifera are made of carbonates, and sink to the seabed to form chalk that may transform into limestone; and skeletons of protozoan radiolarians are made of silica, and sink to fossilize into flints. Some rocks are transformed into others by heat and pressure, and thus does limestone become marble. Rocks, however formed, are sculpted by wind and water. Rivers are carvers of caves and gorges that grow deeper by the century.

Underlying these basic ideas was a new sense of time. The nineteenth-century naturalists could see how quickly sediments formed, and could calculate how much sediment would compress to form a particular rock; and thus they could guess, roughly, how long it took for a particular sedimentary rock of a particular thickness to form. The result was a general sense that the Earth had been around for a very long time. Underlying the basic ideas, too, was the general notion that was consolidated in the seventeenth century and is epitomized in the physics of Newton: the idea that effects have causes, and that natural things have natural causes. Whether or not the scientists believed in God (and most of the early scientists were extremely devout), it simply was not good enough to say that God had made things as they are. Always the question for science was "How?"

Thus, in the eighteenth and early nineteenth centuries the traditional commonsensical view of the world—a landscape created by God according to his mysterious will—was transformed. The Earth came to be seen as a dynamic body, its structure constantly re-formed by volcanoes and sediments, and reshaped by winds and rivers and inundating and retreating seas. The world in short was an evolving thing—and is still evolving—a thing furthermore that took a great deal of time to evolve; but a thing in the end whose transformations could be analyzed and understood by applying the still-emerging laws of physics and chemistry.

On the crest of these ideas Charles Darwin (1809–1882) devised his theory of evolution, summarized in *On the Origin of Species by Means of Natural Selection* of 1859. To be sure, most of the evidence that Darwin brings to bear to support his ideas is based on the distribution and similarities of existing creatures, such as the famous finches of the Galápagos Islands. But he adduced that evidence to demonstrate the specific mechanism of evolution: that is, by means of natural selection. Before he began dealing with the mechanisms, Darwin needed to satisfy himself that the idea of evolution was necessary in the first place—the idea, that is, that living things had indeed altered over time and had not been created in exactly the forms in which they appear now. The geology that was emerging when he was young (Charles Lyell published vol. 1 of the first edition of *Principles of Geology* in 1830) instilled the general notion that the world as a whole has indeed changed, and has taken a great deal of time to do so. The emerging plethora of fossils did not alone demonstrate that animals and plants had evolved, and still less did the fossils show that animals came to be the way they are by natural selection. But the fossils did demonstrate a crucial fact: that in the deep past there existed hosts of creatures that were sometimes very different from those of the present, and sometimes intriguingly similar. Thus, the emerging geology and paleontology of the eighteenth and early nineteenth centuries did not direct but they set the tone of Darwin's inquiry. They showed that the biblical account of the world's past, however powerful and lyrical, was far from complete, and also showed in the broadest way that change, not permanence, was the way of the world.

Yet as these broad ideas unfolded, they were held back by the need, or at least the perceived need, to take account of the Bible. The new geologists envisaged seas advancing and retreating not once but many times; but Genesis spoke only of a single flood. The new paleontologists recognized fossils as the remains of animals long gone; but the traditionalists acknowledged only that some creatures had failed to survive the Flood (even though Noah was supposed to have collected two of each kind). In practice, the attempt to relate the evidence on the ground directly to the Noachian Flood was more or less abandoned after the 1820s; but as we will see, the general idea that floods accounted for almost everything persisted in some quarters until almost the twentieth century.

The transformations envisaged by the new geologists and paleontologists required a huge expanse of time to unfold. The great Scottish geologist James Hutton (1726–1797), ever ahead of the game, said that the Earth had "no vestige of a beginning—no prospect of an end." Yet the idea persisted for a century beyond Hutton that the Earth was comparatively young. Traditionalists continued to look over their shoulders at Archbishop Ussher, who in the seventeenth century calculated from the chronology of the Old Testament that the Earth began about 4,000 B.C. More importantly, the physics of the nineteenth century seemed to support the general notion of Genesis that the age of the Earth must be measured in hundreds of thousands of years, rather than in thousands of millions as is accepted to be the case. Thus, they assumed that when the Earth was formed it contained a fixed amount of heat which, they believed, must constantly be lost; so if the Earth was older than a few hundred thousand years, it would by now be cool right through. Even Darwin was bothered by this idea, late in the nineteenth century, long after he published his ideas on evolution in the *Origin of Species*. Not until the end of the nineteenth century did scientists discover the existence of radioactivity. Only in the early twentieth century did they realize that radioactivity is a constant source of heat in the Earth's interior.

Overall, then, the eighteenth and nineteenth centuries were a time of wonderful boldness, of thinkers who envisaged Earth forces of a magnitude and scale that far transcended common experience, but who needed, as they did so, to fight the deepest traditions of church and science that still prevailed. This was not a time for timid thinkers. By the middle of the nineteenth century, practitioners of the life and earth sciences should have been shockproof.

But even they were astounded by the ideas that are associated with the Swiss geologist Louis Agassiz: the notion that in the past, much of the world had been covered by ice, and that much of the landscape had been fashioned not by Noachian floodwaters but by glaciers.

ICE

Louis Agassiz was not actually the first to formulate the notion that parts of the Earth that now are friendly countryside were once deep under ice. The original insight belongs to his naturalist friend Jean de Charpentier, and the term "ice age" (*Eiszeit*) was coined by another friend of Agassiz's, the botanist Karl Schimper. Indeed, Charpentier had to persuade Agassiz. But once Agassiz grasped the idea, he was the one who expanded and promulgated it.

Three kinds of observation led Charpentier, Schimper, and then

Agassiz to this idea. First, they were Swiss: in their native Alps they could see glaciers at work, and could see firsthand what they can do. Second, they perceived that there is a range of geological phenomena so extraordinary that extraordinary explanations are demanded—and so they provided one. It was perverse of others to object to their ideas on the grounds that they were outlandish, since outlandishness was obviously required. Third (the corollary of the second point), the explanations put forward by the other geologists of the day did not seem up to the task.

To be specific, there are, all around northern Europe (and North America, too, but Agassiz knew mainly of Europe), heaps of debris and isolated rocks that just should not be where they are, and rocks and landscapes so damaged and scarred that they must have been subject to awesomely destructive forces. The former include peremptory piles and streaks of mud and stones, known as moraines; and single rocks known as "erratics," smoothed like pebbles but sometimes big as a hotel, dumped on some hilltop hundreds of kilometers from any bedrock of the same kind. The damage includes great areas scoured of soil, and neat round pits gouged in the open ground, and valleys smoothed in the shape of a U, not in a V like the cuts of rivers; and pavements and faces of rock scratched deep and long as if by malice, and requiring careless but outrageous power.

The geologists of the nineteenth century were keen to apply science and to explain what they saw by natural law. But they were hampered by theology, and the particular desire to show how the landscape of today had resulted from the phenomena described in Genesis. Sometimes, too, their imagination was not quite up to the task. And sometimes (surprisingly often, actually) even good scientists allowed themselves to fudge the physics and to suppose, wishfully, that known phenomena could perform feats that in their bones they must have known were impossible. In short: the standard explanation for the displaced boulders, the smoothed-out valleys and the deep-scored pavements was flood. Of course, flood can move stones; the pebbles on the beach chirp with each wave. But it cannot dump a boulder half a kilometer up a mountain, as some erratics have been dumped. It throws rock at rock: but it does not scour rock against rock, pressing the two together for meters at a stretch, in the manner required to produce the etchings seen in nature.

At least, liquid water cannot do this. But frozen water can, and can be seen to do so in the places where glaciers still exist. Here there is no problem with the physics. Agassiz was the principal proselytizer, and he needed principally to overcome his contemporaries' incredulity. He needed to convince them that the ice rivers in his native Alps had once spread to the deep south of lowland France and into God's own garden of England's Surrey; and to show that that ice could exist in truly outrageous quantities. The physics was fine. It was just imagination that others were lacking: the ability to see, in their mind's eye, a mountainous sea of ice where now there were green fields and olive trees. Others went to extraordinary lengths to find alternative explanations. One proposed that the erratic boulders had at some time sunk into caverns, compressed the air within, and thence been expelled like champagne corks, thence apparently to fly across the valleys and land light as sparrows on some distant hilltop.

But as the nineteenth century wore on, Agassiz's idea took hold, and scientists of many different kinds have since enlarged upon it. The evidence is difficult to interpret, but overall Agassiz has been vindicated: the landscape of the north and the plants and animals of all the world, including us, have indeed been shaped to a very large extent by the widespread presence of ice in the not so distant past. But the story is also proving to be endlessly involved.

It is now clear, for example, that there was not a single Ice Age. Over the past million years there have been about ten discrete ice ages, or glacial periods, occurring roughly at 100,000-year intervals. The latest began some 70,000 years ago, reached its height—its glacial maximum—around 18,000 years ago, and effectively ended around 8,000 years ago (although some argue that it has yet to end). The ice ages were separated by warmer interglacials, which were sometimes warmer than the present day, and allowed a brief reflowering of tropical animals in northern Europe; and within each ice age there were briefer periods of warming known as interstadials. In many places the landscape clearly shows the influence of more than one ice age. Sometimes, for example, there are successive waves of moraines, each one dumped by successive waves of glaciers. But such evidence is bound to be limited, because the glaciers and ice mountains of any one ice age are liable to obliterate the geological evidence of the one before.

Classical geology, however—the study of landscape and the chemistry of rocks—is now assisted by a battery of techniques that can be astonishingly informative.

Conceptually straightforward, though far from simple in practice, is to extract cores of rock or mud or ice from the bottoms of ancient lakes, or the seabed, or the ice fields of Antarctica or Greenland. Mud and ice builds up over time—the deeper you go, the older the sample.

With luck, in some cases, the depth alone tells you the age. In other cases the age of a given sample can be inferred by comparison with samples from other places that seem the same and can be dated. Examining the degree of decay of various radioactive isotopes—potassium, carbon—provides exact dates for at least some of the samples and allows the age of others that are similar to be inferred.

The fossils or subfossils of various organisms within the samples show what was going on at the time. Pollen is extremely resilient and extremely informative because the type of plant can be identified, and from the type of plant we can infer the climate. Alder pollen has been trawled from deep in the mud in Indonesia. Yet alder nowadays grows in dank, impoverished mud by temperate rivers: clearly, Indonesia has been a very different place in the past. Cores from the seabed contain the extremely informative skeletons of foraminifera—"forams." As with pollen, the family or even the species can be identified. We know their modern relatives and how they live, and so can infer from the ancient types whether the sea that harbored them was warm or cold.

These relatively simple techniques are now further augmented by extraordinary insights by Nick Shackleton at Cambridge and his colleagues. They base their work on the fact that oxygen exists in two atomic forms, or isotopes. The more common isotope is the lighter one, with an atomic weight of 16. The rarer one is heavy, with an atomic weight of 18. Water, of course, contains oxygen, so water molecules in their turn are relatively lighter or heavier depending on which isotope they contain. Most water molecules obviously contain the more common, lighter isotope, but a few contain the heavier one.

Anyway, Shackleton points out that the lighter water molecules, containing oxygen 16, evaporate more readily than the heavier ones. For once, physical theory seems to accord fairly straightforwardly with common sense. In normal times the water that evaporates from the surface of the sea simply condenses and runs back in again in rain or river water, so nothing changes. But in an ice age, much of the water that evaporates off the sea is trapped in ice and does not return. So the proportion of oxygen 18 in the sea increases, since the oxygen 16 that has evaporated away is entrapped on land. The skeletons of forams also contain oxygen; and the proportion of isotopes within them reflects that of the surrounding sea. Hence, the isotope ratio of the foram skeletons, revealed in cores from the seabed, show whether or not an ice age was in progress at the time they were formed. Cores from the Pacific now show the coming and going of ice ages throughout the Pleistocene with remarkable precision. In general, they support the classical geological data, but they also reveal details that on land have long been obliterated.

The amount of ice that accumulated on Antarctica and on the northern continents during glacial periods was huge. In the latest ice age, the ice in the north was focused in two vast patches: the Laurentide ice sheet of North America stretched down to what are now the southern states, while the Fennoscandian ice sheet was focused on Finland, Norway and Sweden but extended south deep into France and southern England. At the height of the last glacial maximum, 18,000 years ago, the ice in the Laurentide sheet lay to a depth of almost 4 kilometers—the height of the present-day Rockies; while the Fennoscandian ice sheet rose to about 3 kilometers, the height of the Alps.

Now we see why the sea level falls during ice ages. Eighteen thousand years ago the ice in the north and at Antarctica contained 5.5 percent of all the world's water. That results in only a small percentage drop in ocean level; but since the world's oceans have an average depth of around 3.8 kilometers, that small percentage drop translates into 150 meters, or 500 feet. Because the continental shelves are gently sloped, that drop in turn translates into a huge land area; so at the last glacial maximum there was 40 percent more dry land than there is today. The continental shelf of the eastern United States now extends 70 to 150 kilometers into the Atlantic, but in the fairly recent past it was dry land. Only 1.7 percent of the world's water is now locked up in ice, so presumably we need not fear a further rise of anything like 150 meters if the rest of it were to melt. Nonetheless, some of the world's most populated, productive, and beautiful areas barely have their eaves above the water: Bangladesh, the Florida Everglades, Holland, and the county of Norfolk in England.

The mass of ice, the cooling, and the locking up of water begin a chain reaction. The weight of the ice causes the continents themselves to sink into the mantle below. But the mantle rock will not be crushed, so sinking in one place must be compensated by a rise elsewhere. As the ice melts, the continents spring back to their natural position. Clearly, the sinking and rise of rock influence the level of the sea relative to the height of the land. Thus, the actual change in the sea level in any one place is under two main influences: the amount of water left in the oceans (which depends on how much is locked up in ice on land), and whether and by how much the land itself is depressed by the ice upon it.

Because global temperature is low during an ice age, and because the area of the sea is reduced, the amount of water evaporating from the ocean surface is reduced. It used to be thought that ice ages were pluvial—cold and wet, like an English winter. Now the opposite is known to be the case. Ice ages are cold and arid, and therefore dusty; and the dust itself, revealed in seabed cores, again shows whether and when in the past the Earth was cold or warm. So, too, does the presence of loess on land, loess being soft rock formed from dust. China has a huge area of loess that was formed during the ice ages from Gobi dust.

Taken all in all, paleoecologists now argue that the drying up of the world during ice ages—or periods of global cooling in general—is of more importance, biologically, than the ice itself. Of course, the spreading ice drives animals away from the high latitudes and toward the Equator. Because of the fall in sea level, too, land masses are joined during ice ages that in interglacial times are separated: Siberia with Alaska; Britain with mainland Europe; New Guinea with Australia.

But the dryness of ice ages affects the tropics, and the tropics are where most species live. Mainly because of the drying, the tropical forests retreat in periods of global cooling. The endless sea of trees gives way to savanna dotted by islands of forest. Many species must die, but many others, for various reasons, are prompted to come into being. Our own species was one of them—not in an ice age, in fact, but almost certainly at a time of global cooling that was not far from being an ice age.

When the ice was in place, it was far from passive. Continuing research is showing that the sheets (or at least the Laurentide sheet) had a life of their own, apparently producing surges of giant ice floes in the North Atlantic. Furthermore, these floes did not simply break from the edges of the ice sheets like splinters from an iceberg.

Particles of rock contained within the floes, which are now to be found at the bottom of the ocean, suggest that they emanated from the center of the Laurentide ice sheet, and crossed half a continent before reaching the sea. How come? Well, the earth itself is heated by radioactive energy. The huge thickness of ice at the center of the Laurentide ice sheet provided insulation, and allowed the earth beneath it to warm up. So the ice at the earth's surface melted. Thus, the huge mountains of ice found themselves sitting on a surface that was lubricated. Their weight then rendered them unstable, and they collapsed and split, sending islands of ice out across the land and into the ocean, where they floated as giant bergs until they melted and deposited their accompanying loads of continental rock into the ocean, where they have now been picked up and tell us what must have happened. As more and more ice was lost, the insulation was reduced and the soil could freeze again, to restore a little stability. Computer models show that a complete cycle, from thaw to thaw or from buildup to buildup, takes around 7,000 years.

These ice floes from the center of the Laurentide ice sheet are named after Hartmut Heinrich of the German Hydrographic Institute in Hamburg, who began their discovery in 1988 (see *New Scientist,* September 4, 1993, pp. 36–41). Collectively, the Heinrich floes were huge—equivalent to half of all the ice that is now to be found in Greenland. Because of this they would have affected the ocean currents, and in particular would have interrupted the normal flow of warm water from the tropics to the poles, which keeps them somewhat warmer than they would otherwise be. Geoff Boulton of the University of Edinburgh has made an intriguing extrapolation from that observation. Thus, he points out that if the Earth does indeed begin to warm during the next few decades and centuries because of the greenhouse effect (more of this later), then the Antarctic ice would start to melt. This could cause surges of melting ice, comparable with Heinrich floes of the north. These could interrupt the warm currents flowing south and so make the South Pole even colder. Thus—paradox though it seems—in times of greenhouse warming the ice of the Antarctic might actually increase.

Indeed, the ramifications seem endless. Even the outline of it is salutary, however. It shows again how complex the world really is—and how immensely difficult it is to predict what might happen, how common sense can apparently be defied, when warming the world might conceivably cause more ice to form. It shows again how extraordinary the world is—that there is no place for incredulity, that the least plausible explanations so often turn out to be the most likely. It demonstrates the phenomenon known to mathematicians as catastrophe: that an apparently static mountain of ice can suddenly break up, and produce radical and global changes in climate perhaps within months. The general point is that the Earth may exist in any one of many different stable states, but the stable states do not form a continuum; so if the Earth is caused to alter, then it may fall from one stable state to another in as little time as the physics allows. In the same way, a rock may be at the top of a cliff or at the bottom, but not in the middle. Ice ages seem to end catastrophically. Evidence from Greenland ice cores now suggests that the ice may disappear in as little as twenty years.

This, then, is where the insight of Louis Agassiz has led us. Ice has emerged as a shaper both of landscape and of world climate and a key determinant of sea level, which in turn may crucially decide whether it is possible to walk from one continent to another.

But there are other forces at work, even more momentous, that did not occur to anyone in the nineteenth century—or at best only fleetingly. Indeed, it occurred only to a very few people in the twentieth century; and when it did, it was resisted vigorously, not to say insultingly, for almost four decades. For in the eighteenth and nineteenth centuries geologists acknowledged, indeed they reveled in the fact, that masses of land may rise and fall. But they never supposed that entire continents moved sideways. Yet this is the notion which, more than any other, explains everything else. It has become the founding stone of late-twentieth-century earth science.

FROM CONTINENTAL DRIFT TO PLATE TECTONICS

We cannot blame the scientists of previous ages for failing to grasp that continents move laterally. On the face of things the idea seems ridiculous. In fact, it did not become fully accepted in the twentieth century until well into the 1960s, postdating the invention of the laser and the unraveling of the genetic code. It is indeed a pleasant irony that the "dusty" and gentlemanly science of geology has emerged as the most modern of all: the fundamental idea on which the modern discipline is founded is the most recent to be established.

It is a pity, though, that nineteenth-century geologists could not bring themselves to believe that the continents move, for this explains many of the features of the world that bothered them and led them down false paths. For example, although they perceived with wonderful boldness that mountains had once arisen, and were not simply present from the start, they never proposed a convincing mechanism. What, in fact, had pushed them up? At one point it was suggested that the Earth had cooled, and shrunk, and wrinkled like a prune, but few were entirely convinced by this.

There was also the paleontological puzzle pointed out not least by the Austrian geologist Eduard Suess (1831–1914): that widely separate continents such as Africa and India sometimes contained nearly identical fossils. Yet Charles Darwin had posited in his *Origin of Species* of 1859 that creatures that were similar must have evolved from the same ancestor; they were not simply created separately, as Genesis proposed. It followed that different continents with similar fossils must once have been connected, and that the ancestors had flowed from one to the other. Suess accordingly suggested that southern Africa, Madagascar, and peninsular India must once have formed part of a common continent, which he called Gondwanaland—more appropriately called simply Gondwana, since Gondwana means "land." As we will see, Gondwana—now taken to embrace Australia, South America, and Antarctica, as well as Suess's candidates—now plays a big part in modern thinking, albeit not in the way that Suess envisaged.

But in the nineteenth century (and well into the twentieth) scientists went to the most outrageous lengths to explain the difficulty that Suess had pointed out. They proposed that the now distant continents had once been connected by corridors. They even stretched the myth of Atlantis to embrace a vast hypothetical continent between North America and Europe, which supposedly had disappeared beneath what is now the Atlantic Ocean. The existence of such corridors, flung like etiolated causeways across entire oceans, would have set a geological precedent for which there is no modern counterpart, and is, in fact, absurd. Even more ridiculous—straining the most fundamental laws of physics, and defying almost all the painstaking observations of geology—was the notion of the vast, sinking Atlantic continent. But serious scientists entertained such ideas, showing what enormous intellectual lengths they would go to to resist the idea that continents move.

But then of course, the idea that continents do move defies common sense. It just happens to be the case, as science continues to demonstrate, that common sense, though indispensable, is not to be trusted.

Credit for the idea belongs to Alfred Wegener (1880–1930), who was recognized primarily as a meteorologist and explorer and in passing as a record-breaking balloonist. He aired the notion first in a lecture in 1912 and published it in book form—*Die Entsehung der Kontinente und Ozeane*—in 1915, with several more editions in the 1920s.

Wegener supported his thesis with various lines of argument, some of which now seem stronger than others. His observation that the modern continents can be fitted together in a jigsaw to make one big continent still stands in principle—though modern geologists arrange the pieces differently. More to the point, he argued that the general level of the bottom of the sea, and the general level of the surface of the continents, fall into two distinct bands. It looked, indeed, as if the continents were discrete blocks that sat on the Earth's surface, towering above the surrounding seabed. Traditional theories supposed merely that the surface bumps and depressions were effectively arbitrary. But if that were so, then there would be an entire range of intermediate heights, which in fact there are not.

Wegener also picked up on some nineteenth-century physical notions that have proved valid: notably, the suggestion by an English cleric, the Reverend Osmond Fisher, that the Earth had a relatively fluid interior that swirled by convection like water in a kettle. More of this later, for it is an important idea. In fact, the substratum—better called the mantle—that lies beneath the outer crust is not fluid in the sense of liquid, but in the way that glass is fluid. It looks solid, and feels solid (assuming that the deep layers of the Earth were accessible, and cool enough to feel), but in practice it behaves like an extremely viscous fluid. It does indeed flow, as glass does: the windows in a medieval cathedral are discernibly thicker at the bottom than at the top. Lead also seems solid enough, but it "creeps" along roofs, and old lead pipes are distinctly droopy. But however viscous glass or metal or rock may be, if it flows it is fluid; and the continents, said Wegener, could slide across the fluid mantle like ocean foam.

Wegener also argued, cogently, that rocks on continents on opposite sides of the Atlantic—the east side of South America and the west side of Africa—were in line with each other. It was as if, he said, a newspaper had been torn, leaving the newsprint still aligned to show how one piece had once been joined to the other. Similarly, many of the fossils from the Mesozoic (the Age of Reptiles; see page 77) and the preceding Paleozoic were similar on the southern continents, though these now are widely separated. He was impressed particularly by the Permian aquatic reptile *Mesosaurus* and the seed fern *Glossopteris*, which are found only in the southern continents and India. Suess had used the same examples as evidence for Gondwana. But Suess, in common with the generality of biologists, assumed that there must simply have been some land bridge between the continents that now are separate. Wegener simply pointed out that a land bridge would have been composed of relatively light continental rock, and could not for simple reasons of physics have sunk into the denser substratum below. If there had been no land bridge, said Wegener, then the continents must once have been joined and had simply drifted apart.

At first, Wegener's ideas seem to have had a surprisingly easy passage. He certainly received some brickbats—such an astonishing idea was bound to meet incredulity—but he also found excellent support. Notably, the great Milutin Milankovich (one of the boldest thinkers in a bold field) told Wegener after an early meeting, "I am still totally under the influence of your brilliant lecture." But he added a cogent philosophical aside: "You should not be discouraged if you find it more difficult to persuade empiricists [that is, scientists who rely primarily on observation, in the manner of naturalists] than students of the exact sciences [i.e., theoretical physicists]. Quite the contrary, in fact." Absolutely. It is clear in the late twentieth century, in the age of quantum mechanics and cosmology, that theorists who define what is actually possible often get closer to what actually happens in the Universe than those who merely look and see.

On a more practical front, too, in 1929 the British geologist Arthur Holmes proposed a model of the Earth that resembled that of Fisher in the nineteenth century, although more precise. Holmes knew by now that the Earth's interior was heated by radioactivity, which explained the swirling by convection. Thus, at least in part, he provided the component that Wegener's hypothesis was lacking: a mechanism and a source of energy to drive the continents along.

Nevertheless, as the twentieth century wore on, opinion hardened against Wegener until by World War II his support had all but dried up. Some of the world's greatest scientists were simply dismissive. The American paleontologist and mammal specialist George Gaylord Simpson declared that "the known past and present distribution of land mammals cannot be explained by the hypothesis of drifting continents . . . the distribution of mammals definitely supports the hypothesis that the continents were essentially stable throughout the whole time involved in mammalian history." In fact, as we will discuss in the next chapter, the "whole time involved in mammalian history" is at least 200 million years. The continents have moved spectacularly in that time.

The tide began to flow toward Wegener after World War II, when new geophysical techniques came on line and provided the essential hard data. Of crucial importance was the study of rock magnetism. Some kinds of rock that are formed by cooling from a molten state become magnetized as they cool; and the direction of their magnetism is in line with the Earth's own magnetic field. But a series of key studies throughout the 1950s showed that not all magnetized rocks had the same direction of magnetism as the present-day Earth. Yet the variations from the present were not random. They varied consistently, depending upon the age and geographical location of the rocks.

The conclusion was inescapable. When those ancient rocks were formed, the landmasses of which they are a part were oriented differently with respect to the Earth's magnetic field than is now the case. Either the world had the same geography in the past as it has now, but the magnetic poles had moved, or else the poles had stayed roughly in the same place, and the continents had moved.

Two sets of key observations answer that point. In the mid-1950s, S. K. Runcorn of Cambridge University and his colleagues showed that the magnetic orientation of rocks from Europe had changed steadily over time. They further concluded that in the Precambrian the North Pole must have been where Hawaii is now—either that, or Europe had moved relative to the North Pole. Then in 1960, P. M. S. Blackett from Imperial College, London, and his colleagues showed from observations in the southern continents and India that it must be the continents that moved, rather than the pole that wandered. In short, they could explain their results beautifully if they arranged the big land masses of the south in one giant continent—a rethink of Suess's Gondwana—and then envisaged them sliding away from each other. But their results just did not fit the idea of a wandering pole.

There was another complication, again dating from the early 1950s. It became clear, initially from work in Iceland, that every now and again in the past, the Earth's North Pole became the South Pole, and then flipped back again. This does not materially affect the points I have just been making because the reversal was apparently a quick flip through 180°. That is, the polar axis remains unchanged during these times of flip; there is no question here of poles going on walkabout around the surface of the globe. In the 1960s it became clear that the last reversal took place about 1 million years ago, and the one before that, around 2.5 million years ago. In general, these reversals of the Earth's magnetic polarity provide another rich seam of data from which all manner of insights can be inferred.

Also from the 1950s on, this time using the newly developed techniques of echo sounding and deep-sea drilling, oceanographers began to study the ocean floor more thoroughly than ever before. Thus, they showed that the centers of the world's oceans were traversed by ridges of mountains as high as those on land, and those mountain ranges were in turn cut through by deep faults. The Mid-Atlantic Ridge was particularly dramatic. It transpired, too, that the rock of the mid-ocean ridges was almost invariably of the same type as that of the underlying mantle, or was similar to it. Finally, and startlingly, the rock of the mid-ocean ridges was invariably new—at least, none was older than the mid-Cretaceous. Yet conventional wisdom assumed that the seabed was ancient, and should therefore have contained ancient rocks.

So then, in 1960 (published 1962), Harry Hess of Princeton University put forward what became known as "the spreading sea-floor hypothesis." He proposed that new seafloor was constantly being created in the ridges in the middle of the oceans, as rock from the radioactively heated mantle welled up from below. The ocean floor, in fact, was "frozen" mantle. The newly-formed floor then spread out in both directions, outward from the ridge and toward the continents on either side. When it reached the continental margins, it was shoved down into the mantle again. The surface of the Earth is indeed mobile.

Powerful support for this startling idea—indeed virtual confirmation—was provided by the phenomenon of polar reversal. If Hess's idea was true, after all, then the rock of the seabed (below the sediment, that is!) should get steadily older from the center to the ocean's edge. The direction of the magnetism of any one piece of the ocean floor would depend on the orientation of the Earth's own magnetic field at the time that piece of floor first exuded from the mantle beneath, and cooled into solidity. Since the Earth's magnetic axis reverses every few million years, we should expect to find that the sea floor, magnetically speaking, was striped. The floor nearest the ridge, newly formed, should have the same magnetic orientation as the present Earth; the floor slightly farther from the ridge should be reversed in orientation; and so on. Broadly speaking (the observations are far from straightforward), this does indeed seem to be the case. Modern studies of the ocean floor continue to add to the picture. In particular, the mid-ocean ridges have emerged as truly prodigious mountain ranges. Sometimes they even reach the surface: Iceland, the Azores, and Tristan da Cunha are part of the Mid-Atlantic Ridge. Every now and again they remind us of their provenance, as magma spurts forth from them.

The complete modern picture began to emerge in the mid-1960s, beginning (we may say) with the Canadian geophysicist Tuzo Wilson. It was Wilson who suggested that the Earth's crust is in fact divided into a series of "plates"; he was the first to coin the word "plate" in this context. Thus, he also gave rise to the modern term "plate tectonics," where the word "tectonics" is derived from the Greek *tecton*, meaning "building," or "construction," and is applied by geologists to the Earth's crust in general.

Nowadays geologists recognize about twenty discrete plates, some very large, some small. Some plates simply form ocean floor, and others, known as continental plates, have continents floating upon them. Each ocean plate, as Harry Hess first intimated, acts like a conveyor belt. The back end is constantly re-formed from underlying mantle as it wells to the Earth's surface, generally along a mid-ocean ridge, and the front end of each ocean plate is recycled as the plate dives back into the mantle again, typically along the edge of a continent, where an ocean plate dives beneath a continental plate.

All the features of the Earth's basic structure, and much of its behavior, is explained by this plate-tectonics model. We now see clearly the distinction between the ocean beds and the continents. The plates themselves consist of solidified magma, which is very heavy. The continents are indeed discrete entities—blocks of lighter rock that sit on top of some (but by no means all) of the shifting plates. Where the plates go, the continents must go, too. They are like foam on rough water: thrust hither and thither, pushed together and torn apart, but always floating on the heavier material beneath.

Unsurprisingly, the contact and shearing of plates may result in earthquakes—which indeed occur mainly in the zones where the plates are known to meet, such as California, Japan, and Azerbaijan. Where plates are shoved together, mountains are thrown up, like ripples in silk. The energy of the convection currents in the underlying mantle is prodigious, the mass of the plates is enormous, and the force of the collision, slow though it is, is awesome. With such power, it is not at all difficult to see how mountains can be so casually tossed up, and the strata folded and overthrust. Thus have the Rockies and Andes grown; thus do the Himalayas continue to buckle as the plate that carries India is thrust into the underside of Asia, although the mountains are no longer growing taller as their rise is countered by erosion. But parts of the Himalayas are notoriously unstable, as the people of Nepal are only too aware. The continuing upthrust leads to landslides— helped by the erosion that is encouraged by deforestation.

Volcanoes emerge as special cases of the general process that occurs constantly at the mid-ocean ridges as magma escapes to the surface. There are places in the hot mantle known as plumes, where convection currents produce particularly fierce upthrust; and where plumes coincide with weaknesses in the overlying crust, the magma beneath bursts through as lava. Traditional textbooks depict this outflow as "molten" rock surging from deep within the Earth, but as we have seen, the mantle is not molten in the sense of being liquid. The physics of volcanoes is more subtle than that, as now revealed by Dan Mackenzie of Cambridge University. The point is that the rock of the underlying mantle would be molten—it is certainly hot enough—but it remains in a semisolid, glasslike state because it is under such enormous pressure from the weight of the overlying crust. When the pressure is released—that is, when the crust is breached—then the rock changes phase, from semisolid to liquid. Professor Mackenzie compares the phenomenon to the fizz of carbon dioxide gas when you take the top off a soda bottle. As the pressure is released, the dissolved gas is able at last to stretch itself, and comes swooshing out. Not surprisingly, volcanoes occur in areas where the crust is being pulled about. Thus, there are clear zones of volcanoes where the crust is under stress, just as there are zones of earthquakes where the plates make contact.

More broadly, the layout of the present continents has resulted entirely (or almost entirely) from the movements of the plates since the world began, and over the past few million years, they have moved athletically. Exactly where and how they have moved is crucial to our story, for the positions of the different landmasses at different times have largely determined the peregrinations of animals and plants around the world, and therefore have largely determined their evolution. So where have they been?

WHERE THE CONTINENTS HAVE BEEN

The plates that bear the continents upon them move only slowly by human standards: from less than a centimeter a year to several centimeters. But the Earth is about 4.5 billion years old and over that vast time most of the stretches of land that now combine to form the modern continents have traveled over most of the world, splitting from old neighbors as they go and joining to others as if in a global square dance, sometimes far under the sea and sometimes forced upward in long-gone mountain ranges that at times were higher than the modern Himalayas. Animals of a modern kind, with hard shells and skeletons forming solid fossils, began to appear about 570 million years ago in the eon known as the Phanerozoic, which begins with the period known as the Cambrian. The Cambrian seems extraordinarily ancient; yet when it began, the Earth was already 4 billion years old.

In the Cambrian the landmasses of the world had a somewhat scattered look. For example, the mass known as Laurentia now forms much of northern North America, but then it was an island that straddled the Equator in the midst of a huge ocean, and it was oriented 90° relative to its present position. Most of present-day North America was then under water, forming part of Laurentia's continental shelf.

Siberia, now the epitome of land-locked northernness, was a subtropical island in the Southern Hemisphere; and again, most of it was under water. The area that now is northern Europe was even farther south than Siberia, and again was largely submerged. A piece of present-day northern China also floated free, again straddling the Equator, again oriented 90° differently than today, again largely under water. Almost all the rest was fused into the huge continent of Gondwana.

Throughout the next four periods of the Paleozoic—the Ordovcian, Silurian, Devonian, and Carboniferous—the landmasses drifted together, and there was a net movement northward; until by the time of the mid-Permian, around 265 million years ago, almost all of the land of the world was joined into one huge continent known as Pangaea—"All Earth"—which was focused mainly around the South Pole but stretched practically to the North Pole. Pangaea was hook-shaped; and within the eye of the hook was the then giant sea known as Tethys. Modern-day Africa lay to the bottom of the hook; modern-day Europe lay to the top of it. As the two came together and the eye of the hook closed, Tethys was squeezed to become the Mediterranean Sea.

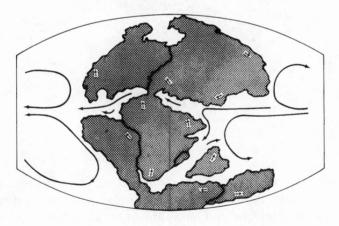

Jurassic

By the early Jurassic period, say 200 million years ago, Pangaea began to break up, and appeared roughly as above. For a time, roughly speaking, there were two landmasses—Laurasia to the north, and Gondwana to the south. Laurasia contained the landmasses that now are North America, Greenland, Europe, and most of Asia; and Gondwana contained present-day Antarctica, South America, Africa, Arabia, Madagascar, India, Australia, and New Zealand. The two great supercontinents remained in contact for a time but only by a narrow isthmus between what is now North Africa and what is now southern Europe; in fact, the isthmus was roughly in the region of present-day Gibraltar. Tethys filled the space to the east of that isthmus. As the Jurassic wore on, say by about 175 million years ago, India broke free from Gondwana at a place that now forms the coast of Antarctica and began its long existence as an island. It also began its long drift north toward Asia.

Mesozoic India was, indeed, like modern-day Australia: an island continent, traveling north. We do not know what India's island creatures were like, however, or how they might have turned out, because its surface was obliterated around 60 million years ago by the Deccan volcanoes, the biggest known on Earth. India's present land and its continental shelf is formed from Deccan lava.

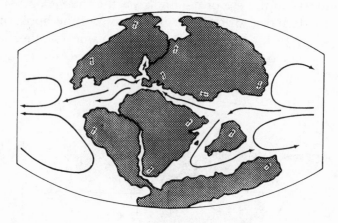

Early Cretaceous

By the early Cretaceous period, about 125 million years ago, the outlines of some of the modern continents can be made out. Laurasia was now split completely from Gondwana. Greenland still provided a broad link between North America and Scandinavia, so that North America, Greenland, and Eurasia were still one continent. South America was still joined to Africa to its east, along the coast of Brazil, but was clearly beginning to break free; and it was still joined to Antarctica to the south, though only by a narrow isthmus. India was now fairly well advanced in its journey toward the south of Asia.

Laurasia remained intact through the Cretaceous, but very sig-

nificant breaks began in Gondwana: the start of the southern lands
of today. By the time the Cretaceous came to an end around 65 million years ago, the landmass that was to become Australia and New
Zealand had broken from Antarctica, and South America remained
only feebly attached, while Madagascar had split from the east coast of
Africa. Note India: still an island, still floating north like modern-day
Australia.

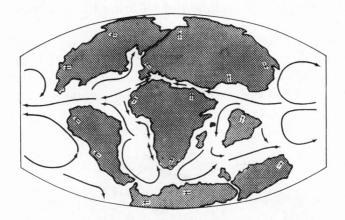

Late Cretaceous

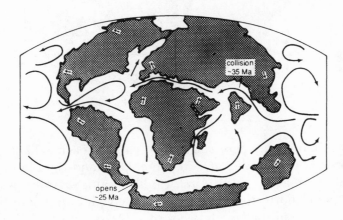

Late Eocene

And then, on into the Cenozoic: the time of dominant mammals. In
the Eocene and Oligocene the ancient landmasses finally began to assume their modern layout, although North America, Greenland, and

Eurasia continued to form one landmass throughout the Eocene. India finally made contact with Asia around 35 million years ago, soon after the start of the Oligocene. We must assume, however, that the continental shelves of the Indian and Asian plates met long before the land itself, and so initiated the rise of the Tibetan Plateau. It rose, quite simply, and is still rising, because there was nowhere else to go but up. As we will see, this ripple in the surface of southern Asia has affected the ecology and evolution of all living things throughout the Cenozoic.

South America finally split from Antarctica around 25 million years ago, virtually at the start of the Miocene, to begin its long and crucial period as an island continent in which, in the manner of islands—like modern-day Australia—it evolved its own unique suites of animals and plants.

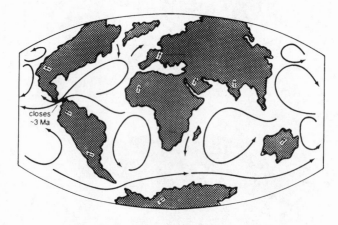

Early Miocene

By 20 million years ago, still in the early Miocene, Laurasia had finally broken up. North America, Greenland, and Eurasia were now separate land masses. North America, like South America, was an island—though shortly to make contact again with Eurasia not through its eastern side but through its western, as Alaska was linked intermittently with Siberia via the vast but transient land bridge of Beringia.

Africa had almost split once more along the Red Sea, and the fragment that was almost broken off—Arabia—had joined with southwest Asia. India was now very clearly a part of Asia, and Australia had moved closer to the tropics.

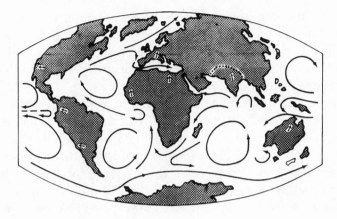

Today

The greatest change of a primarily tectonic nature, and one of huge significance, occurred 3 million years ago in the Pliocene. North and South America, two very different islands with very different origins, were joined via the isthmus of Panama. This must have had huge climatic consequences worldwide as one huge ocean, shaped for some time like a dividing amoeba, finally became two. Whatever currents had flowed between them now had to find another route. The collision also brought two very different faunas into contact for the first time—and as we will see, animals from each of the continents, though particularly from the North, were anxious to migrate into the other. This most recent continental liaison took place within our own history. The family Hominidae was well established when it took place, though living, at that time, in a quite different continent.

In practice, however, the relatively simple picture of joining and splitting landmasses has been complicated by rises and falls in sea level. Thus, the continental shelves of two shifting continents may butt together long before the bits that appear above the water are contiguous. After the shelves are in contact, they may be covered by sea and uncovered again many times in the years that follow. These rises and falls have been particularly significant over the past million years or so because of the ice ages.

In particular—a matter of crucial ecological importance—a land bridge has opened from time to time between North America and Eurasia—not via Greenland, as in the days before the Oligocene, but through the exposure of continental shelf between Alaska and Siberia. It was by this bridge—the hefty chunk of land known as Beringia—that mammals of many kinds, including human beings, flowed both ways between the two continents during the Pleistocene. Very significantly, too, New Guinea and Australia have been joined and divided from time to time as water has flowed and retreated over the shallow shelf between them; and the present-day islands of Indonesia and Malaysia have been joined to Southeast Asia.

In Europe the Strait of Gibraltar through the past few million years have sometimes been just that (a narrow channel of water) and have sometimes been dry land. When the latter prevailed, the Mediterranean has all but dried up; and at the end of those arid periods the ocean has surged back in through the straits to form what must have been a truly stupendous waterfall that flowed until the whole vast sea was full again. Further south, in Africa, the Great Lakes have dwindled time and again as the tropics dried in the ice ages, and grown again, and dwindled. Populations of fish have correspondingly been isolated (as the waters fell and left them divided by pools too shallow to cross) and then reunited. The periods of isolation are ideal for creating new species, as we will see again in later chapters. In Lake Victoria there were until recent years about three hundred species of fish from the single subgroup of cichlids ("sick-lids") known as haplochromines, while Lake Malawi, for much the same reason, now has more than five hundred recognizably different cichlids.

The movements of continents will presumably continue until the Earth finally dies. Exactly what will happen is impossible to predict, at least with present knowledge, though some generalities seem certain enough. Africa will one day split from end to end along the fault line that is now the Rift Valley, and Australia will crash into the base of Asia or perhaps slide along Asia's eastern edge, just as it has been girding itself to do these past 80 million years or so. If it crashes into Asia, mountain ranges big enough to dwarf the Himalayas will appear along what is now the southern rim of China, just as the Himalayas themselves arose 40 million years ago when India first drifted in from the south. But if Australia avoids China it will sweep the islands of Japan before it and shove them steadily northward until they hit Siberia.

So the understanding has grown: first, from the eighteenth century on, the general realization that the Earth is dynamic; then in the nineteenth century came the acknowledgment of ice, although its full significance has become apparent only in the last few years—and it will yield more surprises. Then the extraordinary notion has grown that the seafloor is constantly flowing and continents are always on the move.

Once we add the modern understanding of the atmosphere, the whole picture truly starts to come together.

THE ATMOSPHERE

The outermost reaches of the atmosphere are now known to stretch deep into space and to include a recondite assortment of unearthly ethers: single atoms, too energetic to form molecules; atoms disrupted to form electrically charged ions; particles totally detached from each other and forming inchoate, electrically charged plasmas. They are all formed from atmospheric gases, as their molecules are bombarded and torn apart by high-energy radiations from the cosmos in general and the Sun in particular.

But closer to the Earth the atmospheric gases are largely cushioned from these damaging rays and for the most part remain intact. Closest to the Earth and extending 10 to 15 kilometers up into the sky is the troposphere: this is the "air," the bit we breathe, whose perturbations account for most of our weather. Above that, stretching outwards to about 50 kilometers, is the stratosphere, which is the principal bearer of debris from volcanoes and from nuclear bombs. The stratosphere does contain some disrupted particles, particularly in its upper reaches, but for the most part contains conventional molecular gases.

It is tempting to assume that whatever is present in the greatest amounts is the most important. It is also very deceptive. In the troposphere and stratosphere some of the crucial roles are played by gases that are present in amounts so small that the highest of technologies are needed to detect them at all, and by unstable ingredients that persist for only a twinkling before they are converted to something else; and indeed at present by exotic ingredients produced by human activities. Our own effect on the atmosphere, and therefore on our own and our fellow creatures' lives, is immense, yet is largely out of our control. It is also very tempting to assume that the fundamental features of the Earth are a given, that they are as they were destined to be, and shall be ever more. We have already seen that this is not so. Applied to the atmosphere, it is a desperate and dangerous untruth.

Thus almost 80 percent of the modern-day atmosphere consists of molecular nitrogen, N_2. This has immense direct importance for life on Earth because nitrogen is an essential ingredient of proteins and nucleic acids; and microorganisms living in the soil and the sea are able to trap, or fix, this atmospheric nitrogen by turning it into soluble ammonia, which then, after further chemical conversion of a very simple kind, becomes food for plants to turn into proteins. Thus, the vast mass of atmospheric nitrogen (there are around a quadrillion tons of it—that's about a thousand million million tons) is a huge and vital source of nutrient.

Almost all the rest of the atmosphere—about 20 percent of the whole—consists of oxygen. Each molecule of gaseous oxygen contains two atoms, and so has the chemical formula O_2. That the atmosphere should contain gaseous, "free" oxygen is chemically peculiar, for oxygen is extremely reactive and ought simply to disappear as it combines with whatever else is around, notably the world's apportionment of rocks. Indeed, until about 3 billion years ago, the Earth's atmosphere contained very little oxygen, and the only organisms on Earth were those known as anaerobic, which respired without the benefit of oxygen. Indeed they would generally have been poisoned by it; although, in its virtual absence, they thrived in vast numbers and variety. Free oxygen gas could not exist in the atmosphere as more than a rapidly passing trace until bacteria, and then plants, developed the skills of photosynthesis. For in photosynthesis, the energy of the Sun is employed to split molecules of water, which releases the oxygen that water molecules contain. The hydrogen in the water molecules is then attached to atmospheric carbon dioxide to create carbohydrates, from which (with nitrogen and other elements added here and there) all the flesh of plants, animals, and fungi is subsequently constructed. Note in passing that the atmosphere is the greatest food resource of life on Earth: carbon from carbon dioxide as the basis of all organic molecules, and nitrogen the principal extra ingredient of proteins and nucleic acids. Plants and bacteria expropriate these gases; and we and other animals eat what they produce.

As the atmosphere first began to acquire oxygen from the water that was severed by photosynthesis, the oxygen-hating anaerobic organisms retreated to the airless marshes where their descendants can still be found, and left the surface of the world to the oxygen-dependent aerobes, among which the plants, animals, and fungi are now the most conspicuous. But oxygen is able to persist as a constant component of the atmosphere only because plants and photosynthetic microorganisms produce it in such vast amounts; and they in turn can survive in the midst of such a reactive and destructive gas only because they have evolved a battery of chemical mechanisms for diverting and suppressing it (including the protective ascorbic acid, otherwise known as vitamin C). But although the majority of present-day creatures are adapted to oxygen—we cannot live without it—an atmosphere containing free oxygen cannot be taken for granted. It is no more inevitable than is a world with ice at its poles and forests along its Equator. Indeed, we may note in passing that the oxygen content of the atmosphere is at present diminishing slightly, though measurably, as we employ it to burn fossil fuels (see *New Scientist*, September 12, 1992, p. 16), and that in times past the concentration has been far higher.

Every other component, save these two, is present in small amounts. But "small" does not mean "unimportant," and "amount" is not the same as "turnover": "amount" means the quantity present at any one time, and "turnover" refers to the quantity that passes through over time. Water vapor and carbon dioxide are present only in minute proportions, but both are extremely important in more than one way, and their turnover is vast. Billions of tons of each are thrust into the atmosphere every day, but billions also disappear, leaving only a small amount present at any one time.

Then there is a whole catalogue of ingredients of a somewhat exotic nature, perhaps to be regarded as pollutants. Some of them are produced naturally, and some by human industry. Some are highly unstable but still have a very high turnover, while some remain in the atmosphere for many years. Ammonia and methane are among the gases produced naturally in huge amounts, but are turned over rapidly and so are present at any one time only in trace. Methyl chloride, which arises naturally from sea salt, is always passing through. The vapor of chlorofluorocarbons (CFCs) is man-made: CFCs have been manufactured in vast amounts in recent years for various purposes, of which the chief is refrigeration. They are being pushed into the atmosphere at a steady pace and remain there, because they are extremely long-lived; they are like lead in the body of an urban child, building up slowly over time, with very little elimination. Then there is ozone, which is a special form of oxygen: oxygen in which the molecules contain three atoms rather than the usual two, and so is given the chemical formula O_3.

Three of these trace gases are especially important because they play an enormous role in regulating climate, and therefore to a large extent determine which creatures live where—and indeed, whether life is possible at all. The most important by far, in the past, now, and forever in the future, is carbon dioxide. But two others whose fates are intimately linked are having an impact at least for the time being, and their story has many instructive features. These are ozone and the CFCs.

OZONE AND CFCS

The CFC-ozone story provides three powerful lessons that are highly pertinent to our theme. First, extremely small quantities of pollutant can have a huge effect. Second, even the very best science, brilliantly applied, cannot predict the unexpected; nature invariably proves to be more complex than anticipated. Third, although the trouble is now diagnosed (we hope!), and legislators for once have acted promptly, there is nothing we can do to stop the rot over the next few decades. The world has to bite the bullet. The processes of CFC infiltration and ozone destruction will take more than a century from now to run their course.

Ozone, O_3, is the Jekyll and Hyde of the atmosphere. As Hyde, it is a pollutant at ground level; the acrid by-product of sparking electric motors, and a hideous and aromatic component of traffic fumes. But ozone is also generated naturally through the action of solar radiation upon oxygen in the stratosphere, where it is concentrated at 20 to 25 km above the Earth's surface. This is the ozone layer. Even at its densest, the layer contains only 10 molecules of ozone per million total molecules. But without the ozone layer, all living things would be very different, and it is hard to see how any could have ventured onto the land.

Ozone is vital to modern life forms because it is extremely good at absorbing and screening out ultraviolet radiation from the Sun. Ultraviolet (UV) is light of short wavelength and correspondingly high frequency; light is a form of electromagnetic radiation; and the higher the frequency of electromagnetic radiation, the higher the energy. X-rays and gamma rays are electromagnetic radiations of extremely high frequency and are notoriously damaging. Ultraviolet is somewhat less energetic than X-rays, and far less than gamma rays, and so is not quite so destructive. Yet it can still be lethal. In particular, it is absorbed by molecules of DNA, which it thus disrupts, and so it induces genetic mutation that can lead to skin cancer. Even with the ozone layer intact, a small proportion of UV reaches ground level, and an estimated 5 percent decrease in stratospheric ozone would lead to an additional 100,000 cases of cancer per year in the United States alone. Smaller creatures are simply killed by ultraviolet; sunshine is a great antiseptic, and the allegedly lethal waters of the Bay of Naples contain fewer live bacteria than some of the bracing seas around cloudy Britain.

Creatures that live in the light of the Sun have various mechanisms to protect them from UV. Human beings have skin pigments. Corals contain algae within their cells that contribute to their nutrition by photosynthesizing. So corals must sit in the Sun near the surface of the sea if they are to survive, because otherwise their essential algal residents would die. Corals have now been shown to contain extremely effective chemical UV screens, which manufacturers of suntan lotions are now seeking to imitate. So, too, as a matter of no small interest, do slime molds. But as ozone is simply a form of oxygen, it is clear that the ozone layer could not have existed until after the atmosphere began to contain significant amounts of oxygen, which was after living things evolved photosynthesis. Clearly, the very first photosynthetic organisms must have been in something of a dilemma—needing to expose themselves to the Sun, but in mortal danger from ultraviolet streaming from an ozone-free sky. I have no idea how the trick was pulled. Presumably, UV protectors were among the first pigments to evolve. It does seem clear, though, that plants and animals could not have ventured onto the land, exposing themselves to the full blast of UV, until the ozone layer had formed.

But now, as all the world knows, the ozone layer is being punctured; and the principal causes are gases created by human industry, and in particular by the chlorofluorocarbons, or CFCs. In the long term—that is, after another century or so—this may not matter. This is one pollutant that could be cleared up; indeed, the latest legislation to control CFC production was adopted in Copenhagen in 1992 and could well do the trick. But pollutants like CFCs cannot be purged from the atmosphere overnight, and in the next few decades the degradation of the ozone layer could matter a great deal, for the holes could spread to parts of the world where crops are grown, for example in Canada; and if the world lost the Canadian wheat crop, there would be trouble indeed. Already there have been reports that the lichen on which the reindeer of Scandinavia feed ("reindeer moss") is being killed by excess UV.

So what exactly is going on? What and where is ozone, and what for that matter are CFCs, and how do they work their evil way? Well, ozone is formed naturally in the stratosphere in a two-stage process: first, radiation from the Sun breaks down molecules of oxygen into two isolated oxygen atoms (monatomic oxygen); then each single atom is liable to combine with another oxygen molecule to form ozone. In irreducibly simple chemical terms, $O_2 = 2O$; and $O + O_2 = O_3$. As we have seen, the resulting ozone absorbs UV, and as it does so, it is warmed. Thus does the stratosphere as a whole grow warmer. Because the stratosphere is heated from above, it is warmer than the upper layers of the troposphere below. Thus the stratosphere is less dense than the upper troposphere and has no tendency to sink into it. In fact, it remains happily on top of the troposphere and is stable, much to the delight of pilots and passengers of high-flying aircraft. Thus, too, the thin and fragile ozone layer stays where it is.

A huge amount of energy is required to create the total ozone layer around the world: in fact, about 10 trillion watts, which is about three times the total energy consumed by human beings. Thus, if the ozone layer was destroyed, we could not even begin to replace it. Only the Sun can do this. But as Joe W. Waters of the Jet Propulsion Laboratory in Pasadena comments in *Engineering and Science* (Summer 1993, p. 3), the ozone once formed is very easy to break down; indeed, in concentrated form it is "about as explosive as dynamite." We should be grateful that in the stratosphere it remains dilute.

Ozone does disappear naturally from the stratosphere. For example, it reacts directly with ordinary molecular oxygen. But this process is slow, and the ozone lost by this means is constantly replaced as solar radiation generates more. It has become clear since the 1950s, however, that the breakdown can be accelerated by various materials in trace amounts: materials including hydrogen, nitrogen (in certain chemical forms), and the two related elements, bromine and chlorine. Each of these materials acts as a catalyst. It is the feature of catalysts that very small amounts can have huge effects. And when the breakdown is catalyzed, the loss of ozone can exceed the rate of formation.

Of the four catalytic destroyers of ozone, chlorine has in practice proved to be the most significant. It occurs naturally in the atmosphere in very low concentrations—namely, 0.6 parts per billion by volume (0.6 ppb). Its principal natural form is methyl chloride (CH_3Cl) from sea salt. But the present level of chlorine in the atmosphere is 3.6 ppb. The additional 3.0 ppb—five sixths of the whole—comes from CFCs, produced industrially.

That the chlorine in CFCs could damage the ozone layer was first pointed out by American scientists in 1974, but the initial calculations, based on the laboratory knowledge at the time, suggested that the ozone would be depleted only by about 15 percent over a hundred years. As time passed, the calculations grew even more optimistic; by 1980 the depletion was estimated at a mere 5 to 7 percent. But the British Antarctic Survey Team had been measuring the ozone layer over Antarctica since 1957 (which was the International Geophysical Year) and their measurements of what was really happening showed a very different picture. The ozone over Antarctica dropped dramatically after the mid-1970s. By 1985 it was down by a hundred times.

But chlorine from CFCs was not thought to be the cause, for laboratory studies suggested on theoretical grounds that chlorine could not remain in the stratosphere in sufficient amounts to do the damage. Other causes were invoked: a change in solar activity, leading to an increase in oxides of nitrogen, which also catalyze breakdown of ozone; or an upwelling of air from below—air that was low in ozone. But then Susan Solomon of the U.S. National Oceanic and Atmospheric Administration organized an Antarctic expedition to test the various ideas and found that neither a change in solar activity nor an upwelling of ozone-poor air could be responsible. Chlorine—CFCs—were indeed the cause. So why had the laboratory chemists not predicted this? The answer is simple: because nature is more complicated than the laboratory. The basic reason why CFCs break down ozone was clear enough. First, CFCs enter the atmosphere, and CFCs are extremely stable. They are not broken down by the sunlight of the troposphere, they do not react with other atmospheric components, and they are not soluble in water and so are not washed out by rain. Instead, over a decade or so, they slowly percolate up to the atmosphere: through the ozone layer and above it.

But once the CFCs are above the ozone layer, they are no longer protected from UV; and UV, in upper stratospheric quantities, does destroy CFCs. Each CFC molecule consists of one atom of carbon, plus one of fluorine and three of chlorine: $CFCl_3$. Ultraviolet releases one of the chlorines. This single chlorine atom is extremely reactive, and reacts with one of the equally reactive ozone molecules. In practice, the chlorine wrests an oxygen atom from the ozone to form chlorine monoxide, leaving behind a molecule of ordinary molecular oxygen. In short: $Cl + O_3 = ClO + O_2$.

But the stratosphere, in the region of the ozone layer, also contains single atoms of oxygen: single atoms formed as solar radiation breaks up molecules of ordinary, bimolecular oxygen. Single oxygen atoms are desperately reactive.

Chlorine monoxide, for its part, is desperately unstable. So where ClO meets O, there is reaction: the two O's join to form molecular oxygen, and the Cl is released once again. In chemical terms: $ClO + O = Cl + O_2$.

The monatomic Cl, released afresh, is now free to react with another ozone molecule.

Thus, once the Cl is released from CFC, it is endlessly and rapidly recycled in the stratosphere. I say "endless"; but fortunately the process does come to an end, for eventually the Cl may react not with ozone but with methane, CH_4, to form hydrochloric acid, HCl. HCl diffuses downward into the troposphere, meets the clouds, and is washed out of the atmosphere. But before this happens, the CFC may stay aloft for a century or more. ClO may also react with nitrogen dioxide, NO_2, to form chlorine nitrate, $ClNO_3$; so here is another theoretical way for it to be got rid of.

All this, however, was known to the chemists who predicted that CFCs would not have a hugely destructive effect on the ozone layer. So where did they go wrong? Well, in 1987, James Anderson and his colleagues from Harvard University flew a converted U-2 spy plane into the ozone layer over Antarctica and found that when they hit the hole in the ozone layer the concentration of the notoriously unstable chlorine monoxide, ClO, rose dramatically; rose, in fact, to levels a hundred times greater than theory predicted was possible. Something else was going on, therefore: something that ground-level chemistry had not predicted.

That something else was the presence over Antarctica of stratospheric clouds of ice. Within those clouds, using the particles as a substrate, the HCl and the $ClNO_3$ that should have transported the chlorine safely out of the atmosphere stopped to react with each other. Thus, they released bimolecular chlorine gas, Cl_2, plus nitric acid, which remained locked in the ice clouds. Solar radiation at that height is easily powerful enough to break the Cl_2 molecules apart to form monatomic chlorine, Cl. So the Cl is given a new lease on life. The ozone holes occur in Antarctica in September—springtime in Antarctica—when the first rays of the Sun start to break up the accumulated Cl_2. The ozone over Antarctica bears the brunt of the damage because that is where the ice clouds form, which frustrate the processes that should remove the chlorine. The stratosphere of the Arctic is not so damaged because it is warmer, since it is close to warming land: the continents of North America and Eurasia. So we see again that the climates of the two hemispheres are not symmetrical.

The first regulations intended to reduce CFCs, passed in Montreal in 1987, would have allowed levels of atmospheric chlorine to rise to almost 9 ppb by 2100 A.D. The tighter Copenhagen guidelines of 1992 will still allow a rise to more than 4 ppb by 2010 A.D.—less than twenty years' time—falling to about 1 ppb by 2100 A.D. If we stopped producing all CFCs tomorrow, we would not halt ozone destruction, because the last decade's worth that was released has still to work its way up to the stratosphere.

It is possible, then, that present regulations have stopped the rot of ozone in the nick of time. But just think what might have happened if CFCs had been manufactured in quantity in the nineteenth century. That was a great age of industrial chemistry—all those tanneries and factories for matches and dynamite—but the science of the day was conceptual eons from the modern age of high-atmosphere chemistry. Inter alia, the moderns rely heavily upon very high technology, including probes fixed to high-flying aircraft that can detect ridiculously tiny quantities of ClO. The scientists of the late nineteenth century did not even know that the ozone layer existed, or that it filtered out UV; nor did they appreciate the destructiveness of UV. Victorian CFC factories, had they existed, leaky as all Victorian factories were, could have destroyed the ozone layer decades before the science and technology existed to show what was happening. Once in the atmosphere, the CFCs could not have been recalled any more than they can now—even assuming that anyone had guessed that they mattered.

The people of the early twentieth century would just have stood by and watched each other die, along with all the plants and other terrestrial creatures. This would have been the mass extinction to end all mass extinctions, and if intelligent creatures had ever evolved again— after the CFCs had finally dissipated, and the ozone layer reformed, and animals crept back onto land—then they could never have guessed the reason. Well, perhaps they would have got close; for they would have seen that human beings had once become anomalously populous, and then had suddenly stopped; and might have found enough of the stinking remains of nineteenth-century factories to have inferred some form of self-destruction.

Of course, that did not happen. The invention of CFCs was delayed. Just in time, atmospheric chemists developed the skills required to see what was going on. No one who owns a 'fridge can feel pious about this, for we have all made use of CFCs. But it seems apt to quote the Duke of Wellington's comment on the Battle of Waterloo: "It was a damn'd close-run thing."

However, the loss of ozone could be cured within another century or so, and the story of CFCs may remain as a curious morality tale for the history books. Carbon dioxide is unlikely to prove so tractable.

CARBON DIOXIDE: HERE AND NOW

Carbon dioxide, CO_2, is just a gas, present in the atmosphere in minute amounts, and yet it is the principal food of plants: taken up in the course of photosynthesis and combined with hydrogen, split from water, to produce organic molecules that are the source of plant and animal flesh. It is streamed into the atmosphere as plants themselves respire, and as they die and decompose, and as animals breathe out, and decompose in their turn. The immediate biological importance of CO_2 is therefore obvious. But CO_2 also plays another, vital, and quite separate role. It is the principal short-term regulator of earthly temperature. Like ozone, CO_2 acts as a filter, or rather as a trapper, of electromagnetic radiation: not of ultraviolet this time, but of infrared.

Infrared has a much longer wavelength than ultraviolet, and hence lower frequency, and therefore a much lower energy content. When it strikes molecules in material bodies, it is not inclined to break them apart, as UV does. Instead, it merely vibrates them. The vibration of molecules is experienced as heat. In other words, infrared is warming.

Infrared is called infrared because its wavelength is somewhat longer than that of red light; in fact, it is just too long for human beings to see. The wavelength of ultraviolet by contrast is just shorter than violet: just too short for human beings to see (though bees and some other animals can see UV well enough, and many flowers are UV-colored, however that may look to the creatures who can see it). Most of the radiation from the Sun that actually reaches the surface of the Earth is in the visible spectrum: between infrared and ultraviolet (which, as we have discussed, is filtered out by ozone en route). So the surface of the Earth is warmed mostly by the energy in the visible light.

But once the surface of the Earth is warmed, it immediately starts losing its heat again; indeed, there is a constant flow of heat energy between all bodies, and the temperature at any one time is merely the net effect of heat in versus heat out. But the molecules of the surface of the Earth jettison their heat again in the form of infrared.

Carbon dioxide intercepts infrared, just as ozone intercepts ultraviolet, though not in the same way. It does not stop much of the Sun's energy from hitting the Earth's surface, because most of the energy coming from the Sun is not in the form of infrared but of visible light. But CO_2 does stop radiation from leaving the Earth's surface, because most of the energy leaving the Earth is in the form of infrared. Thus, the net effect of CO_2 in the atmosphere is to trap escaping energy, but not to exclude incoming energy. Thus, carbon dioxide in the atmosphere tends to warm the Earth's surface. Greenhouses work in exactly the same way. Hence, carbon dioxide is said to be a greenhouse gas.

At present, as most know, the burning of fossil fuels is causing atmospheric CO_2 to rise. This is the greenhouse effect. It will almost certainly cause a rise of sea level, just as occurred at the end of previous ice ages; although, as we have seen, there are reasons why such an outcome is not certain, and we need not expect such a dramatic rise as the one that followed the last ice age since most of the ice that remained from that time has already melted. Nevertheless, the rise could be dramatic.

In practice, we could do quite a lot to decelerate the rise in atmospheric CO_2, just as we can do a lot to reduce the rise in CFCs. But, as is the case with CFCs, our control is imprecise, and the most dramatic feasible action will not prevent further changes over the next few decades. In other words, as with CFCs, we are having an enormous impact, but we are not in control. There are two possible ways to achieve reduction (or at least to slow the increase). The first is to burn less fossil fuel, and the second is to plant more trees. One way to reduce the use of fossil fuel is to burn more firewood. Such burning of course releases CO_2; but the CO_2 released is equal only to the CO_2 that the tree that provided the wood took up during its own lifetime, so the net effect of burning firewood is zero. Note, by contrast, that when we burn fossil fuel we are releasing carbon into the atmosphere that was trapped initially by the plants and marine animals whose fossilized and undecayed bodies compose that fuel. Hence we are releasing within a few decades carbon that those creatures trapped over many millions of years.

In addition, if the world had a policy of burning firewood, then for the next few decades at least we should be planting forest specifically for that purpose; and so long as the planting of trees outstrips the burning, there would be a net reduction in atmospheric CO_2, as the growing trees photosynthesized. Indeed, it has been estimated that an extra forest the size of Arizona could absorb all the surplus CO_2 that the world is liable to produce in the foreseeable future. There would be no danger of precipitating another ice age by this means, because if the extra trees removed too much CO_2, then they would stop growing, because the world would be too cold for them and, more to the point, the loss of CO_2 would rob them of their food supply.

Overall, then, living things—or the dead bodies of living things manifesting as fossil fuels—play a crucial role in regulating the amount of CO_2 in the atmosphere; and this in turn is a critical determinant of earthly temperature and therefore of climate.

Finally, there are jokers in the modern greenhouse pack. Water vapor is itself a powerful greenhouse gas. There is not a great deal we can do about this, except to point out that as the world grows warmer, evaporation increases, and as the world enters an ice age, the air grows drier, so here we have an example of positive feedback of the kind that helps to determine, and complicate, climate. Methane is also an important greenhouse gas. It is produced by the actions of anaerobic bacteria—in marshes, and from the gut contents of animals. Oddly, human action is increasing methane, too. Cattle are a major source, because as they ruminate they belch, and their belches are a significant source of methane, and the world's population of cattle is anomalously high because we raise vast herds of them for beef, milk, and transport. The flatulence of termites is also a major source, and they, too, are increasing because termites prefer grassland to forest, and we are felling the forest to make way for grass so that we can keep more cattle. CFCs are also significant greenhouse gases, but perhaps that is being put right.

But I am sure that this general picture is now very familiar. Even politicians tend to write about it. Some in Britain seem to regard the pending greenhouse effect as a bit of a lark: what fun it will be, they suggest, when dour northern constituencies are warmed like the Mediterranean. They clearly do not appreciate the possible magnitude: what a rise in sea level even of a few meters may mean; what effect a novel climate—changing temperature and rainfall with no corresponding alteration of day length—will have on wild creatures or on crops; or of the fact that resounds through all modern descriptions of earthly behavior, of sheer unpredictability. For example, they clearly do not realize that the direction of sea currents depends upon the distribution of energy around the Earth. This can have grand effects, as demonstrated by Hartmut Heinrich's and Geoff Boulton's ideas on the behavior of ice. Of particular relevance to Britain is the behavior of the Gulf Stream, which now brings warmth from the semitropics. With a shift in the world's distribution of energy, it could go into reverse. Then Britain could be as cold as present-day Baffin Land while the rest of the world—or most of it—basks in the new tropics. The point is not to spread gloom. The point is simply to realize that the Earth does indeed behave in extraordinary ways—and incredulity is no defense.

Over the past decade or so, however, an even grander picture of carbon dioxide has been emerging.

CARBON DIOXIDE: THE GRAND PICTURE

The picture of the pending greenhouse effect that has now become well known can be dramatic if we take all the implications into account. Yet it does not convey the full significance of carbon dioxide in shaping the behavior of the world. Two more components need to be added. First, a knowledge of climate history: the extent to which atmospheric carbon dioxide has determined the climates of the past and all that depends upon climate. Second, the extraordinary interactions of carbon dioxide with the rocks of the Earth's surface.

For one thing, it is now clear that the pending (or present) greenhouse effect is not an isolated phenomenon. Theory now suggests that the climate of the past for as far back in time as can be measured has largely been determined by the concentration of carbon dioxide in the atmosphere. Thus, Robert Berner of Yale University has produced a computer model to show that over the past 600 million years the concentration of atmospheric CO_2 has correlated at least roughly with the Earth's climate. He infers what the climate was like from the fossil record: knowledge, or at least informed guesswork, about the kind of conditions each class of creature would have preferred. The model shows that 500 million years ago atmospheric CO_2 was eighteen times the present value; and 500 million years ago the fossils show that the world was warm. Then the level of CO_2 dropped until the Permian to reach levels similar to today's—and we know that the Permian was cold. CO_2 dipped again in the early Cretaceous, around 120 million years ago; another cold time. Overall, simple physics tells us that if the atmosphere contained no CO_2 at all then the Earth would freeze almost instantly, and irrevocably. If it contained much too much, we would boil like our neighboring planet Venus.

The grander insight, however, also largely from Berner, is that the carbon dioxide interacts on a massive scale with the oceans and with the rocks of the Earth's surface, and that these geological interactions, rather than the respirations of living things, in the end determine how much carbon dioxide is in the atmosphere at any one time. Indeed, we may envisage carbon dioxide like currency: money circulating in a society. Living things shuffle CO_2 between them in the way that consumers exchange cash. But the amount of cash that is available for circulation is determined by the government and by the banks. In this metaphor the oceans and ultimately the rocks are the bankers of CO_2. They ultimately determine how much is in the atmosphere, for plants and animals to play with, and how much remains under wraps.

Thus, in 1983, Berner suggested that at any one time there are earthly processes at work to remove carbon dioxide from the atmosphere, and others that tend to increase it. Carbon dioxide is removed from the atmosphere—or scrubbed, as industrial chemists might say—by rain. Rain dissolves atmospheric CO_2 to produce a weak solution of carbonic acid. Carbonic acid in turn reacts with the common rocks of the land, such as granite, which are silicon-based and contain various metals, such as calcium and magnesium. The reaction produces bicarbonates of those metals, which flow away from the land and into the sea. What is left on land is oxide of silicon—alias silica, alias sand. The bicarbonates that wash away become integrated with the seafloor, and as the seafloor finally dives back into the mantle at its leading edge, those bicarbonates are temporarily lost from circulation. Thus is CO_2 gas leached from the atmosphere, and the carbon it contained is buried deep within the Earth.

The CO_2 is restored again to the atmosphere as mantle wells up at the ocean ridges to form seafloor, and by volcanoes, and as rocks are put under stress and pressure as the sidling of plates forces continents together. Thus tectonic movements have a profound effect on the CO_2 content of the atmosphere, and hence on the climate of the Earth.

I find insights like this quite wonderful. They are conceptually simple, but the scale is awesome. It takes yet another great leap of the imagination to see that such a process is taking place.

The final synthesis, though, comes from elsewhere: from Maureen Raymo of the Massachusetts Institute of Technology, and Bill Ruddiman of the University of Virginia. They have explained what for our lineage and for our species is the key issue of Earth history: why the world has cooled over the past 50 million years, during the time that the mammals—including us—were finally taking their modern forms. Raymo and Ruddiman's idea brings together all the grand notions we have been discussing: the movement of continents; the huge circulations of energy by ocean currents; the interactions of atmosphere and rocks. The key to the whole issue is the Plateau of Tibet. This, I suggest, is one of the great insights of the late twentieth century.

TIBET AND THE COOLING OF THE WORLD

The computer model of Robert Berner that described the effect of carbon dioxide upon climate had one shortcoming. The fossils tell us that the Cenozoic began on a tropical note and reached a climax of warmth in the Eocene but began to cool as the Eocene wore on, and that the cooling has continued and culminated in the cool Pleistocene, which flickered in and out of glaciations. If falling CO_2 is the cause of the Cenozoic cooling, then the model should show a dip in atmospheric concentration some time after 50 million years ago. But it does not.

The missing component in Berner's model, according to Maureen Raymo, is the Tibetan Plateau, with the Kumlun Mountains to the north and the Himalayas to the south. The key is India, which broke free from Gondwana more than 180 million years ago, floated north, collided with southern Asia around 40 million years ago, and has been ploughing into Asia ever since like a very slow but immensely powerful locomotive that has overshot its terminus. The land in the path of India has risen to form the Tibetan Plateau; and this plateau is, says Raymo, like "a giant boulder thrust into the atmosphere." It has absorbed CO_2 from the atmosphere by the processes that Berner described. Indeed, it has been big and active enough to scrub much of the CO_2 from the atmosphere of the whole world. This in turn has subjected the entire planet to an antigreenhouse effect—an icebox effect.

This, then—the rise of the Himalayas—has been the fundamental cause of earthly cooling since the Eocene, and of all its ecological and evolutionary consequences. The jerkiness of the fall in temperature, as in the Miocene and in the late Pliocene, 2.5 million years ago, is easily explained by the complexity of the planet itself and by the phenomenon of catastrophe. Most obviously, when the temperature cools below a particular threshold, the distribution of energy around the world is changed, ocean currents may reverse their flow, and the world takes a sudden, catastrophic lurch into a new stable state. The constant split and reunion of continents also has its effect; as we have seen, the liaison of North America and South America 3 million years ago would have stopped all flow between what are now the mid-Atlantic and the mid-Pacific. The lurches and reversals of climate since the Eocene is not surprising at all. The overall cooling has been inexorable.

Why did Berner not arrive at this conclusion? Why did he not take the Tibetan Plateau into account in his own model? Well, he did, of course; but he did not acknowledge its full potential. He assumed that the absorptive effect of rock depended more or less exclusively upon its area. But Raymo and her erstwhile academic supervisor Bill Ruddiman, now at the University of Virginia, concluded from studies that other scientists had carried out in the Amazon that the topography of rock, and in particular its slope, is very important, too. The edges of the Tibetan Plateau, and in particular the Himalayas to the south, are steep, and that improves their efficacy as extractors of atmospheric CO_2.

Even so, the Raymo model would not work unless the Tibetan slopes were washed with truly prodigious quantities of rain, to bring carbonic acid to the steep rock. But of course they are. The wind that drives against the plateau from the southeast has crossed the Indian Ocean and is heavy with water. As it strikes the mountains, it is forced high into the sky. As it rises, it cools, and as it cools, the water that it carries precipitates. It falls as the rain of the Asian monsoons, which in turn determine what and who lives where. The Tibetan Plateau as a whole covers less than 5 percent of the world's land, but it is drained by eight of the world's mightiest rivers, including the Ganges, the Brahmaputra, the Indus, the Yangtze, and the Mekong, and between them they carry 25 percent of all the dissolved material that reaches the world's oceans. On its lee side, the plateau of Tibet throws a rain shadow that embraces the Gobi to the north and the Mediterranean and the Sahara to the west.

Raymo and Ruddiman thus present a beautifully simple picture that again beggars credence only because of its scale. But there is independent evidence for their idea—for example, in the amounts of various forms of metals such as strontium and osmium that have been washed from the Tibetan Plateau over the past 50 million years and are now to be found on the ocean floor. Support comes also from other climate modelers. In 1992, for example, models run at the National Center for Atmospheric Research at Boulder, Colorado, suggested that if southern Asia did not have mountains to the north then the region would be an astonishing 12°C warmer than in fact is the case. The model also shows that the temperature would fall and the rainfall would increase if the land of Tibet was raised even further.

Indeed, the main theoretical problem with Raymo's proposed mechanism is that it might be too effective. Some calculations show that a CO_2 extractor as efficient as the Tibetan Plateau would scrub all the CO_2 from the atmosphere in a few hundred thousand years. Perhaps as CO_2 is removed and the world grows cooler, the rate of the chemical reaction between CO_2 and rock is reduced. Here, then, is another possible negative feedback, comparable with the feedback that prevents the overefficient removal of carbon dioxide by photosynthesis. Perhaps a new balance has now been struck between the Himalayas and the atmosphere, albeit leaving less free CO_2 than before.

Some scientists do not like the Raymo-Ruddiman model. They argue that it is too ad hoc. For my part, for what it is worth, I like it very much indeed. In particular, it is in the right spirit of Earth science as it has emerged these past two hundred years; just the right blend of boldness and discipline. The math is there, but so, too, is that wild streak of romanticism. I love the notion of the rising Himalayas, thrust up by the awesome power of migrating India, soaking half a continent in its monsoon rains, starving another half-continent of rain altogether, and cooling the whole world through a trick of esoteric chemistry which, just a couple of decades ago, was unsuspected. Perhaps it is wrong to judge the truth of science by its emotional impact; but that in practice is how truth is ultimately judged.

Still, though, we have left a gap in understanding. The Raymo-Ruddiman model explains how the world has cooled since the Eocene, culminating in the bleakness of the Pleistocene. It does not explain why that bleakness has flickered in and out of ice ages. To understand this we need to look away from the Earth.

THE SUN

We should be grateful that the Earth is just the right distance from the Sun. Much closer, and we would boil, like Mercury; much farther away, and the ice might stretch from the poles to the Equator. We should be grateful for its spin: without it, one side would cook and the other would freeze. Our position vis à vis the Sun is critical. It is not so surprising, then, that small and cyclic alterations in that position critically affect our climate.

That this might be so was first suggested in 1867 and again in 1875 by James Croll, but then the thought was put to one side, as is so often the fate of bold ideas in science. Milutin Milankovich then elaborated the notion in the early twentieth century, but again it was put aside.

Milankovich's ideas were then refined mathematically in the 1970s by Anandu D. Vernekar and A. L. Berger, and since the 1980s they have become the orthodoxy. Changes in the Earth's position and attitude relative to the Sun are now believed to have had a profound effect on climate, and indeed perhaps to have provided the final spur to the ice ages, which have so dominated the past million years. In fact, to anticipate, we should probably envisage the Pleistocene Earth on a climatic cusp: uncomfortably cool to begin with, and nudged into glaciation every now and again by fluctuations in the path and attitude of the Earth.

In practice, the Earth's relationship to the Sun changes in three ways: in its orbit, or eccentricity—the path it follows around the Sun in the course of a year; in its tilt, inclination, or obliquity—the angle of its axis of spin, relative to the Sun; and in its precession, which can be loosely defined as wobble. Each of these alterations has its own kinds of effects; and because all three kinds of change take place simultaneously, the different effects are superimposed, sometimes building upon each other, and sometimes tending to cancel each other out. Each kind of change works to its own time scale, each of which is quite independent of the others. Milankovich's brave contribution was to show (or begin to show) how these interacting cycles of the Earth's position and attitude would have affected the Earth's climate, and in fact to relate the Milankovich cycles to the ice ages of the past few hundred thousand years. We should look at the three kinds of change in turn.

The Earth's orbit around the Sun, like that of all the planets, is not a circle, as Pythagoras supposed, but an ellipse. Ellipses are like circles with two hubs, or foci, instead of one. The Sun is positioned in one of the foci of the ellipse, so that at one end of the ellipse (the perihelion) the Earth is closer to the Sun than at the other (the aphelion).

The shape of the ellipse changes over time. Sometimes the ellipse is almost circular, and at other times it is much flatter, more dish-shaped. The total amount of energy reaching the Earth from the Sun in the course of an entire year is the same whether the orbit is (almost) round or is flat; the mean distance from the Sun must remain the same. But the way the energy is apportioned throughout the year is different. With the orbit at its flattest, there can be a 30 percent difference in the input of solar radiation between perihelion and aphelion. The total cycle of orbital change takes 95,800 years: that is, it takes 95,800 years for the orbit to progress from nearly circular, to dish-shaped, to nearly circular again.

The inclination of the Earth changes, too. The current tilt away from vertical is 23.44°, but it varies between 21.39° and 24.36°. At present, therefore, the Sun is directly overhead at 23.44° latitude north during the northern summer (the Tropic of Cancer) and at 23.44° latitude south during the southern summer (Tropic of Capricorn). The total width of the tropics is therefore $2 \times 23.44°$. Clearly, the tropics grow wider or narrower depending on the degree of tilt—though this, in practice, is not of huge consequence, and the tropics are not much affected by change of tilt. But the poles—or high latitudes in general—are greatly affected. Greater tilt means longer winter darkness, and longer midnight Sun. Contrast between winter and summer is thus enhanced even more when the Earth is more tilted (and it is extreme at the best of times); and so, too—which is very important—is the contrast between the poles and the tropics. The periodicity in the change in inclination is 41,000 years: that is, it takes 41,000 years to go from a tilt of 21.39° to 24.36° and back again.

Finally, the Earth wobbles as it spins, the phenomenon of precession. The effect is generally explained by analogy with a spinning top, which never spins quite upright, for its axis of spin always describes a cone around the vertical. Precession matters, because it means that different parts of the Earth are pointed toward the Sun at different points on the orbit. Thus, there will be times when the Northern Hemisphere is enjoying its summer—that is, is inclined toward the Sun—when the Earth is at perihelion, and times when the north is tilted to the Sun when the Earth is far from the Sun, at aphelion. At present, northern summer occurs with the Earth at aphelion, but 11,000 years ago (at the time known as the Younger Dryas) northern summer occurred during perihelion. The periodicity of precession (the time the axis takes to work through a complete cone) is 21,700 years.

Just as a theoretical exercise, consider the kinds of things that might happen when we start to put these three sources of variation together. Note first that the three periods of change—of eccentricity, inclination, and precession—have nothing to do with each other: 95,800 years is not a simple multiple of 41,000 years, or of 21,700. This means that in theory any combination of orbital position and tilt is possible. Note, too, that the various changes affect different parts of the Earth differently. Low latitudes (near the Equator) are affected mainly by variations in the Earth's orbit, and by precession: whether a given hemisphere is pointing at the Sun during perihelion, or during aphelion. High latitudes, by contrast (toward the poles) are affected mainly by changes in obliquity (tilt). Note also—again—that the two hemispheres are not mirror images of each other. They can be affected differently. Depending on which hemisphere is inclined toward the Sun at which point in the orbit, it can transpire that seasonality (the contrast between summer and winter) will be exaggerated in one hemisphere but reduced in the other.

The Milankovich cycles also clearly affect the degree of difference between the amount of radiation falling at the poles and at the Equator; and this in turn affects the buildup of polar ice. For whether or not ice can form at high latitudes depends only in part on how cold it is. It also depends on how much water (precipitation) is falling at high latitudes.

In short, when the effects of the three Milankovich cycles are superimposed upon each other, as in reality is the case, then the ramifications and the possibilities seem endless. Yet reality is even more complicated, because the effects of the Milankovich cycles are superimposed on other events: changes brought about by tectonics, and changes in the atmosphere. It does seem, however, that the constant flip from ice age to interglacial throughout the Pleistocene is related to

the Milankovich cycles. In particular, the roughly 100,000-year periodicity of those cycles corresponds with the roughly 96,000-year cycle of
orbital change; and the fluctuations of the past 130,000 years (the late
Quaternary, which includes the Holocene and the late Pleistocene)
may also reflect the 41,000-year oscillation in tilt.

We may ask why the Milankovich swings did not apparently trigger
ice ages at intervals since the beginning of time. The answer seems to
be that before the Pleistocene the world was usually too warm to be
pushed easily into ice age. But it has cooled throughout the Cenozoic
and by the time of the Pleistocene was in a hair-trigger state, poised to
freeze. From time to time, Milankovich cycles pushed it into glaciation;
not (to repeat) by changing the total amount of radiation reaching the
entire Earth in the course of a year (for that would be impossible), but
more subtly by changing the places on the Earth's surface where the radiation has impinged, and the timing of that impingement, thus varying the contrasts between winter and summer, tropics and poles, and
Northern and Southern Hemispheres.

So that is how the world works most of the time. There is, however,
one more unworldly influence which, from time to time, throws everything into turmoil and on at least five occasions has wiped out huge
swathes of animals at a stroke.

THE JOKER IN THE PACK: ASTEROIDS

That the Earth has been struck by asteroids in the past is not in doubt.
No planet or moon can escape their impact. At the very least, they remind us that the Earth is not an island, but part of a hostile or at least
indifferent Universe. The questions are whether those asteroids have
really been as influential as some scientists maintain, and whether, as
some maintain, they strike at regular intervals; and if so, when we can
expect the next.

The physics of asteroids, like that of plate tectonics, is awesome. Asteroids are in effect large meteors, and "large" might easily mean a
solid mass with the same diameter as, say, an English county. Movement in space is relative. The Earth hurtles through its orbit at around
12,000 km an hour. Speed has little meaning until different bodies
come together. So imagine a body that was the size, say, of the Isle of
Wight or Manhattan hitting us with a speed relative to us of around
20,000 km per hour, a perfectly plausible scenario. We might envisage
the softest possible landing—it would have a good chance of hitting

the Pacific Ocean—but that would make no difference. It would hit the Pacific like a brick in a puddle; or, rather, it would bore through, vaporizing as it went.

Such an impact is bound to have a significant, indeed a tremendous, impact on climate. Even if a large asteroid struck the deepest part of the deepest ocean it would throw up more debris than can be imagined, like a truly super volcano. Such debris would surely create the mother of all nuclear winters, blotting out the Sun for decades. On the other hand, the energy of impact has to be dissipated somehow, and in practice would be transposed into heat. So perhaps the world is cooled after such impacts, and perhaps warmed, and perhaps hovers between uncomfortable extremes as contradictory forces fight it out.

That asteroids might have such an influence was first suggested informally in 1980 by the Californian physicist Luis Alvarez. He looked at clay dating from around 65 million years ago, when the Cretaceous was ending and the Tertiary was beginning. This of course represents the transition between two great eras, for the Cretaceous is the last period of the Mesozoic, and the Tertiary* is the first of the Cenozoic. This time of transition is known as the K/T boundary, K standing for *Kreide*, which is the German for Cretaceous, and T standing for Tertiary. Anyway, this clay—a layer that had been laid down within only about a thousand years—proved to contain about three hundred times more of the rare metal iridium than did the surrounding limestone. The uninitiated might say, "So what?" but in practice this kind of thing is not too easy to explain away. Indeed, said Alvarez, the most likely explanation is that the iridium was brought in by an asteroid.

It was of course at the end of the Cretaceous that the dinosaurs came to an end. The most likely cause of their decline—in fact the only one that is truly plausible—is a shift in climate. The question is, was an asteroid responsible for that shift? There are problems with this idea. If, for example, such an asteroid had simply caused a giant nuclear winter, then we might have expected the dinosaurs to die out within a few years—effectively, over one long weekend. In practice, they clearly declined over several thousand years and many were living after the time that the iridium was laid down. But there is much support for the no-

---

*Please note that the Cenozoic era is divided into two periods: the Tertiary and the Quatenary, each of which is equivalent to Mesozoic periods such as Cretaceous and Jurassic. The intervals of time that are more generally referred to in this book—Eocene, Pliocene, Pleistocene, etc.—are epochs: subdivisions of the Cenozoic period. All is made clear on page 77.

tion that an asteroid was at least involved, and there is even a possible site, in the Caribbean. David Raup and Jack Sepkoski of the University of Chicago suggest that asteroids have been a regular feature of Earth history and were probably responsible in large part not only for the demise of the dinosaurs but for many other episodes of widespread extinction, including the five great mass extinctions. In fact, say Raup and Sepkoski, if you look closer at the fossil record, then you see "minor" mass extinctions occurring at intervals of 26 million years. For it happens, they say, that every 26 million years, the Earth's orbit crosses that of a belt of asteroids which, like the planets, circulates around the Sun. If Raup and Sepkoski are right, then we are due for another impact, and another mass extinction of unpredictable extent, in 13 million years' time. We will just have to wait and see.

That, in outline, is how scientists now explain the behavior of our planet and its history. I think it is interesting, but why does it matter? What can we learn from history? In practice, all lessons drawn from history must be general, for the first, ironical lesson of history is that nothing, including climate, ever happens in quite the same way twice. The world was often warm in the past and it will be warm again in the future, but the weather nonetheless will be different, because the continents are not in quite the same place as they were before, and the mountains are not quite the same height, and the currents of the sea have found a new position, and the Earth is tilted slightly differently, and there are bound to be new combinations of temperature and day length and temperature and rainfall, while the contrasts will be greater or lesser between winter and summer and between the Northern Hemisphere and the Southern. In short, whatever is happening at any one time is unique.

But the broad lessons are cogent nonetheless. History tells us that the world is capable of extremes, and it tells us, too, that those extremes are inevitable. Though we may in practice have plenty of time to live our lives and spread our civilizations and our species, we all live in the gaps between disasters. History tells us, over and over, not to be too incredulous. It is difficult to believe that the mountains of today must once have been beneath the sea, yet this has been commonplace knowledge for at least two hundred years. It seems astonishing to cover a continent with ice and lower the sea by 150 meters, but this has clearly happened and could happen again. The shift of continents is ridiculous and the consequent rise of the Rockies and the Himalayas defies belief, and yet the evidence is there: and so, too, is the evidence

GEOLOGICAL TIME SCALE

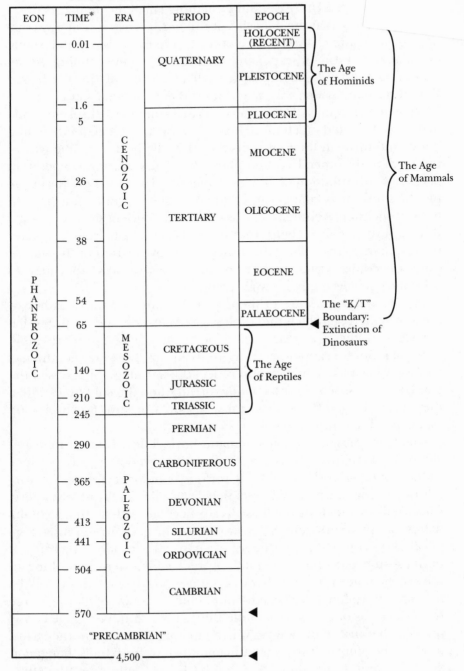

EON	TIME*	ERA	PERIOD	EPOCH	
P H A N E R O Z O I C	0.01	C E N O Z O I C	QUATERNARY	HOLOCENE (RECENT)	⎫ The Age of Hominids
				PLEISTOCENE	
	1.6				
	5		TERTIARY	PLIOCENE	⎬ The Age of Mammals
				MIOCENE	
	26				
				OLIGOCENE	
	38				
				EOCENE	
	54				The "K/T" Boundary: Extinction of Dinosaurs
	65			PALAEOCENE	
		M E S O Z O I C	CRETACEOUS		⎫ The Age of Reptiles
	140		JURASSIC		
	210		TRIASSIC		
	245		PERMIAN		
	290	P A L E O Z O I C	CARBONIFEROUS		
	365		DEVONIAN		
	413		SILURIAN		
	441		ORDOVICIAN		
	504		CAMBRIAN		
	570				◄
			"PRECAMBRIAN"		
	4,500				◄

*Note: Numbers refer to millions of years since the beginning of each period.

for the icebox world that the Himalayas have created. Asteroids are a fact of living in the Universe; and something killed the dinosaurs.

We learn, too, that the behavior of the world is extraordinarily complex. In this brief chapter I have outlined the major forces that shape our planet and its landscape, and the list is not overwhelming. Atmospheric chemistry, topography, the tilt and orbit of the Earth—we should be able to cope with that. But as the components interact, the complexities multiply. The oceans are the year-by-year brokers of climate, for they hold much more heat than the labile rocks of the continents, and through their currents they shift the heat from one part of the Earth to the other. How and where those currents move depends in part on the distribution of energy around the Earth—where for example the sunshine is falling—and also on where the continents happen to be at the time. Another factor is whether the ocean surface is being invaded by vast cold icebergs that sit on the surface because the water they contain is fresh. They can upset the whole rhythm of the sea because the coldest water ought to be underneath, and so they may reverse the currents on their own account.

The heat and moisture of the oceans is translated into weather by the atmosphere, in which the whole saga of currents and upsurges that are seen in the ocean are acted out again on a far more rapid timescale. But we can add even more to all this; for example, the fact that ice will act like a mirror, and reflect the warming radiations of the Sun back into space. So as ice fields spread when the cold times come, they reject the source of heat and so create a positive feedback loop that makes the world even colder.

Finally, we must add the mathematical principles of catastrophe and chaos. Catastrophe tells us that forces acting smoothly may produce sudden changes. A slowly cooling world may suddenly freeze. A warming ocean current may suddenly decide to take its favors elsewhere. Chaos tells us one thing that has already become evident: that a combination of apparently simple forces can produce a horrendously complicated outcome. But the theory of chaos adds a rider which is not obvious—the outcome is literally unpredictable: there is nothing in the system at any one time that can tell us precisely how things will be in some future time. Perhaps the most chilling lesson of all to emerge from the ideas of chaos is that little things really can have huge consequences, because of the ways in which one thing leads to another. Thus it is that the puny industries of humankind—our fossil fuels that generate CO_2, and refrigerants that send CFCs to infiltrate the atmosphere—are affecting the prospects of all life on Earth.

The awesome complexity and the innate unpredictability does not stop climatologists worldwide from trying to understand and forecast the weather, both in the short term and the long. It is worth doing, for there is nothing more important to the economy of human beings or the ecology and evolution of all our fellow creatures; and it is not entirely hopeless because the predictions, though never cast-iron, can in theory and practice approach greater and greater degrees of certainty. Without the climatologists, we would know nothing of the pending greenhouse effect; and that knowledge alone justifies all the effort. Without the paleoclimatologists, the moderns would have nothing solid to go on.

Now we should see how the climates and the continental movements of the past have influenced living things, especially the ones who, as things turned out, proved to be our ancestors.

CHAPTER 3

THE DANCE THROUGH TIME

Old-style paleoanthropologists liked to suppose that the present-day human species, *Homo sapiens,* represents some kind of ultimate. Thus, in 1933, Robert Broom, who discovered the early hominid now known as *Australopithecus robustus,* suggested that after the appearance of human beings "there was no further need for evolution" and felt it was "very doubtful if a single new genus had appeared on earth in the last two million years." Yet as Roger Lewin comments in *Bones of Contention* (London: Penguin Books, 1987; p. 42) Broom's opinion "was unusual only in the very forthright way in which it was stated." Other leading authorities of the day, including Henry Fairfield Osborn in America and Sir Grafton Elliot Smith in Britain, were "close spiritual fellow travelers with Broom." Such a view of evolution is in the tradition that Misia Landau described as mythmaking—and not of the kind that is acceptably heuristic. Broom's belief in evolution was founded in mysticism—a conviction that evolution must move toward some destiny, that this destiny is us, made in the image of God, and that God, having achieved this much, can now retire with a clear conscience.

To modern biologists such ideas are terrible nonsense, and I am inclined to suggest that they are mildly blasphemous. If there is a God, then I am sure He would find such conceit most irritating, and also the implication that He can now put Himself out to grass. In practice, other species have appeared since us, including quite a few plants that can produce instant novelty, of true species status, simply by a doubling of chromosome number. More to the point, our own species almost certainly will not last forever. We may ourselves be transmogrified into other forms or, if we simply go extinct, then we can expect that whatever remains will radiate (diverge to form new species) in all kinds of directions. If the only animals left by the time we disappear are rats, cows, chickens, and cockroaches, then they for several million years will have a field day, their descendants refilling at least some of the niches now occupied by tigers, elephants, parrots, and wasps.

In short, we should see evolution not as a mechanism that was designed to lead to us or even—more broadly—as something that happened in the past. It is happening now. We are part of it. We ourselves have evolved and, like all living things, have been affected by the evolution of other creatures and are affecting them in our turn. To understand how and what we are and what effect we are having, we should see ourselves as part of a process; or, more graphically, as latecomers to a global party that has been in progress for at least 3.5 billion years, since life first began, and will continue until the death of the planet itself. It is a fabulous party, with billions of participants from all walks of life; with riotous games that have rules that lie deep beneath the melee; with giant quadrilles that span entire continents and take tens of millions of years to unfold—what Elisabeth Vrba of Yale University has called "the dance through time." It is a party, furthermore, that takes place upon a shifting stage, as outlined in chapter 2—a stage that is itself crucial to the action.

In this chapter and the next I want to explore the party that human beings have joined, and at which we have become such noisy guests; to see who else is there, and what they do, and to seek the rules that underpin the turmoil. To do this properly we need to go back farther than our allotted 5 million years. We should ask what was happening before the hominids arrived, for only then can we judge our effect, and indeed understand how and why we came to the party at all. We do not need to review the whole 3.5 billion years. But it is appropriate to look at the emergence of the primates—the group, or order, to which the Hominidae belong—and, even more broadly, at the rise of the class of mammals of which the primates in their turn are a part. There are historical lessons here that are pertinent now. This exploration takes us back to the end of the dinosaurs, which 65 million years ago suddenly disappeared from the Earth after 150 million years of world dominance.

THE IDEA OF EVOLUTION

Modern understanding of the party's rules began with Charles Darwin, who in 1859 published his seminal *On the Origin of Species by Means of*

election. There he showed, once and for all, that present-day
  had evolved into their present form, and provided the first
plausible explanation of the means by which they did so. Darwin was
extremely astute, but many of his immediate followers were less so and
they, rather than Darwin himself, developed a version of his ideas that
now seems crude but is widely perceived as traditional "Darwinism."
But as the twentieth century has progressed, much of that crude super-
structure has been stripped away while refinements have been added
to produce a new but still essentially Darwinian view of nature (which
in fact is in many ways closer to Darwin's own, pure perception). I get
the impression that many nonbiologists still equate Darwinism with the
somewhat coarse overstatements of the immediate post-Darwinians,
and have come to believe that the moderns have in large part rejected
Darwin. That simply is not the case. What might be called the mainstream of modern biology is still emphatically Darwinian. To illustrate
the difference between the old and the new—and, more to the point,
to advance our exploration of our own history—it is worth looking at
the demise of the dinosaurs and at the rise of the mammals through
the eyes of both the "traditional" post-Darwinians and the moderns.

DINOSAURS, MAMMALS, AND EVOLUTION:
THE OLD VIEW

Many biologists who followed Darwin felt that the rise of human beings
to the top of the heap virtually needed no explanation. Darwin's great
contribution, after all, was, first, to establish that living things do indeed evolve, that they were not simply placed on Earth in their present
form, and also, crucially, to establish the mechanism of evolution—"by
means of natural selection." His concept of natural selection was
rooted in one piece of late-eighteenth-century economic theory, and
two simple observations. The economic theory was due to the English
cleric and economist Thomas Malthus (1766–1834) who in 1798 first
published *An Essay on the Principle of Population.* Malthus argued that
the human population increases exponentially, but that the human
food supply can be increased only arithmetically. "Exponentially"
means increasing by a fixed proportion of what was there before; and
since what was there before gets larger and larger, steady proportionate increase implies that the growth becomes faster and faster.
"Arithmetically" implies growth by a fixed absolute amount; there is no
increase in the speed of growth as time passes. Since population does

grow faster and faster, and resource does not, population must eventually outstrip resource.

Darwin perceived that this generalization of Malthus must apply to all species. All creatures, even the ones that breed most slowly, such as elephants, must tend by this process to outstrip their resources. Therefore, there must be competition between them to survive, or in Darwin's words, a "struggle for life." But he also observed that among any creatures of the same kind—elephants, cats, oak trees—there is variation. It follows that some of the variants must be better adapted to the prevailing conditions than others—or are fitter, "fit" here having the Victorian meaning of "apt." The ones that were most apt, or fit, were the ones that were most likely to survive and leave offspring of their own. Hence, in the words of post-Darwinian biologists (though not of Darwin himself), there is "survival of the fittest." In other words, said Darwin, the prevailing conditions—the environment—effectively select the creatures that will survive and reproduce, in the same way that a pigeon-breeder favors the birds that conform to the ideals of the show bench. The pigeon-fancier's art is called "artificial selection." The analogous process in nature may hence be called "natural selection."

With this idea firmly in place post-Darwinians found no difficulty in explaining the "ascent" of human beings. First, effortlessly, they conflated the concept of "fit" with the idea of "superior," and thus they were able to argue that human beings have prevailed simply because we are better. So, more generally, is the primate order to which we belong; indeed, the term "primate" was coined by Carolus Linnaeus in the eighteenth century when he first set out to classify all living things, and it means "first." Linnaeus called other animals "secundates." More generally still, the fossil record showed that the dinosaurs disappeared suddenly around 65 million years ago, and that the mammals then very rapidly filled all the niches that they had hitherto occupied, and a few more besides. Indeed, the period since the dinosaurs, known as the Cenozoic, has clearly been the Age of Mammals (with the birds playing a very conspicuous supporting role). Again, the reason seemed obvious. The dinosaurs were dolts who had outgrown both their strength and their wits, given to lolling in shallow seas since they could no longer support their own weight and freezing passively to death when the climate took a turn for the worse. The mammals, by contrast, are fast-moving, quick-witted creatures who make their own body heat and could chew the dinosaurs to death as they lay torpid under cloudy skies. The mammals, in short, are a superior caste. Natural selection was bound to favor them; and among the mammals, selection was bound to favor the primates; and among the primates, it was bound to encourage the magnificent creature who in those days was invariably called Man (which indeed is the literal meaning of *Homo*).

Later in this book, very unfashionably, I will argue that this view of things is not quite so gross as most biologists now suppose. I will suggest that some animals are indeed superior to others, and that this superiority does correlate to some extent with survival, and hence that there is progress in evolution. But I will also argue that "superiority" has to be defined extremely carefully, and that "progress" absolutely does not mean "destiny." Thus, I will maintain that some evolutionary outcomes really are more likely than others (and it is idle to suggest otherwise), but no historical path traced by evolution can be considered inevitable. Hence, despite my defense of old-fashioned notions of progress and superiority, I agree absolutely with all those modern biologists who argue that it was not the mammals who ousted the dinosaurs, but time and chance, and indeed that the dinosaurs could be with us still and might have dominated the Earth until the Earth itself finally died. In short, we are lucky to be here. So let us look again through modern eyes.

DINOSAURS, MAMMALS, AND EVOLUTION: THE MODERN VIEW

The traditional view of the death of the dinosaurs and the subsequent rise of the mammals contains a number of flaws. First, there is an awkward detail of fact that was known perfectly well to the Victorians: that the lineage of reptiles that evolved into mammals (known as synapsids or simply as mammal-like reptiles) is at least as old as the lineage that led to the dinosaurs. Indeed, those reptiles show the first mammalian tendencies as long ago as the Carboniferous, which ended 290 million years ago; and from the time of the Permian—more than 245 million years ago—the mammal-like lineage can be clearly discerned and indeed became numerous in the Permian and Triassic. So the mammal-like reptiles flowered before the dinosaurs were established.

Mammals differ from reptiles largely in details that are not preserved as fossils—in their warm-bloodedness, their possession of hair, in suckling their young and, I would say, in their heavy reliance upon olfaction (which means they are particularly smelly, and responsive to smells). But some features that are diagnostic of mammals are preserved in their bones; in particular, the lower jaws of reptiles and other vertebrates are compounded from several bones, whereas the lower jaws of mammals are built from only one. The one-bone structure is stronger than the multiboned arrangement (and may enable the mammals to eat more, with less wear and tear), and the extra bones that are no longer needed to form the jaw were evidently modified to form the chain of bones that transmits sound through the mammalian middle ear. Here in passing is a marvelous example of evolution's opportunism.

The point, though, is that the peculiar jaw structure of the mammals enables us to trace their transition from mammal-like reptiles through the fossil record, and we can be sure that bona fide mammals had appeared by the mid-Jurassic, say 170 million years ago; a time when the dinosaurs were only just reaching their peak. Then for more than 100 million years these mammals—true mammals—played second fiddle to the dinosaurs. During that time they must have evolved enormous genetic diversity, and, as we will see, the main branches of modern mammals were clearly discernible long before the dinosaurs disappeared. But so long as the dinosaurs remained, the mammals were ecological also-rans, none bigger than a polecat. They emerged and radiated only after the dinosaurs disappeared, like genies wakened from a thousand-year sleep. Their initial radiation in the early Cenozoic was spectacular—there were huge mammalian carnivores and herbivores as big as modern rhinos by the time of the Eocene—but that radiation had to wait until the dinosaurs unscrewed the lid. The notion that the emergent mammals inevitably outwitted the dullard reptiles as soon as they got into their stride simply does not stand up.

The birds put up a slightly better showing. They clearly evolved from a line of dinosaurs and the first true bird, *Archaeopteryx*, dates from the late Jurassic (somewhat after the first mammals). Some—notably the ancestors of modern gulls and auks—flourished in the Cretaceous. Modern gulls are prone to harass anything that flies, and we can imagine Cretaceous protogulls hassling the pterodactyls (although we cannot assume that they got the better of it!).

We may also question whether mammals are indeed superior to reptiles in general or to dinosaurs in particular. Thus, a key difference between modern reptiles on the one hand and mammals and birds on the other lies in their control of body temperature. Modern reptiles rely for body warmth on external energy (the Sun) and on heat generated as a by-product of muscular activity. They are sometimes said to be cold-blooded—although a better term is "poikilothermic," which emphasizes simply that their temperature is variable. In contrast, mammals and birds can generate heat internally just to keep themselves warm and so maintain a core body temperature of around 40°C, which is close to the optimum for biological systems. They are colloquially called "warm-blooded," but "homoiothermic" is the preferred term, stressing the constancy of body heat rather than absolute temperature.

When homoiotherms and poikilotherms are in competition, it seems obvious that the homoiotherms have the advantage; after all, homoiotherms can be active when it is cold, while poikilotherms often tend to be torpid. But all is not so simple. For a start, many poikilotherms clearly can be active in cold conditions—like the fish and krill which operate happily in Arctic seas, or the lizardlike tuataras that run about on their New Zealand offshore islands on freezing nights, leaving the biologists who study them in severe danger of frostbite. True, most of the modern poikilotherms that we are now familiar with, from tortoises to snakes, do prefer warmth; but this is clearly not invariable.

Then again, so-called poikilotherms in general employ a range of strategies for maintaining constant temperature, whatever is happening outside: from basking in the sun, like tortoises and butterflies, to whirring the wings like bees, or lying with the tail immersed in a warm tropical river, like crocodiles.

What really counts, however, is that poikilotherms for the most part heat themselves up for free. A combination of sunshine and normal movement gives them all the warmth they need. By contrast, animals committed to homoiothermy commonly use 90 percent of their food energy simply keeping warm. Indeed, many homoiotherms are obliged to compromise or abandon their warm-bloodedness just to save energy. Some, like armadillos and sloths, allow their temperatures to vary in a somewhat reptilian fashion, while hibernating mammals tend simply to cool down, as bats do by day and hummingbirds do by night. What truly matters, however, is that poikilotherms get by on far less food—so a crocodile that can bring down a wildebeest may eat no more than a goose may do. The problem for homoiotherms is not simply to find enough food for one individual, but enough to support an entire, viable, *self-sustaining* population.

Thus, we find that in habitats where food is limited, such as deserts and small islands, poikilotherms have often prevailed and in many cases still do. You will find no leopards on small islands, but you will find flourishing populations of big lizards and sometimes of snakes. Thus, the largest living lizard of all is a monitor that lives on the small island of Komodo in Indonesia and is known as the Komodo dragon. It is big enough and fast enough to ambush and eat a goat—or a human being. Australia, which is both an island and is largely desert, also has huge monitor lizards known as goannas; and its most formidable predator ever was the giant goanna known as *Megalania,* which was up to seven meters long and weighed about 1,460 pounds—about eight times the weight of a modern Komodo. In the Pleistocene, Australia also had land crocodiles, like *Quinkana,* which lived in Queensland. That *Quinkana* lived on land is suggested by its rounded tail—not flattened to form an oar, as in aquatic forms—while its claws were modified into hoofs. At three meters, *Quinkana* was not huge by crocodilian standards, but it was quite big enough for most prey. Indeed, it seems highly likely that such reptiles inhibited the evolution of truly impressive marsupial carnivores in Cenozoic Australia just as the dinosaurs apparently suppressed mammalian development worldwide through the whole of the Mesozoic.

As a twist in the story, Robert Bakker of the University of Colorado argues that the dinosaurs were not cold-blooded at all. They were perfectly good homoiotherms, like mammals and birds. Specifically, he points out that dinosaur bones show the close-packed pattern of blood vessels that denote rapid bone growth—and such a structure is typical not of poikilotherms but of homoiotherms. Probably, too, flying reptiles like the pterodactyls had insulating fur.

But even if the dinosaurs were poikilotherms, they would not have slumped into cold-blooded torpor overnight. For above all, they were large, and big animals are like storage heaters: they take a long time to cool down. In any case, it is clear now that the dinosaurs are not uncomplicated layabouts. Some that were once supposed to be too heavy for unsupported life on land are now known to have trekked happily across deserts. Some were obvious sprinters. Many lived in herds like cattle or elephants. Many were excellent parents.

In short, modern biologists tend to argue that the dinosaurs were not innately inferior and they were not simply brushed aside by the mammals. Instead, modern theory has it that around the end of the Cretaceous, at the K/T boundary, there was a dramatic change of climate. The precise cause is unknown, but the leading candidate is an asteroid, perhaps of around 10 km diameter; one of those which, as Dave Raup and Jack Sepkoski suggest, are liable to strike the Earth every 26 million years. There are problems with the impact theory, notably that the dinosaurs clearly died out over a period of at least several thousand years, whereas a single crash would presumably have led to a sudden die-off. But mounting evidence supports such an event, including a possible site in the Gulf of Mexico.

At present, however, it is not clear whether this hypothetical but likely impact brought about a dramatic cooling, or a warming. Common sense suggests cooling, as debris would have been thrown into the stratosphere. It also fits in with the traditional notion that dinosaurs would not have coped with cooling as well as mammals would, because of their alleged poikilothermy. Others argue, however, that the world is not so simple and inter alia that the massive energy of the asteroid's impact would have "burned" surface rock and thus released carbon dioxide and so provoked a period of greenhouse warming. So perhaps sudden cold was followed after a few years by egregious warmth. Certainly, the first epoch of the Cenozoic, the Paleocene, seems to have been warm, while the subsequent Eocene seems to have been as hot as the world has ever been, with tropical forests pressing toward the poles. Indeed, I wonder whimsically whether the real problem of the dinosaurs was not an inability to keep warm, but to stay cool. Perhaps the irony is that dinosaurs were homoiotherms, as Robert Bakker suggests, but that they lacked the methods of cooling that mammals have, and in particular the ability that many of us have to sweat, which is a profligate and potentially dangerous way to cool because it squanders so much body water, but is highly effective nonetheless.

If it is the case that excessive warmth rather than extreme cold killed off the bulk of Mesozoic reptiles, then this would account for the lack of cold-adapted reptiles in the modern world. Only the ones that were best adapted to warmth survived the events of the K/T boundary. If this were true, then the present-day lack of cold-adapted reptiles would not be a matter of physiology—an innate inability of poikilothermic reptiles to survive the cold—but simply of history. Neither need we be surprised that Australian dinosaurs survived Cretaceous ice ages. Why not? In fact, the whimsical notion that dinosaurs disliked heat more than they feared cold may well have some mileage.

But whatever the precise cause of their passing, it is clear that the extinction of the dinosaurs has profound implications for our understanding of evolution. Some feel that because the crude, late-Victorian explanation of dinosaur extinction does not work, then this (and other such examples) show that "Darwinism" itself does not work. After all, if mammals replaced dinosaurs just because a meteor and a change of climate wiped the dinosaurs out, then what price natural selection, and the idea that the environment chips away to produce the best adapted lineages? Other biologists, however (including me), argue that Darwin's view of evolution by means of natural selection is alive and well, and absolutely deserves its status right at the heart of modern biology. All that is wrong are the overzealous and simplistic applications of Darwin's ideas. To get a true feel for what has gone on—for the rules of the evolutionary party—we must tell the Darwinian story with greater subtlety, as indeed Darwin himself did; and also add a few twentieth-century refinements. Chief of these has been the transformation of "Darwinism" into "neo-Darwinism," a term which in this context implies the fusion of Darwin's ideas with those of Gregor Mendel.

NEO-DARWINISM

Darwin himself knew full well—it bothered him until his dying day—that as it stood, his idea of evolution by means of natural selection was incomplete. It fails completely unless creatures of any particular kind give birth to offspring who are like themselves. After all, there would be no point in nature selecting horses that were particularly swift unless those swift horses produced offspring which, like themselves, were swifter than the rest. On the other hand, his theory depends upon the notion that the offspring of any one creature are variable; without variety, there is nothing to select between. These observations seem incompatible—for what mechanism of heredity could on the one hand produce offspring that closely resemble their parents and on the other produce offspring that demonstrably differ both from their parents and from each other? For a long time, indeed, Darwin toyed with the mechanism proposed by the earlier French biologist Jean Baptiste Lamarck (1744–1829)—the notion that offspring inherited characters that their parents had acquired in their own lifetimes. He even began a book about it. But he seemed to know in his bones that it was not right and was easily talked out of it by his friend and adviser Thomas Henry Huxley.

Did he but know it, however, his problem was solved in principle in his own lifetime by the Austrian monk Gregor Mendel (1822–1884), whose work on heredity in garden peas and other plants began the science now known as genetics.

But alas, Mendel's work was effectively ignored, and despite speculation to the contrary, there is no evidence that Darwin knew of it. Indeed, the force of Mendel's ideas was not realized until the early twentieth century; "genetics" is a twentieth-century word. The fusion of

Darwin's ideas with Mendel's was achieved from the 1920s to the 1940s, a fusion commonly known as the Modern Synthesis, which produced Neo-Darwinism. Neo-Darwinism is one of the intellectual triumphs of the twentieth century and represents the essence of modern biology.

Mendel's explanation of heredity, looked at in retrospect, is delight-fully simple. It is easy to forget how confusing the facts of hereditary re-ally are when that explanation is lacking—how it is, for example, that a family of brown-eyed people can suddenly produce a blue-eyed child (but never vice versa); or how it is that some extremely unpleasant dis-orders, like cystic fibrosis, tend to run in families (but only erratically) while some (like hemophilia) affect one sex far more than the other. Yet this and much else is explained just by applying the very simplest of Mendel's notions. If you are *au fait* in basic genetics, then please skip the next few paragraphs.

Mendel's first great and crucial insight was that each feature (or "character") of each creature was "determined" by some discrete "fac-tor"; a "factor" which in twentieth-century parlance is called a "gene." In practice, each individual possesses a pair of genes for each charac-ter, one inherited from the mother and one from the father. Contrari-wise, each parent passes on only one of each pair of genes to each offspring, in either egg or sperm. The complete set of genes in any one individual is its genome. Mendel's second crucial insight was that each gene within each species could exist in more than one version; and each version nowadays is called an "allele." Allele is a very useful and much-underused word that ought to be in common parlance and is not. A pity. Third, Mendel perceived that some alleles are dominant over others. Put the three notions together (with just a few extra refine-ments) and almost everything you need to know about heredity is ex-plained.

Thus, we can say that human beings are different from carrots because they have different genes. Carrots have genes to make chloro-phyll, and human beings do not. Human beings have genes to deter-mine eye color, and carrots do not. Humans and carrots do in fact have some genes in common, which for example make various fundamental enzymes involved in respiration. But on the whole, the genomes of car-rots and those of humans are compounded from quite separate sets of genetic ingredients.

On the other hand, it is clear that my genes are different from yours. If it were not so, then we would be identical, give or take a few battle scars. But we are both human, so we both have the same kind of genome. What is the difference? Simply that although we have basically the same set of genes (those that befit a human), we have different versions of those genes. Where you may have an allele that gives you blue eyes, I have an allele for brown eyes. And so on. In practice, biologists speak of the genetic variation within species, or within populations. It would be far less confusing to speak of the allelic variation within populations or species, and the genetic difference (or similarities) between species.

Those, then, are the basic Mendelian ideas. Neo-Darwinism now seeks to impose Darwin's notion of "evolution by means of natural selection" onto a Mendelian framework. The result, again, is delightfully simple.

We have merely to envisage the reality of life: that almost all animals, and most organisms of all kinds, practice sex, which means that they swap genes with each other. In fact, a species is commonly defined as a group of creatures that can mate successfully with each—which means they can swap genes—and thereby produce "fully viable" offspring. For example, asses and horses are considered separate species because their offspring—mules—are sexually sterile, and hence are not fully viable. A population is simply a subset of a species, a group of creatures from any one species that happen to be in contact with one another.

All of us are members of sexually reproducing populations, and we can therefore say that each of us partakes of, and contributes to, the collective gene pool of that population (though it should more accurately be called the allele pool). We can also say that we are members of lineages, a lineage being a population (or a gene pool) that unfolds through time, generation by generation. Since all human beings are members of the same species, our genomes all have the same basic structure, the same basic array of genes. In general, none of us can contain more than two allelic variations of any one gene; yet any one gene in any one population may in theory exist in the form of dozens or even hundreds of different alleles. The alleles in the population pool are like books in a central library. Each of us is allowed to possess a selection of the books at any one time—but only a selection. And it is a private library. To be a member of a species is to be a member of a club, able to borrow alleles of the genes that collectively belong to the club, but forbidden or at least inhibited from borrowing genes from other clubs.

On the whole (and within limits), a species (or population) benefits if the alleles within its gene pool are as various as possible. First, if there is little variation in the population as a whole, then each individual will tend to inherit more or less the same alleles for each gene from its mother as from its father. Individuals who inherit similar alleles from each parent are said to be highly homozygous, which in effect means they are inbred. Inbred individuals tend in general to be more feeble than heterozygous individuals who inherit different alleles from each parent and benefit from a version of what Darwin called "hybrid vigor." Second, populations that contain a great deal of allelic variation are much more able to adapt to changing circumstance than populations that are more uniform—obviously, because as Darwin said, variation is the raw material on which natural selection can work. The amount of allelic variation within a population (or in the species as a whole) largely determines its chances of long-term survival in a changing world.

In the Neo-Darwinian view, the net effect of natural selection is perceived in the gene pool as a whole.* Thus, natural selection favors individuals who carry particular alleles that confer advantage. The individuals who have such an advantage are more likely to live long enough to reproduce, and they will leave offspring who also carry that advantageous allele. Thus, over time, as natural selection continues to operate, the advantageous allele will become more commonly represented within the population—or more frequent, as biologists say. Contrariwise—unless the population is simply increasing in size—individuals who do not carry the advantageous allele will be squeezed out. Hence, the effect of natural selection over time is to change the composition of the gene pool. Over time, some genes become more frequent, and some become less frequent, and some will be squeezed out altogether. After many generations of such change the population will be altered so much that members of it would no longer be able to interbreed with members from the original population, even supposing that the original ones were still alive. Hence, the population would have changed so much that it had given rise to a new species. The idea that lineages can transmogrify into new species was one of Darwin's key contributions—and one which, at the time, gave rise to some of the

---

*Nowadays most biologists agree that natural selection operates "at the level" of the gene; genes spread through populations if they have the means to further their own replication. This is the notion behind Richard Dawkins's "selfish gene." In practice, however, genes tend to survive best if they promote. They bring advantage to the individual of which they are a part; and the net effect of natural selection, albeit working on "selfish genes," is to change the frequency of the alleles in a gene pool, just as Neo-Darwinism suggests.

fiercest controversy.* But the mechanism of Neo-Darwinism shows how this is possible.

In practice, the basic pool of alleles (the gene pool) is constantly added to. First, alleles themselves alter; that is, they mutate to form novel alleles. Most of these variations are disadvantageous (mutations are random—like kicking a television set), but some of the novelties remain in the pool and sooner or later some of them become useful. In fact, such mutations are the prime source of the "variation" Darwin required as a basic strut in his original theory. Gene pools are also added to by introgression. That is, populations that may have been separated for some time and have evolved along different lines may sometimes—I believe more often than biologists often suppose—be brought together again. Then there will be matings between individuals from each population, to produce a flow of genes from one gene pool to the other, with each pool then enriching (or polluting) the other. Ruddy ducks introduced into Spain from North America are currently showing introgression in action as they mate (hybridize) with the native white-headed ducks. I also think it is highly likely, as many paleoanthropologists do, that at least some introgression occurred in Europe between the first modern European people—our own ancestors—and the Neanderthals, whom the moderns replaced.

But in addition, of course, alleles are constantly lost from the gene pool. Some are eliminated by natural selection, but many are simply lost by chance. Thus, as Mendel said, each parent possesses two alleles of each gene but passes on only one of each pair to each offspring. Unless each parent has a huge number of offspring, there is a good chance that any one parent will fail to pass on at least some of its alleles to any of its children. Most alleles are possessed by many individuals in any one generation, so most of them will be passed on, despite that possible source of loss. But rare alleles, by definition, are possessed by only a few individuals, and there is a good chance that they will not be passed on at all. Loss by such means is called genetic drift. Clearly, the smaller the population becomes, the rarer any one allele will become,

---

*Ernst Mayr points out that the heresy was not against God but against Plato, who had suggested that each species represents an "ideal," and that ideals were sacrosanct. In practice, some who accepted the general idea of evolution by means of natural selection rejected the idea that species barriers could be broken; and some who accepted that the barriers could be broken rejected the mechanism of natural selection. In short, responses were complicated. See Ernst Mayr, *Toward a New Philosophy of Biology* (Cambridge, Mass.: Harvard University Press, 1988).

and the greater the chance that the rarest alleles will not be passed on at all. Thus, in small populations genetic drift is a major force, and small populations rapidly lose genetic (or allelic) variation. But we have commented that it is good for a population to be genetically varied. Small populations are more vulnerable than large populations at the best of times and the rapid loss of genetic variation within them, leading to inbreeding in the short term and to loss of flexibility in the long term, can provide the *coup de grâce*.

Note finally that evolution by natural selection is inveterately opportunist. Natural selection is obliged to act simply upon the gene pool as it finds it at the time: on a pool of alleles that has been shaped by the selective forces of the past, and has been enriched (or polluted) by chance mutations and introgressions and depleted by the hazards of drift. Natural selection does not create; it does not conjure appropriate mutations to order.* It is like a sculptor chiseling a block of marble, unable to improve the substance of the stone, able only to remove and never to add; stuck, in short, with what the quarry was able to provide. This is one reason why nature can produce some bizarre forms, and some apparently bizarre solutions to problems—for example, using spare jawbones to make bones for the mammalian middle ear. There is also an important philosophical connotation: that natural selection can shape the lineage only to cope with the pressures of the moment. It does not look ahead. It cannot create halfway, apprentice creatures with an eye to greater things in the future. In the Miocene there were small, three-toed horses, grazing the newly spreading grass. They coped with their surroundings as well as natural selection could manage with the gene pool available, and they did what the environment required of them. They were not a rehearsal for the Arabs and thoroughbreds of today. By the same token, the ancient *Homo erectus* were their own people. They were not a prototype for us.

The fact that mutations-to-order are not typical is, then, the essence of Neo-Darwinism; and that, so most biologists agree, is the essence of modern biology. Natural selection operates to change the frequency of the genes in each gene pool, and hence in the end to produce qualitative change, although, at the same time, the forces of chance act both to provide new alleles (by mutation and introgression) and to remove others (by genetic drift).

---

*In fact, there is evidence that some bacteria are able to undergo mutations in a directional way, producing genes appropriate to the task in hand. But this is a very special ability, not typical of living things as a whole and certainly not of animals and plants.

The theory is excellent. But to understand what really happens in nature we have to look at the component notions more closely. Key among these is the much misunderstood concept of competition.

THE CRUCIAL BUT COMPLEX ISSUE OF COMPETITION

To Darwin the notion of competition was crucial. He implied that constant competition—unremitting pressure from other creatures—led inexorably to change. But many modern biologists question this. First, they say that when you look at nature you do not often see overt competition. Most of the time the various creatures seem to rub along well enough—like people in a football crowd, who interact up to a point, but apart from a little passing abuse have little impact on each other. There is little to justify Tennyson's pre-Darwinian vision of "nature red in tooth and claw." These modern biologists also doubt whether the competition that is discernible actually leads to adaptive change. Sometimes the fossil record shows that animals remain unchanged for millions of years even though they are presumably engaged in constant contest. Sometimes, on the contrary, competition seems simply to lead to obliteration. So what is really going on?

THE NATURE OF COMPETITION

What, first of all, do creatures actually compete with? Well, consider that beautiful and versatile African antelope, the impala. In the dry season it eats mainly grass, in the wet season mainly browse (the leaves of bushes and trees). All year round it may be preyed upon by leopards. In the breeding season both sexes must find mates. Some biologists apply the term "competition" sparingly, and regard predator-prey relationships as a special and separate case. But it seems neater simply to admit that in practice impalas (like all animals) must compete with a whole range of other organisms in order to stay alive. The competition in part is interspecific—that is, between species. So impalas compete with grasses and trees, which on the whole would prefer not to be eaten, and adorn their leaves with spikes and toxins; with other herbivores, from gazelles to rhinos, that also eat grass and browse; and with leopards who seek to eat them. They also compete with their own kind—intraspecific. The males in particular may fight for mates; and all impalas must compete with others both to find food and to escape from leopards. The leopards for their part have their own suite of competitors.

Note, in passing, that the competition takes two completely different forms. Sometimes impalas must indulge in out-and-out fighting—when males battle for mates or females attempt to drive predators away from their young; and sometimes they merely strive to outshine the others—for example, as they flee from leopards. The first kind of contest is comparable with, say, a boxing or tennis match, where you win by undermining the opponent; and the second is more like a race or a game of golf, where neither competitor should interfere directly with the other, but both "compete" with the environment. The two are conceptually different. In short, competition in nature is a complicated business.

Yet it seems undeniable that impalas and leopards do compete with each other. So if Darwin is right, and competition really does drive evolution, then we should expect the fossil record to show that leopards have grown steadily stronger and more agile, and impalas ever more swift. In practice, however, as Elisabeth Vrba points out, the record shows no such thing. The two animals have evidently remained more or less unchanged for the past 3 million years. This is a long time: time enough for our line, the hominids, to change from virtual apedom into us. So what is going on? There are two points. The first we have noted: that leopards are only one of many problems faced by impalas, and vice versa. Every creature in the wild has to keep many balls in the air. Thus, an impala might in theory evolve tremendous fleetness of foot, but if it did, it might thereby sacrifice some of the strength of shoulder and weight of horn needed to beat off rivals in the fight for mates. It might on the other hand develop an enviable ability to digest coarse vegetation. But if it did, it would come into competition with zebras, or acquire a huge belly that would compromise its fleetness. In short, the real impala in the real world has to compromise, and its present size and shape allows it to balance one need against another. Indeed, because the pressures are so various, different pressures are liable to prevail at different times. Thus, at times impalas may not be pressured by lack of food, because leopards may check their numbers before this becomes a problem. But at other times the lack of food may indeed be keeping them in check—and then the leopards merely take the surfeit. Whichever way you look at it, the two beasts clearly get along well enough. In this case natural selection is not operating to produce change.

However, the example of the leopard and the impala does not tell the whole story. Africa has changed in the past 3 million years, and the fact that leopards and impalas have not is in part a tribute to their individual versatility. Impalas are indeed more adaptable than most herbivores, and leopards can live anywhere. But apart from climate, their habitat has been more or less left alone—except for the steady encroachment of evolving hominids. But no place on Earth can ever be left alone indefinitely. Sooner or later something changes. Sometime or other there is bound to be a change of climate that is too extreme even for impalas; or, for one reason or another, faunas and floras that hitherto have been separated will be brought together. Then there is a new ball game. Then the phenomenon of competition takes on a whole new aspect as the various creatures vie to exploit the new ecological opportunities and to come to terms with new neighbors.

COMPETITION AND INVASION

When a foreign species or group of species is thrust into a novel environment, this is called biological invasion. Since members of the genus *Homo* have been the greatest invaders of all—they, plus the many creatures they have carried around the world with them—it is salutary to look at some examples. I have chosen three: two brought about by human beings and one by the tectonic movement of continents.

My first example is very straightforward: the flood of creatures that Europeans have taken to Australia over the past three hundred years or so and are now playing havoc with the native creatures. Domestic cats jumped from Dutch ships in the seventeenth century and are still working their way across the continent, laying waste the small marsupials as they go. Foxes, brought in for sport, joined them in the nineteenth century. In the twentieth century, the huge cane toad was brought from Central America to clear pests from the sugarcane fields of Queensland and is now chewing up native reptiles and amphibians state by state. A different kind of competition is provided by rabbits, imported for food and sport in the nineteenth century and now transforming landscapes, and water buffalo, now obliterating delicate wetlands in the Northern Territory by their wallowing and trampling. The impact of the cat on the native marsupials has sometimes been taken to illustrate the innate superiority of a placental mammal over mere marsupials. But it seems much more sensible to observe that Australia, for all its vastness, is an island, and island creatures commonly are not adapted to the depredations of continental-style predators. Moreover, marsupials themselves have sometimes been successful invaders in lands dominated by placentals—like the opossums in North America.

The second example, also generated by human beings, seems more complicated. Britain's native squirrel is the bushy-tailed red; it used to be called the common squirrel. Then in the 1820s gray squirrels were introduced from North America. Decade by decade the gray squirrels spread and the "common" reds retreated, and are now found only in a few special retreats in England, although they still do well in Scotland. But whereas cats in Australia are known to kill small marsupials, there are no convincing records of gray squirrels attacking reds. So why have the newcomers flourished while the natives have faded? Well, two British biologists have finally provided the answer, and it is strikingly undramatic. It simply seems that gray squirrels enjoy a slightly more catholic diet. Red squirrels prefer the seeds of conifers and are reluctant to eat very much else. But grays will eat other things and in particular have a greater predilection for acorns, perhaps because they cope more easily with tannins (though this is speculation, and not part of the original study). So in any one place, at any one time, gray squirrels have a greater chance of surviving. That is, they have a better chance of building a viable population and therefore—crucially—of pulling through in difficult times. Thus, over time, the grays inexorably spread, and the reds inexorably retreat.

The third and final example is of one of the greatest invasions of all time: the exchange of creatures between North America and South America when those two continents came together in the Pliocene. This is the Great American Interchange.

In outline (for more details, see chapter 4), South America evolved several huge suites of unique species during its tens of millions of years as an island continent. These included an amazing suite of marsupials, of which the present-day creatures are but a shadow; and the edentates (now represented by armadillos, sloths, and anteaters), which produced some truly fabulous forms, including the tortoiselike glyptodonts and the giant sloths. South America also produced several entire orders of ungulates (hoofed animals) that are only very distantly related to the ungulates we know today, although they imitated most of the body forms of the moderns, including those of rhinos and elephants. A few of the ancient South Americans are still with us—like the armadillos—but the only ones that would truly seem familiar today are the New World monkeys, which must have lived on the South American landmass even before it finally broke away from Africa.

Following the clash with North America, some of the South Americans went north—including the armadillos, glyptodonts, giant sloths, and the ancestors of the modern marsupial opossums, which are the only wild marsupials in modern-day North America. But a great many more passed from north to south, including cats (the ancestors of modern jaguars and pumas), dogs, bears, horses, camelids (the ancestors of the modern llamas and vicuñas), elephants, and deer. North America's rhinos had already disappeared during the great cooling of the Miocene.

After the northerners invaded, many of the southerners—including most of the marsupials and most of the unique South American ungulates—disappeared. It looks as if they simply could not compete with the swifter and brainier northerners, as if the shortcomings of the cozy southern islanders were exposed by the streetwise and more continental invaders from the north, just as modern bandicoots are now succumbing to cats. Thus, the Great American Interchange has long been presented as the classic example of competition in general, and of the superiority of the modern mammals in particular.

But Elisabeth Vrba has provided a brilliant alternative explanation of the events that followed the interchange. To be sure, the northerners flourished in the south, while the southerners largely disappeared. But there is no reason at all to suppose that the northerners were innately superior. Neither, she says, need we suppose that there was any overt competition between the northern immigrants and the southern natives.

For consider, says Dr. Vrba, what else was happening in the period after 3 million years ago, when the two continents collided. The world as a whole was growing cooler and drier—a fact that is crucial to our own evolution in Africa. Indeed, the cooling helped to bring the two Americas together, because as global temperature goes down, polar ice builds up and sea level falls. But the animals of North America were not adapted to the cold. They had evolved in warmer times. So what did they do? They moved toward the Equator. And where is the Equator? In South America: roughly along the Amazon. On the other hand, it would have made no sense at all for the southerners to go north, away from the tropics to which they were adapted.

But why did the southerners, who stayed where they were, so often go extinct after the clash? Well, tropical Brazil was itself severely affected by the cooling, drying climate. The rain forest retreated and grass encroached to produce a grassland-woodland mosaic. So the southerners found that their favored rain forest was depleted, but they had to stay where they were because there was nowhere else to run. So they died out. But such a mosaic was precisely the kind of country that the northerners had previously experienced. So they were happy enough with the new-style Brazil. They had simply followed their favored habitat southward.

Thus, the northern immigrants survived while the southern stay-at-homes died out, yet we need suppose no direct conflict between them. What brought the southerners to an end was a change in climate. The hypothesis is that they would have died out—or at least suffered severe setback—even without the northern invasion. The northerners merely occupied the spaces left by the retreating southerners. It looks as if the northerners swept the southerners aside; but that is an illusion. By the same token, to revert to our earlier example, it looks as if the mammals pushed the dinosaurs out, but all they did was jump into the void that the dinosaurs had left. Both the Pliocene South American mammals and the dinosaurs worldwide were eliminated by outside forces that had nothing to do with the creatures that succeeded them. They were actually eliminated by a change of climate. But the animals that did survive the American interchange evolved to form new suites of creatures; and this is just as true of the edentate glyptodonts and giant sloths who emigrated to the north as it is of the northerners who made new homes in the south.

If the Equator ran through Massachusetts, says Dr. Vrba, then the entire scenario might have been reversed. The northerners would have stayed at home and died in their beds, while the southerners marched north. Then, perhaps, biologists would now be concluding that the South Americans were biologically superior to the northerners.

These examples, I suggest, show how competition really works; and the lessons that emerge are directly applicable to our own species, and the effects we have had upon others.

Note, first of all, that competition takes different forms at different times. When animals are given a chance to settle down, then they do indeed come to terms with each other, just as the leopard and the impala have so obviously done. Then indeed they may remain side by side, unchanged. But when the system is shaken up—when the climate changes, or different suites of creatures are brought together—there is a new tension, as the animals strive to come to terms with the new conditions and with each other. Such periods may, in practice, last forever, as the fortunes of different groups continue to fluctuate over time. But in general the periods in which competition is obvious and bloody are short-lived. Soon the Australian marsupials that are simply unable to cope with cats will have gone, and those that are left will adapt to cats just as impalas are adapted to leopards.

But the example of the red and gray squirrels shows that competition can be devastating, and yet be covert. American grays wiped out the British reds so subtly that it has taken biologists more than a hundred years to find the reason. From this example two principles emerge that illustrate mechanisms of competition which, I believe, are far more typical than those of the depredations of cats in Australia. For to explain the replacement of reds by grays we merely need to invoke a general principle that may be seen as a component of game theory (discussed further in chapter 6). This principle is that of preemption: once a creature has taken over a particular territory or ecological niche, then it is extremely difficult for another creature to turf it out. This may seem to conflict with our examples of biological invasion, but in fact successful invasions are rare compared with the failures; it is simply that when they do succeed, they are often spectacular.

So, in the squirrel example, we can imagine a scenario in which there are plenty of pine trees and a few oaks, in which the red squirrels compete very well with any grays that are present. But in one particular year the pines may fail to set much seed—it happens to all populations of trees at some times. Then the reds are forced to move out. But the grays stay on, sustained by the acorns. While the reds diminish, the grays multiply. The following season the pines may set seed again and the reds might seek to return, but in practice they cannot because the grays never left and are already in residence. Their presence preempts the reinvasion of the reds. There is no overt conflict—the reds do not even try to get back, because they perceive that the territory is no longer available. Over time, through this ratchet effect, the reds retreat and the grays expand, but never a blow is struck. I am sure that this same mechanism—simply staying in residence when other creatures are obliged to move out—largely accounts for the observed ability of human beings to spread around the world while other creatures retreat. Our versatility gives us the same kind of preemptive edge, for we can occupy places at times when other species cannot.

Indeed, this same principle of preemption, applied on a grander scale, explains why the mammals remained as also-rans for 100 million years while the dinosaurs held the stage. To be sure, a modern lion dropped by some meddlesome time traveler into dinosaur country might have done very well. There were plenty of efficient dinosaur predators, but lions could surely have held their own. But because the dinosaurs already prevailed, there was no opportunity for lions to evolve. For modern lions were preceded by ancestral lions; and before ancestral lions there were big, clumsy predators that simply lacked the agility, cunning, and social organization of the modern beast. Those clumsy predecessors could not have competed with the dinosaurs.

Indeed, they never even had a chance to appear. No big mammalian predator of any kind could get a look in until the dinosaurs had gone and the niches were left vacant. By the same token, as already noted, big lizards and crocodiles may have inhibited the evolution of big marsupial predators in Australia in the Cenozoic. Taken all in all, then, competition is a complex set of phenomena to which animals respond in a variety of ways.

In practice, however, only four options are open to animals and plants when they are challenged by outside stresses. Clever or otherwise versatile creatures may simply modify their behavior, as pigs or leopards or human beings may do—or the foxes and crows that are now adjusting to life in cities. Or they may move into new territory, either by active migration or by dying out in one place and spreading into another. This of course depends on opportunity. Thus, the shrews of Europe were pushed south during the cooling Pleistocene until they ran out of land and some species finally disappeared in southern Italy and Greece; but North America's mammals were saved from a similar fate in the cooling Pliocene by the timely arrival of South America. If a behavioral change is insufficient and emigration is impossible, then animals may exercise the option suggested by Darwin: they may undergo adaptive, evolutionary change. If that is impossible for whatever reason, then they may simply go extinct.

But when are animals able to evolve in the face of pressure, and when are they simply wiped out? This question is crucial to our own evolution, and to our impact on other creatures.

STRESS, EVOLUTION, AND THE VITAL MATTER OF POPULATION

As Darwin pointed out, no population can change unless it contains a variety of creatures from which the natural pressures can make their selection. Without variety, there simply are no options. In these Neo-Darwinian days we know that the necessary variety results from a varied gene pool.

To provide such variety, two conditions must be fulfilled. First, the individuals within the population must be genetically varied. Modern studies of DNA show that wild populations of animals may differ hugely in the degree of genetic variation. Thus, it transpires—serendipitously—that the Indian rhinos that live in the Royal Chitwan National Park in Nepal are extremely varied genetically, even though there are only a few hundred individuals. Evidently, they converged on the park from a vast area of Asia and brought their variety with them. By contrast, studies in the 1980s revealed that all the cheetahs of Africa were as similar to each other as peas in a pod. At some time in the past the world's cheetah population must have been extremely reduced, and stayed at a low level for some time. Then there would be rapid loss of variation through genetic drift, until the remaining animals were practically uniform. Animals that lose genetic variation by such means are said to pass through a genetic bottleneck. By good luck the tiny population of cheetahs multiplied again, but the inheritance of genetic uniformity remains.

This brings us to the second factor: quite simply, big populations can contain more allelic variants than small populations. To be sure, some small populations—like the Chitwan rhinos—may be very varied, but if there were tens of thousands of those rhinos instead of a few hundred, we could expect even more variation.

Thus, we see that small populations of animals lose out in all ways. Just because the numbers are low, small populations are more easily wiped out than big ones by purely physical means: an epidemic, a fire, an exuberant hunting party. But in addition, small populations are at a huge genetic disadvantage. First, they cannot contain as much genetic variation as large populations, and so lack the genetic wherewithal to make evolutionary change. But secondly, as already noted, small populations lose genetic variation much more rapidly than large ones do by the chance process of genetic drift. Because of drift, a small but genetically various population (like the Chitwan rhinos) can degrade into a genetically homozygous and uniform population (like modern cheetahs) within a few generations if the population remains at a low level.

In fact, modern genetic theory allows us to define the size of populations that are needed, in theory, to resist various kinds of pressure. In general, any population of animals that remains consistently at 50 individuals or less is almost bound to be wiped out sooner or later simply by accident—anything from epidemic, to a failure in one generation to produce enough offspring of the appropriate sex. At present, the population of both Javan and Northern white rhinoceroses is less than 50. To have a reasonable chance simply of surviving in the wild, populations in general need to be maintained at at least 500. But to contain enough genetic variation to enable evolution to continue in the genetic, Neo-Darwinian sense, a population must generally be sustained at at least 1,000.

The ecological and conservational implications are obvious. Elephants in Asia, for example, generally need about 10 square kilometers each—so 10,000 square kilometers are needed to sustain Asian elephants in truly viable numbers. Tigers, as top predators, need even more space: anything between 10 and 100 square kilometers each. A truly viable population needs up to 100,000 square kilometers, the size of a small country. That kind of area is just not available. At present, tigers in the wild are simply hanging on. But we see, too, how easily we can explain our own impact on other creatures. It is not necessary to wipe out every last one in order to drive them to extinction. We merely have to reduce the population below viable numbers; but viable numbers are surprisingly high and, in the case of large animals, require an extraordinary amount of space.

But population size also affects an animal's evolutionary fate. As we have seen, if a population is very small (less than a few hundred) then all but the smallest pressure is liable simply to wipe it out. If, on the other hand, the population is huge—many millions—then it is hard to see how Neo-Darwinian change can occur at all. After all, we cannot say that a gene pool has altered qualitatively until some of the alleles have been cast out altogether. Any change of frequency that falls short of total elimination is merely a fluctuation, which all lineages undergo as the generations pass. It seems to follow that Neo-Darwinian change can occur only if the population is middle-sized. This, I will suggest later, has great implications for the emergence of our own lineage—and for our present-day evolution, or the lack of it.

It also seems to me (though I have not heard the matter discussed) that variations in size of population help to explain why it is that evolution sometimes seems to proceed slowly, and sometimes quickly. For when a population is doing well, it grows. Eventually, I suggest, it gets to a size at which Neo-Darwinian change is no longer likely, because there are no alleles rare enough to be shuffled off. But when some new pressure builds up, the population may start to drop. Eventually, numbers fall to a point at which the rarer alleles start to be weeded out. Thus, the gene pool undergoes qualitative change, which is the essence of Neo-Darwinian evolution; and natural selection ensures that it changes in ways that increase its "fitness." So then the population expands again, and the process of change again comes to a halt. If, however, the pressure is too great, then the diminution continues until the population is no longer viable, and then it decays rapidly to extinction. Hence, I suggest, a rise and fall in population, which itself is a reflection of fitness, also helps to regulate the pace of evolutionary change. We could say that such a mechanism helps to ensure that lineages change when they need to change (when pressure is reducing their populations) and to stay the same when they are doing well (and their populations are large).

The need for populations to be large also explains why, when habitats are under stress, some species disappear while others expand. A million years ago the world contained at least a dozen species of elephant (and probably many more), and now it contains only two. But one of the two that remains occupied virtually all of Africa until recent decades, and the other spread from Pakistan to China and the islands of Southeast Asia. What has happened, I suggest, is that over the past few million years various pressures (including the rise of *Homo*) reduced most of the elephant species to numbers that were simply inviable, and they disappeared. But the surviving species were then able to spread into the elephant territory that was vacated. Thus, one effect of pressure is to reduce the number of species—though not necessarily to reduce the total biomass, because the remaining ones expand. Population—its reduction below viable numbers—is again the root cause.

In short, over the past 130 years Darwin's original vision of evolution has been modified by a broader view of what competition really entails and by adding the ideas of genetics. The overall effect has been not to threaten but to enrich the original concept. The third and final area of enrichment is in the matter of timing.

TIMING

Darwin never supposed that evolutionary change always occurs at the same pace. Enough fossils were known in his day to show that some lineages have changed quicker than others, and that any one lineage may alter rapidly in some periods and more slowly in others. But he does imply that the struggle for life is a constant, nagging force, which suggests that the creatures themselves must constantly be modified to stay ahead of the game. Furthermore, he supposed that evolutionary change is bound generally to be slow. Hence, Darwin's original view of evolution is generally said to be one of gradualism: slow but inexorable change.

Modern studies of a fossil record that is much more complete than that of Darwin's day suggest that this need not be the case. On the one hand, we now know that evolution can, at times, be astonishingly rapid. Thus, in the interglacial period between the last two ice ages, around 120 million years ago, the North Sea encroached into western Europe to create the English Channel and the Channel Islands (although it retreated again, temporarily, during the last ice age). Red deer were trapped on the island of Jersey and, as is typical of island animals, they became dwarfed. Astonishingly, fossil studies by Adrian Lister at Cambridge University now show that simply by natural selection their body weight was reduced to one-sixth of the original in a mere six thousand years (*Nature*, November 30, 1989, pp. 539–542). Contrariwise, long fossil sequences of marine snails now show that some lineages remained virtually unchanged for hundreds of millions of years.

With such examples in mind, Stephen Jay Gould and Niles Eldredge suggested in the early 1970s that the slow but steady change known as gradualism is not the normal pattern of things. Instead, evolution has a tendency simply to shudder to a halt—until the system is given a jolt and the lineages rapidly evolve into some new form. They called this stop-go pattern "punctuated equilibrium."

In the light of much that has already been said in this chapter, we can see why this might well be the pattern. Often, after all, it pays an animal to stay the same. Impalas and leopards, for example, are coping very well with life, and change could upset their respective applecarts. South African biologist Hugh Patterson has suggested another reason for staying the same. All creatures that breed sexually (which is most of them) must recognize each others' mating signals. Thus, male crab spiders impress other crab spiders by waving their palps, while male bustards jump up and down to display their vigor. At the chemical level, eggs must recognize sperm from the appropriate species and the stigmas of plants must recognize suitable pollen. The fact that all creatures must communicate with others of the same kind if they are to reproduce and spread their genes implies that they should beware of varying too far from the norm, unless there is special reason to do so. The need to be recognized, in short, is another reason why natural selection should favor the creatures who stay the same.

Thus, a generalization emerges: that natural selection favors change only when the advantages of changing outweigh the advantages of staying the same. And often, as we have seen, there are good reasons for staying the same. I have also suggested that the rate of evolutionary change is affected by population size; very large—which implies successful—populations seem resistant to change.

Elisabeth Vrba* has developed the general notion of punctuated equilibrium into a grand vision of evolutionary change that she calls "the turnover pulse hypothesis." She has looked not simply at single lineages but at entire suites of different creatures in whole ecosystems, notably at the fossils of bovids (mainly antelope) in Africa over the past 10 million years or so. She finds that in general all the species tend to stay much the same until provoked by some outside event, when they all tend to change more or less together. "Turnover" implies change; and "pulse" implies a distinct period of alteration imposed on a background that is generally quiet. She also suggests that the outside event that usually produces such general alteration is a global change of climate. Thus, fossil and geological evidence suggests that the whole world cooled dramatically about 2.6 million years ago. The bovids of Africa changed accordingly; in particular, the gazelles appeared. At about that time, too, the genus *Homo* made its first tentative appearance.

I would also like to add a suggestion of my own, which is that plants may subvert these climatic pulses, and this is of prime importance since plants ultimately provide the food and much of the shelter for animals, and largely determine what kinds of animals can live. Of course, plants are equally subject to climate. We see this clearly as the world cooled during the Cenozoic, and Eocene forest gave way to Miocene grassland (with occasional resurgence of forest in periods of temporary warming). But plants also have a significant tendency to do their own thing whatever the climate is doing. In particular, it seems that the flowering plants, the angiosperms, first appeared in the Jurassic, and became widespread and varied during the Cenozoic. Unlike the rise of the mammals, this radiation of angiosperms did not apparently depend on the loss of some previous life form and did not follow any discernible change of climate. The flowering plants simply appeared, and having appeared, they continued to spread. As they did so, they created a quite new kind of environment: in particular, a canopy of angiospermous trees that both in their chemistry and in their geometry (the shape and layout of the branches and leaves) were different from anything that had gone before. It was in this new environment, cut off from the essentially terrestrial world of the dinosaurs, that our own an-

*Elisabeth Vrba's idea of turnover pulse derives in large part from the general notion of punctuated equilibrium as first made explicit by Stephen Gould and Niles Eldredge. The speculations concerning the influence of population size on the rate of evolutionary change are my own (which means that Dr. Vrba should not be blamed for them!).

cestors developed: the first, essentially arboreal primates. The primates, in fact, are one of the few groups of modern mammals (the others being the edentates and the marsupials) whose ancestors must have existed in recognizable form long before the dinosaurs disappeared. I suggest that the rise of the angiospermous trees, occurring independently of global climatic events, allowed this to happen.

These, then, are the ideas of Darwin as modified and enriched by later thinkers. I would like to suggest, however, that Lamarck, whom Darwin superseded, does have something important to say to us, though it is not the thing for which he is generally remembered.

LAMARCK REVISITED

I mentioned earlier that Darwin toyed for a time with the hereditary notions of Jean Baptiste Lamarck, and was then persuaded to reject them. That has been the fate of Lamarck: to be rejected, and indeed often ridiculed. But although Lamarck said some things that were wrong and should be cast aside, he also said other things which, I believe, are of profound significance but are generally overlooked. The babies have been thrown out with the bathwater.

Thus, in 1809—he really was a pioneer—Lamarck published his *Zoological Philosophy* in which he not only promulgated the idea that creatures evolve, but also proposed four laws of evolutionary change. The fourth is the law of "inheritance of acquired characteristics." It proposes that creatures inherit the features developed by their parents during life. So, for example, a blacksmith who developed big biceps would produce children with big biceps, and a giraffe that stretched to reach the uppermost branches would produce offspring with long necks. This is the law for which Lamarck is remembered and is, unfortunately, wrong. Such a mechanism could not operate unless information from the muscle cells could somehow be transmitted to the sperm or eggs; and in the 1880s the German biologist August Weismann (1834–1914) pointed out that there is no such transmission. The germ (reproductive) cells and the somatic (body) cells developed independently. Ergo, "inheritance of acquired characteristics" was a nonstarter. Gregor Mendel's ideas then delivered the *coup de grâce*.

But another of Lamarck's laws—the second—has much to commend it. This proposes that "the production of a new organ in an animal body results from a new need that continues to make itself felt." Thus, most modern biologists accept, as Darwin said, that giraffes acquired their long necks by natural selection. That is, ancestral giraffes had necks of varying length, but the ones that survived the best and produced the most offspring were the ones with the longest necks. There is no reason to suppose that the parents' straining necks influenced the gametes. As Weismann said, this cannot happen.

But why did natural selection favor long necks in giraffes? Why not in hippos? Because—to put the matter anthropomorphically—giraffes had already decided to earn their living by feeding from the tops of trees. So they are indeed shaped by natural selection, just as Darwin said. But natural selection would not have shaped them in the way it did unless they were already doing the things that encouraged that particular body form.

This idea seems to me of crucial importance in understanding evolution. For example, lineage after lineage has shown its ability to shift dramatically from one way of life to another, and back again. Thus it is that all the lines of herbivorous megareptiles in the Mesozoic evolved from carnivorous ancestors. It is clear, too (as discussed in the next chapter), that the modern ungulates, which are the hoofed animals and generally herbivorous, shared a common ancestor with the carnivores: in the Paleocene there were creatures with the potential to go either way. *Andrewsarchus* is one such. It was the biggest terrestrial meat-eating mammal of all time (about four meters long) and had jaws like a crocodile, but it also had hooves.

A look at some modern animals shows how this kind of shift might come about. Thus, bears are basically meat-eaters, within the mammalian order Carnivora. But in practice all bears are somewhat omnivorous, and even polar bears will eat some vegetation if given a chance. At least two bears, the giant panda (which clearly is a bear) and the spectacled bear of South America, have evolved into virtual vegetarians. Indeed, giant pandas subsist almost entirely on bamboo, although they clearly remember their carnivorous roots and can be lured into traps by the smell of roast pork.

On the other hand, many apparently committed herbivores are more omnivorous than they look. Recently, in a nature reserve, red deer acquired notoriety by biting the heads off ground-nesting birds. Apparently, the deer lacked calcium, and most animals display a remarkable ability to identify what it is they lack and to seek it out. I was also told in India by excellent field biologists that elephants sometimes acquire a taste for flesh. "Doc" Krishnamurthy of the Bombay Natural History Society, a veterinarian acknowledged as India's greatest hands-on authority on Asian elephants, tells of an Asian elephant bull he once tried to rescue from a river as it munched contentedly upon a decaying antelope while the waters rose fatally around it. There are well-attested tales, too, of man-killing elephants occasionally becoming man-eaters. One bull reputedly returned to the shallow grave of his victim and dug her up, to continue his interrupted meal.

Thus, we see that many animals are behaviorally versatile, which implies that the same animal is capable of acting many different roles. Among them these roles provide the variety that Darwin perceived as the essential raw ingredient of evolution. Each of these roles is exposed to its own particular set of selective pressures; and if the conditions are right, then any one of the roles, after a few generations of selection, could evolve into a new kind of creature. Thus, if elephants that had a taste for flesh were marooned on some hypothetical island where there was more animal matter than vegetation (on which, for example, passing whales had a particular tendency to strand themselves), then natural selection might eventually transform them into committed scavengers. But the crucial point is that natural selection could not transform elephants into scavengers unless they had a taste for scavenging in the first place. This idea seems to me extremely important; and it derives from Lamarck. It conforms with his second law, which, when rephrased in the language of Darwin, states that creatures tend to put themselves in a position in which natural selection can start to work on them.

This principle seems to me of profound significance in the evolution of human beings. For we are supremely versatile. Our early ancestors in Africa offered to the anvil of natural selection a whole range of possibilities. If our ancestors had been set in their ways like, say, duck-billed platypuses, then they simply would not have provided natural selection with interesting raw materials, and we would still be duck-billed platypuses. Thus, we, like all animals, have to some extent shaped our own evolution. We have not done this consciously. But like giraffes, we have put ourselves in a position in which natural selection could shape us in particular ways.

A SUMMARY: NATURAL SELECTION AND TIME AND CHANCE

So that, briefly, is the way the game of life has been played. Charles Darwin first outlined the rules in a plausible form; and later authors, from Gregor Mendel to the modern ecologists and paleontologists, have added refinements. It is clear now that the evolutionary machine is not a smoothly oiled and self-motivating engine. Still less does it run on any metalled track, toward any stated goal. I will argue in chapter 9 that evolution does, in fact, "progress," which is an extremely unfashionable idea. But "progress" is quite different from "destiny"; the progress is not in any direction that can be foretold.

Overall, indeed, the evolutionary machine is a ramshackle and patchwork affair. Left to itself it is liable simply to shudder to a halt. All the creatures on board would be quite content to stay the same. In general, it lurches into life again only when provoked. But fortunately (if we think evolutionary change is a good thing), the world and the universe as a whole will always provide provocation sooner or later: a change of climate, collision with some formerly alienated continent, or the importunate arrival for whatever reason of some novel suite of animals and plants.

During these periods of change many creatures are lost. For one reason or another they simply lack the wherewithal for suitable adaptation. They may lack the necessary genetic variation, perhaps because there simply are too few of them to embrace the necessary variety. They may lack the versatility of behavior that would enable them to explore new ways of living and put themselves in a position in which natural selection can act upon them. Or the new pressures may simply be too great or too rapidly applied, and snuff them out before they have a chance to change.

But some of the creatures in these times of change will respond by altering in ways that increase their adaptedness—their fitness. These will be the ones that have the right genes and the potential for suitable behavior and are in the right place at the right time.

Overall, there are plenty of conditional clauses. Evolution by natural selection clearly can produce a huge variety of creatures wonderfully adapted to an enormous variety of habitats. The evidence is all around us. But it works to no plan; it just follows its nose, does what the conditions at the time require, and is obliged to build only on what was there before. Time and chance play a huge part. But for a change of climate 65 million years ago, the dinosaurs would be with us yet, and they well may have remained, still as recognizable dinosaurs, until the end of the Earth. Our ancestors, the primates, did appear long before the dinosaurs had gone; but if the dinosaurs had survived, they could forever have been confined to the trees.

So those are the rules of the games at the global party. It is time to introduce our fellow guests.

CHAPTER 4

FELLOW CREATURES

At this moment we could be sharing this planet with 30 million other species. That is a guess, for less than two million have been named, but preliminary studies suggest that there are about 30 species for every 1 that is known, living in the canopies and roots of tropical forests. If all 30 million had been described, and you recited their names at speed—a second for each—then it would take you a year to get through them all, assuming you did not stop to eat or sleep.

Yet all we see today is a soupçon of what has been. For every present-day species, there have been thousands in the past. The mammals of today are a shadow of former glories, even though they include some of the most beautiful and extraordinary creatures that have ever lived. Thus, the fossil record suggests that throughout the Cenozoic each species of mammal has lasted for an average of around a million years before going extinct or evolving into something else. Since the Cenozoic has lasted 65 million years, then commonsensical but slightly esoteric math suggests that for every species that exists at present, there must have been 30 or more in the past. Add the mammals of the Mesozoic, which shared the planet with the dinosaurs, and we may reasonably guess that for each mammalian species of today there have been at least 50 in times gone by. Now there are about 4,300 mammalian species on Earth, but the total that have lived since the group began can hardly be less than 200,000—not as many, but in the same league, as the known inventory of present-day beetles.

So today there are 5 species of rhinoceros, while the fossil record reveals at least 200, in three whole families where today there is only one; and whereas the present-day rhinos are all tanklike and much of a muchness, some of the ancient rhinos ran like small ponies, some were more hippolike than the hippos, and one, *Paraceratherium,* which inhabited an area from Pakistan to Central Europe in the Oligocene, was built like a massive giraffe, stood six meters at the shoulder, and was the biggest land mammal that has ever lived, dwarfing the biggest elephants. Today there are 2 species of elephant, but over time there have been at least 150. Today there are 4 species of hyena, but almost 70 are known from the past. Today there are 200 known species of primate—apes, monkeys, and prosimians such as lemurs and tarsiers—but the total estimated from the past is around 6,500. Today there is only one species of hominid—us—but since our family first appeared in the late Miocene, there have been at least 20.

A NOTE ON NAMES AND CLASSIFICATION

The concept of species is fundamental in all of biology. In general, it refers to any group of creatures that can breed together successfully, to produce fully viable offspring—that is, offspring that are just as fit and fertile as their parents. Modern horses obviously belong to a different species from domestic dogs because the two cannot possibly interbreed. When horses and asses mate, they can produce hybrid offspring, but they still count as separate species because the hybrids, known as mules, are sexually infertile and hence are not fully viable.

When animals are extinct, and known only from their fossils, it is of course impossible to judge directly whether any two types could or could not have mated successfully together. Then the paleontologist just has to define the species more or less exclusively on the basis of appearance. This is generally acceptable, but there is obvious scope for discrepancy, since we know from living examples that some true species can vary greatly in anatomy, while some groups of creatures look very similar but in fact do not breed together and so form different species. For example, some different species of living gazelles, bush babies, possums, mice, and owls are almost impossible to tell apart by eye, and yet do not interbreed.

Groups of species that are obviously very similar but are nonetheless distinct are grouped together in the same genus—plural, genera. Thus, all living horses, zebras, and donkeys are placed together in the genus *Equus*. By the convention devised by the eighteenth-century Swedish biologist Carolus Linnaeus (1707–1778), each creature is given two names: the generic followed by the specific, which is the name of its particular species. Thus, the modern domestic horse is *Equus caballus*, the

continued next page

continued from previous page

African ass (which is the one that is commonly domesticated) is *Equus africanus,* the plains zebra is *Equus burchelli,* and so on. Formal, scientific names are often called Latin names although the Latin is often made up and many of the roots are Greek.

Three conventions are worth noting. First, when the Latin name is presented formally, it is always written in italics, with the generic name in upper case and the specific in lower case. Thus, *burchelli* is spelt with a small *b* even though it derives from somebody's name. Second, when it is clear what genus is being talked about, the generic name can be presented simply as the initial letter: as in *Equus caballus* and *E. africanus.* Third, zoologists sometimes find it convenient to subdivide species into races, otherwise known as subspecies. Then a third name is added, as in *Equus burchelli burchelli.*

Different but related genera are grouped together in families; families are combined in orders; orders are combined to form classes, classes are grouped to form phyla (singular, phylum); and phyla are grouped to form kingdoms. Thus, the genus *Equus* and various related types (such as the three-toed *Hipparion*) are grouped in the family Equidae; Equidae is linked with the rhinoceroses and tapirs (and others) in the order Perissodactyla; the perissodactyls are grouped with other warm-blooded, furry creatures in the class Mammalia; the mammals are joined with reptiles, fish, etc., in the phylum Chordata; and the chordates are combined with worms, snails, etc., in the grand kingdom Animalia, the animals. Any one of these categories can also be further subdivided or grouped together to give, for example, infraclass or superorder or suborder. Thus, the Perissodactyla are combined with the Artiodactyla (the cattle, deer, and antelopes) and others in the superorder Ungulata—the hoofed mammals. Finally, it has sometimes proved necessary to invent even more subcategories, so that the grouping "cohort" may be interposed between infraclass and superorder, while families may be divided first into subfamilies and then into tribes, before further subdivision into genera.

Some more conventions are worth noting. First, the higher-order groups (family, order, etc.) are always written formally in Roman type, but beginning with a capital letter. But they can be referred to informally in lower case, as in "chordates" or "mammals." The names of plant families tend to end in "-aceae" (as in Chenopodiaceae, the family of Good King Henry) or "-ae" (as in Labiatae, the mint family) or "-eae" (as in Gramineae, the grasses). Botanists then refer informally to the

continued next page

continued from previous page

families in various ways, as in "chenopods," "labiates," or just plain "grasses." But the names of animal families always end in "-idae." Zoologists therefore refer informally to members of a particular family just by shortening the name. So "equids" obviously refers to members of the Equidae; "pongids" are the Pongidae—the family of chimpanzees, gorillas, and orangutans; and "hominids" are members of the human family, the Hominidae.

In addition, members of a particular genus can be referred to informally by adding "-ine." So *Homo sapiens* (us) and *Homo neanderthalensis* (the Neanderthals) are both "hominines." The extinct genus *Australopithecus* is almost certainly the ancestor of *Homo,* and is also placed in the Hominidae and is therefore also a hominid. But members of the genus *Australopithecus* are not hominines. They are australopithecines.

One final point. Modern biologists seek to devise a "natural" system of classification. That is, they seek to place different creatures in the same group if, and only if, they can all be presumed to have evolved from a common ancestor. A group that contains such an ancestor plus all of its descendants is called a clade. Thus, the class Aves is a clade because it contains all the birds that have ever lived, including the presumed ancestor of them all, which was *Archaeopteryx.* But the class Reptilia is not, strictly speaking, a clade, because the reptiles gave rise both to the birds and to the mammals. To qualify as a clade, Reptilia would have to be taken to include Aves and Mammalia. This would mean that you, and me, and sparrows, as well as snakes and dinosaurs, are all reptiles. But in practice most biologists recognize the need to compromise, and retain the term "Reptilia" even though it is not properly a clade. The integrity of the group is preserved by introducing the concept of "grade"—that is, an animal is placed in the Reptilia if it is a descendant of the first reptile, and if it also retains obviously reptilian features, such as a scaly skin.

Not just species, but entire groups of mammals have disappeared: entire orders of tanklike creatures from the Eocene, like the relatives of the horse, the uintatheres; or the two unique orders of ungulates from South America, the litopterns and the notoungulates, which between them emulated most of the forms of the ungulates on other continents; or the creodonts, which were the first large specialist mammalian carnivores, before the true carnivores—the cats, dogs, bears, and their ilk—got under way. Orders that now are sadly reduced, like the edentates, which now manifest as small tree sloths and armadillos, once were glorious, with the immensely successful giant sloths and tortoiselike glyptodonts bestriding both Americas. Other modern groups that enjoyed much better times in the past include the Perissodactyla (now represented by rhinos, horses, and tapirs); the camel family, which once included the long-gone oreodonts that for a time were the most numerous large animals of North America, plus the bizarrely horned deerlike protoceratids; the pronghorn antelopes, or antilocaprids, now down to one species but once a superb and varied family—although they have fared better than the protoceratids, which are gone altogether; and the marsupials, now known best for the kangaroos and koala, which among others once included bearlike forms in South America and an array of sabertooths resembling the sabertoothed cats, plus the hornless but rhinolike diprotodonts of Australia.

Neither were these creatures of the past content to do the things that their modern relatives do today. Animals are obliged to obey the laws of physics, and those that live on land are especially prone to gravity, so they cannot be any old size or shape they choose. Pigs will never fly, and although I have heard of an elephant in Sri Lanka who allegedly jumps, elephants will never soar like a gazelle unless they undergo some vigorous evolution and reemerge in a quite different guise. In practice, indeed, animals on land conform to a rather small number of basic engineering designs that work; and the kinds that work are called ecomorphs.

A quick or even a leisurely survey of present-day mammals suggests that each lineage has more or less commandeered its own particular ecomorph, as if it was somehow destined to take that form and no other. Thus, the only existing practitioners of the giraffelike ecomorph, with Eiffel Tower necks for browsing from the heights of acacias, are the giraffes themselves.

But giraffes did not invent the giraffelike ecomorph. Among the very first in fact was the giant rhino, *Paraceratherium*. The giraffe family (Giraffidae) arose from deer (Cervidae), and while *Paraceratherium* was being a giraffe, the giraffids themselves resembled moose, which is the largest of the living deer. The modern okapi shows the general form, though it lacks the bulk and the candelabra antlers of some earlier giraffids. Others who have played at being giraffes include several longnecked camels and at least one of the South American litopterns. In general, a giraffe is a reasonable thing to be and there is no reason why giraffes should have a monopoly on that way of being. By the same token, the hyenas that now specialize in crushing bones (apart from the perversely peg-toothed aardwolf, which contrives to subsist on ter-mites in South Africa) have in their time aspired to be chasers and har-riers—like dogs. So, too, on occasion, have bears. Dogs in their turn have at times played at bone-crushing, like hyenas; and perhaps it was a bone-crushing dog that kept the hyenas out of North America when the opening of Beringia apparently gave them a chance to spread themselves.

In short, the notion that dogs are meant to be like dogs and giraffes were destined to be giraffes is giving way to the realization that any-thing can be anything, or, more precisely, that most big lineages at some time have explored a wide array of ecomorphs and that most eco-morphs have at different times been essayed by a wide array of lin-eages. The arrangement we see now—giraffes with an exclusive claim to steeple necks and hyenas as the world's specialist bone-crushers—is just the way things happen to be at the moment.

It is the case, too, that modern-day teachers of biology and publish-ers of children's books are somewhat obsessed with zoogeography. Zoos at this moment are being redesigned to present "The Asian Cen-ter" or "The African Experience." Well, there is something to be said for zoogeography. The idea of the animal and plant community is not entirely fatuous because many creatures have coevolved—have been shaped by each other's presence—and there are many interdependen-cies. Zoogeography has an aesthetic element, too: antelope, lions, and cheetahs give the African savannah its character while feral camels and water buffalo simply look out of place in Australia, which really has been the land of marsupials (and honey-eaters and goannas and banksias and grevilleas) for 40 million years.

But again, a preoccupation with zoogeography can be misleading. Elephants now are creatures of the tropics, but their relatives have lived almost from pole to pole—from islands north of Siberia to the south of South America. Marsupials today are represented in North America only by the opossums that came in from South America dur-ing the Great Interchange. But marsupials evidently arose in North America about 120 million years ago, when the dinosaurs still had sev-eral chapters to run. The rhinos now seem inveterately African, but they came much later to Africa than to other continents. They, too, arose in North America. Indeed, the large mammals of North America now consist largely of Eurasian immigrants who arrived recently via Beringia, and yet North America is the birthplace of many a fine group who have long since left: the perissodactyls as a whole and the camelids, as well as the marsupials. Elephants may well have arisen in

Africa, but the modern family of elephants, the Elephantidae, apparently arose in Asia. But the Elephantidae spread all over the world (not least in the form of mammoths) and the genus to which the modern Asian elephant belongs, *Elephas,* arose from a group that had migrated back to Africa.

Tapirs today are found only in South America and Southeast Asia, while lions are found in Africa and in one tiny pocket of northwest India. How come? Did God plant them there, like a whimsical gardener? Not at all. Tapirs and lions (and battalions of other creatures, too) have in their time spread throughout Eurasia (and North America), and the populations we see now are remnants. Tapirs left Europe and North America a very long time ago, but lions lived in Europe, Asia Minor, and deep into China in historical times. How else would Androcles have pulled a thorn from a lion's paw, or David have killed the lion that attacked his sheep? Perhaps those stories are not literally true (though there is no good reason to doubt them), but they are based on a fact: the fact that lions are just as able to be European as they are to be African.

The generalization emerges that zoogeography expresses no deep principle but simply reflects the way things happen to be at the moment. Most of the groups of larger animals have in their time been everywhere (except of course Australia, which so far has guarded its island status). They have been motivated to move by changing climate and allowed to move by seas that conveniently drained from time to time as the Mediterranean has done and by land bridges that have obligingly appeared to straddle oceans, like Beringia and the Isthmus of Panama. Their peregrinations stop only when they run out of habitat, as the shrews that fled through Europe in the ice ages were driven south through Spain, Italy, and Greece and were finally snuffed out when there was nowhere further to run. We think that animals belong where we happen to have found them, but usually they are just passing through.

Add the dimension of time, then, and we no longer see a landscape of creatures with each one doing its allotted thing in its allotted space. Lineages like pronghorns and elephants that now contain just one or two creatures can be seen to include entire suites. While some animals like platypuses may be stuck in ecological and genetic grooves, most of the more vigorous types are trying on each others' hats, being giraffes or hippos or sabertooths where today they are content to be rhinos or phalangers. And while these suites of creatures are swapping roles and reclaiming them again, they trek the entire surface of the globe once the opportunities arise: cascades of animals migrating, radiating, moving on or drifting back from whence they came. Over just a few centuries they may flit across a continent as if it were a city park, and in practice they have had millions of years to play with.

I intend to look at "us" in the next chapter and for the rest of this book. But for now I want to look further at the kind of world our family joined in the Miocene—and our genus entered in the Pliocene and began seriously to influence in the Pleistocene. Before introducing our fellow players in more detail, it is worth looking more closely at the concept of the ecomorph. It helps us to see what our fellow creatures are actually trying to do—and why we are different.

THE IDEA OF THE ECOMORPH

For animals that live in the sea, the physical problems of life have to do with the properties of water: with its mass, its viscosity, and its adhesiveness, which is felt in particular by very small animals. But aquatic creatures are spared the full rigours of gravity. Thus, unless they choose the life of a swimmer—in which case they must generally be torpedo shaped—they can take almost any form, and so it is that the free-floating creatures of the plankton can be wonderfully bizarre, with waving feathery gills and spicules and yolk sacs stuck on wherever they happen to fit; or so it may seem to the land-bound naturalist. But on land, the constant reality of gravity ensures that animals must conform to the rules of civil engineering; and the bigger and heavier they become, the more critical are the parameters. After all, anyone can make a garden shed. But only a master craftsman can build a cathedral that will stay standing. For homoiothermic land animals, too, the physics of heat comes into play. Thus, gravity and thermodynamics between them determine that, on land, body size, shape, and lifestyle are bound to be intimately linked.

So it is that each ecological niche tends to require creatures of a particular body form—a particular ecomorph, and every lineage that aspires to exploit that niche must adopt that form. It is because there have been many more lineages than there are feasible ecomorphs that convergence has been such a powerful theme in evolution: like it or not, all creatures that aspire to live in a particular way find themselves subject to particular physical laws that must be respected and that largely determine what form they take.

In broad terms—the broadest possible—some animals have elected to survive by being big, and some by being small. "Big" and "small" seem purely relative terms, but because the laws of physics are as they are, these parameters in fact impose some absolutes. Gravity alone would ensure, for example, that a creature on land that weighed much more than 100 tons could not move—except perhaps as some slithering mass. In practice, then, land mammals have rarely been smaller than present-day pygmy shrews and only a few giraffelike rhinos have been bigger than today's elephants. Within this range, though, what are the advantages and drawbacks of being any particular size?

THE SNAGS AND BONUSES OF BIG AND SMALL

On the face of things, it pays to be big. In general, large size means greater strength and, broadly speaking, more speed over the ground for creatures that need to run. Big predators can in theory catch a greater range of prey, while big prey can outface a greater range of predators. Big males are more likely to defeat their rivals and win more mates. And, for thermodynamic reasons that are discussed below, big homoiothermic animals have a much lower metabolic rate than small homoiotherms, and weight for weight they need much less food than small creatures do or they can make do on food of much lower quality.

Yet there are snags. Large size may be a bonus in winning mates, but it takes a long time to reach a large size; so, for example, male natterjack toads in Britain leave more offspring if they breed at two years old when they are small than they do if they wait another year until they are big, and risk dying along the way. Big homoiotherms do need less food than small ones weight for weight, but in absolute terms they obviously need far more. Elephants eat thousands of times more than mice. Big carnivores can in theory catch a greater range of prey, but in practice it does not pay them to catch anything too small—so a tiger, like a weasel, generally finishes up catching things that are roughly its own size. In general, a whole range of niches that are open to small animals, from the canopy of forest to life underground, are closed to big ones. Worst of all, perhaps, in the long run is that within any given area populations of big animals are bound to be smaller than those of small ones, and this makes them vulnerable to extinction.

The advantages and disadvantages of smallness are simply the converse of the above. The biggest advantage for the small animal is the number of niches available: the bark of trees, a crack between the rocks. A single haystack can contain a viable population of mice while viable numbers of elephants need a large county. The chief disadvantage for the small animal is its extremely high metabolic rate. Small homoiothermic animals like shrews must eat every few hours or die, and all small animals need food of high quality and concentration—either insects or seeds, which is plant food at its most concentrated.

Having chosen their size, however, big or small animals find themselves subject to physical laws that shape their bodies and their lives in particular ways.

LENGTH, VOLUME, AND SURFACE

Size is a matter of geometry and in particular of the relationship between linear dimensions, volume, and surface area. If you double the linear dimensions of a flat plane, you increase its area four times. If you double the linear dimensions of a solid body (length, width, and height), you increase its volume—and therefore its mass—eight times. More generally, surface area increases in proportion to the square of the linear dimension, so a threefold increase in height and width increases area nine times; and volume increases in relation to the cube of linear dimensions, so a threefold increase in linear dimension gives a twenty-seven-fold increase in volume and hence in mass. Thus it is that a fish one foot long may weigh a pound, which is big enough when tailed and gutted to give the angler a pleasant supper. But a fish of the same kind that is two feet long weighs eight pounds and can feed the family for several days. A three-footer is a monster.

Because fish are supported by water, they can increase their length with impunity without altering their body shape. Water can support an eight-pound body as easily as a one-pound one—or indeed can support a whale of 100 tons. But animals on land have no such freedom. When an animal gets above a certain size—above that of a rabbit, say—the multiplication of volume and mass that accompany even a modest increase in height and length begin to make themselves felt. Body forms that serve small animals perfectly are unacceptable in big ones.

Therefore, small mammals in general, whatever their genetic provenance, conform to the "generalized mammalian form." This is seen nowadays in the weasel and the rat: long-bodied, short-legged. The notable point, though, is not the shortness of the legs but their attitude. Small mammals hold their legs permanently bent, a position that gives them permanent spring, the constantly coiled potential to twist, turn, and take flight on the instant. High-class squash players and young boxers similarly stand with bent legs for the same reason. Aging pugilists stand straight-legged, hoping to make up in ring-craft what they no longer possess in surplus strength.

Big animals, with their volumetric increase in weight, are obliged like aging boxers to stand straight-legged. Above the size, say, of a coypu, land mammals have legs like stilts. As long as they remain middle-sized, then the stiltlike legs can be relatively thin, as in a gazelle. But as linear dimensions increase further, the proportions must change. Doubling height and length increases body weight eightfold, but it increases the cross-sectional area of the limbs only fourfold. Tripling height and weight gives a twenty-seven-fold increase in weight but only a ninefold increase in limb thickness. But limb thickness largely determines limb strength. Hence, as linear dimensions and body mass increase, the limbs must thicken disproportionately to keep up with the body mass. So gazelles may indeed have needlelike legs, but the related eland has legs like a bull, while elephants have legs like Doric pillars. Note, incidentally, that coypus have bent legs, while domestic cats, which are no bigger than coypus, have legs like stilts. But then, evolutionarily speaking, coypus are big mice, while cats are miniature lions.

From such general body forms further refinements follow. Very big animals must also have big heads—at least if they are homoiotherms, and need to eat a lot. A big head means big weight: the head of a white rhinoceros weighs as much as an entire Jersey cow. But again, while head weight increases relative to the cube of the linear dimension, the thickness and hence the strength of the neck increases relative only to the square. So a big animal with a big head cannot afford the sinuous neck of the weasel, even if it wanted such a thing. The simple solution is to settle for a short neck, like that of a modern rhino or elephant.

But the short neck raises problems of feeding. If a rhino were tall, it could not reach the ground. Hence, huge mammals often have short pillar legs to go with their short bullish necks. This produces a general body form that might for the sake of brevity be compared to that of a tank; and as we will see, the modern rhinos are only one of many mammalian lineages that, over time, have adopted the tanklike ecomorph. The "small tank," with relatively thinner legs—effectively shaped like a Regency sideboard—is perhaps the commonest large mammalian ecomorph of all, seen in creatures as diverse as pigs, tapirs, oreodonts, and bears. When in doubt, natural selection has it, be a Regency sideboard.

Tanklike animals face problems of defense. Their feet must be planted foursquare to the ground, and their necks are short. Their heads, suspended at convenient predator height, seem terribly exposed. So it is that tanklike animals often have face armor. The rhinos we know, and other armed leviathans we will look at later. The unarmored state of the rhino-like marsupial diprotodonts must have left them particularly vulnerable. This might have hastened their demise after the arrival of the aborigines.

There are other ways out of tankish problems, however. One is to leave the neck short but extend the nose and upper lip to create a trunk—essentially a fifth limb—providing dexterity in an ad hoc manner after the rest of the body has been converted into a kind of gun carriage. With a proboscis, tankish animals may grow tall—and trunk and height together bring the canopy into reach. This is the approach of the proboscideans, which include the elephants; but the elephant ecomorph has also been adopted by several long-gone nonproboscideans from South America, and it is at least possible that the proboscideans reinvented trunks more than once. An ancient proboscidean known as *Moeritherium* was not nearly as tall as the later elephants but had the beginnings of a trunk. Modern tapirs show the same trend, and so do black rhinos with their prehensile upper lips.

Another solution for big mammals is to produce a long neck after all but to hold it more or less upright like the Eiffel Tower—or, more accurately, like a flag pole, supporting it with guys (tendons) firmly anchored to huge protrusions of the thoracic vertebrae. A long neck for a large mammal is clearly a big commitment, but the Eiffel Tower morph has been adopted nonetheless by many different creatures: camels, rhinos, and South American litopterns, as well as giraffes.

Many middle-sized and large herbivorous mammals, particularly those that live in open country, need to be able to cover large distances; and since such rewarding prey inevitably attracts its suite of predators, it often pays them to run at high speed. Hence, many have adopted the cursorial mode. The relevant physics here has to do with centrifugal forces and momentum. Animals have legs rather than wheels, and as they run, they need to reverse the direction in which the leg moves: power stroke backward, recovery stroke forward. If the feet and forelimbs were heavy, the reverse of swing would render them unstable, or would at least waste energy. In practice, then, in cursorial herbivores the bones that are farthest from the body—equivalent to human forearms, hands, and fingers—are elongated. The accompanying muscle is reduced to a minimum, and the muscles that move the limb, on what is equivalent to the human upper arm, thighs, and buttocks, are packed as closely to the body as possible. Thus, the weight, primarily that of the driving muscles, is as close to the body as possible. Again, the compact, light-limbed cursorial form has been reinvented many times: by deer, antelope, antilocaprids, the protoceratids, some ancient rhinos, and of course by horses who, with their mighty buttocks and ultrareduced feet, show the form to perfection.

Small animals, as already outlined, tend to be much of a muchness. The mammals of the Mesozoic must have been extremely varied genetically, which presumably is one reason why they were able to diversify so rapidly and dramatically once the dinosaurs left them to it; but in general shape they all looked much like modern tree shrews which, broadly speaking, resemble large mice (although they are quite unrelated to mice, and may be more closely related to primates than to shrews). Nonetheless, the huge strength-to-weight ratio of small animals gives them latitude for experiment that is denied to the bigger ones. Thus, a range of small mammals have rediscovered the ecomorph of the glider, culminating in the powered flight of the bats: and it may be that the microbats (like pipistrelles and vampires) are unrelated to the megabats (like fruit bats), which would mean that mammals invented powered flight more than once.

In general, if you want to be big, you really do have to make concessions, which in the case of rhinos include the need to keep all four feet on the ground most of the time and in horses include the loss of four toes from the primitive pentadactyl limb that mammals originally shared with all the other tetrapods, right back to the amphibians. Little animals by comparison are overengineered. Relative to their body weight they have strength to burn. Crucially, they have been free to retain the standard five-toed limb, for they did not need to evolve feet like pegs, or plinths that could withstand huge forces. And of course there was one particular group of small mammals that simply made a virtue of its generality, abandoning few or none of the primitive features but simply building upon them. These small generalized mammals bit by bit produced a form whose five-fingered limbs could grip, a creature that exploited its smallness to explore the trees, and particularly the angiospermous trees that came into prominence during the Cretaceous. These small generalists, which eschewed the dramatic solutions of the big and diverse ground dwellers, became the primates. Some of those primates did grow bigger later, and invented a range of quite new ecomorphs—including the brachiating apes (that is, creatures that progress by swinging arm over arm, as exemplified by modern gibbons). The brachiating apes in turn produced the only truly convincing bipeds that the mammals have yet provided, which is the family Hominidae.

But we will come to this. For now we should look at the other great ramification of body size: that of metabolic rate. As a generalization, it is supremely relevant. In the context of the steadily cooling Cenozoic, which in the Miocene caused huge areas of forest to give way to grassland, it is crucial.

METABOLIC RATE AND THE PARTICULAR RELEVANCE OF GRASS

The basic issue here is again the relationship between linear dimension, surface area, and volume. Homoiothermic animals like birds and mammals generate body warmth for the sake of it—they burn energy for no other purpose but to provide heat. Poikilotherms like fish and reptiles also generate body heat but only as a side product of muscular activity. That, at least, is a reasonable working generalization. In either case the amount of heat generated is related to the volume of the animal: more flesh can generate more heat. But the amount of heat lost is related to the surface area. Body volume increases to a much greater extent, relative to linear dimensions, than does surface area. So small animals have a relatively high ratio of surface area to body volume, and big animals have a very low ratio of surface area to body volume. It follows that little animals cool down much more easily than big ones. It is also the case, of course, that the center of a mouse is far nearer to the surface than the center of an elephant. All this, of course, everyone knows instinctively: tiny social cups of coffee cool down far quicker than big working mugs.

For small poikilotherms, the relatively high surface area can work to advantage. A lizard well placed in the sun can heat up in seconds. But for small homoiotherms, seeking to retain a body temperature above the ambient, the high surface area is a nightmare. They need to generate huge amounts of heat relative to their body weight. To do this, they need an extremely high metabolic rate. Indeed, the metabolic rate of a mouse is about a thousand times greater than that of an elephant. Dr. Michael Brambell, director of Chester Zoo in the north of England, calculates that if an elephant had the metabolic rate of a mouse you could fry an egg on its back.

Because small homoiotherms need to generate so much heat, they need to eat high-energy food. Thus, they eat insects or worms, as insec-tivores do, or seeds, as small rodents typically do. If they ate low-energy foods, they simply could not consume or process enough in a day to keep themselves going. It happens to be the case, however, that by far the most abundant food on land is the green leaf. It also happens to be the case that green leaves, which in general are watery and shot through with fiber, contain very little energy relative to their bulk.

Hence, in homoiothermic herbivores we see a very clear ecological cutoff between the small creatures and the large ones. The small ones cannot subsist on leaves, and the large ones can. The smallest specialist leaf-eating mammals are rabbits. The smallest specialist leaf-eating birds are grouse, which eat the young leaves of heather. Smaller mam-mals and birds will of course eat some leaves, but if they tried to subsist on them, they would starve.

The factor that for small homoiotherms is a negative becomes, for large homoiotherms, a tremendous positive. For creatures bigger than a rabbit have access to the greatest and most ubiquitous food source of all—the green leaf. We have already noted some of the benefits of being large. But for a homoiothermic herbivore, the ability to make use of the greatest of all terrestrial bonanzas is surely the greatest. The potential rewards for a large mammalian herbivore are huge; and of course, where there are large herbivores, there is likely to emerge a commensurate suite of large predators.

But subsistence on leaves is not straightforward. The point is not simply that the ratio of energy to bulk is low. Equally troublesome is that most of the energy in leaves is in the form of fiber, the most ubiq-uitous component of which is cellulose. But no animal is known to pro-duce enzymes that are able to digest cellulose. Rumors that snails can do so seem to have been exaggerated. In general, animals that aspire to digest cellulose and hence to release the energy that it contains must employ the services of bacteria and other microbes that live in their guts and do the digesting for them.

Thus it is that mammals that contrive to subsist on leaves fall into three broad categories. First, there is the giant panda, a formerly car-nivorous bear that has evolved to live mainly on bamboo but, per-versely, still has the small gut of a carnivore and retains no worthwhile colonies of gut microbes. Therefore it cannot digest cellulose and has to make do on the sugar in the bamboo leaves and has virtually no en-ergy to spare. Hence, a giant panda is among the most spectacularly lethargic of all mammals.

The second and third categories are the bona fide leaf-eaters that do retain microbes in their guts to break down cellulose. The more obvious strategy, adopted by a wide range of lineages, is to house these microbes in the hindgut: either in the colon or in a diverticulum at the start of the colon known as the cecum. Elephants are hindgut digesters. So too are rabbits, koalas, and all the perissodactyls, like rhinos and horses.

The last possibility is to retain the digestive microbes high up in the gut—in fact, in a modified stomach. Such foregut digestion is a harder trick to pull evolutionarily (not least because conditions in most animals' foreguts do not generally favor bacteria), but mammals have nonetheless achieved it several times. Kangaroos are foregut digesters, with stomachs modified to contain appropriate microbes. But the best-known and ecologically most influential foregut digesters are among the artiodactyls. The camels and their relatives use a form of foregut digestion. So, too, do the ruminants, which have evolved a stomach with four chambers, one of which, the rumen, is a huge fermenting vat; an in situ compost heap. Rumination has proved the most successful mechanism for digesting herbiage. Two families of ruminants in particular—the bovids (antelope, cattle, sheep, and goats) and the cervids (deer)—have become the most diverse and widespread large mammals on Earth.

In general, foregut digestion is somewhat more efficient than hindgut digestion. The basic reason is simple: in foregut digestion the nutrients that are released from the cellulose then have to pass through the small intestine, where there is plenty of opportunity for further digestion and absorption. Hindgut digesters have already bypassed the small intestine by the time the cellulose is broken down. Rabbits overcome this problem by eating their own droppings, and giving the semidigested vegetation a rerun. The droppings of horses are similarly nutrient-rich, but horses leave them for others to consume. Jonathan Swift was notoriously coprophobic and in *Gulliver's Travels* he elevated horses—"Houyhnhnms"—to the highest status precisely because their droppings are inoffensive. Compared with most, they have suffered little decay.

In general, too, as we have seen, big homoiothermic animals are more efficient than small ones: that is, they need less food per unit of body weight. Thus, it seems that although big hindgut digesters compete well enough with big foregut digesters, in the middle-sized range the foregut digesters have the edge. So in the modern world we find that the biggest mammalian herbivores of all are in fact hindgut digesters—like the elephants, rhinos, hippos, and horses. Yet I can think of no living ungulate hindgut digester that is smaller than a tapir, which even in its smallest wild forms is a big animal. But of course there is a huge range of small-to-middle-sized ruminants, including most of the antelope, sheep, goats, and deer. There is, however, a range of small but specialist hindgut digesters, including the koala, whose remarkable cecum and equally accommodating liver enable it to cope with the abundant but uncompromisingly vicious leaves of eucalyptus, shot through as they are with fiber and toxic oils.

Finally, and crucially, there is a hierarchy of digestibility among leaves that corresponds roughly to the difference between the leaves that grow on trees and shrubs on the one hand, and grass on the other. Animals that eat the former are browsers, while grass-eaters are grazers.

In general, in the wild, good browse is the most nutritious food; specialist browsers like giraffes do extremely well on the tender leaves of acacia. Wild grass on the other hand tends to be fiercely fibrous, and indeed contains spicules of silica that are like needles of glass. On the other hand, grass has become awfully abundant; and where it grows, it is easy to consume. Hence, there is a classic contrast between the persnickety gourmet browsers, like giraffes, and the mass gourmandizers of grass, epitomized by the barrel-bodied horses and the grazing ruminants, especially cattle, sheep, and a range of big antelopes such as the eland and the gnu.

As we saw in chapter 2, the world has cooled throughout the Cenozoic as the Himalayas rose and leached the carbon dioxide from the atmosphere. The cooling happened in jerks; and even when the cooling was steady, the vegetation responded in jerks as successive suites of plants were pushed beyond their thresholds of tolerance. The Miocene was the watershed, as the warm, wet forest of the early Cenozoic gave way, over huge and rapidly increasing areas, to cool, dry grassland. This was when the grazers truly took over from the browsers. This is when the ruminants of various sizes and the big grazing perissodactyls came into their own and the browsing forest generalists went into relative decline. Our own family, the Hominidae, entered this newly grass-covered world in the late Miocene, although it did take them several million years to graduate from forest, through open woodland, into truly open country.

Being a grazer on the ever more extensive open plain carries further implications. In such spaces it helps to be able to cover large areas, and to do so quickly, especially when harassed by the specialist predators evolving at the same time.

Hence the cursorial mode. Food is abundant, but the dangers are great, and both are factors that favor sociability. So the grazers of the plains typically roam in herds, while their forest relatives, like roe deer and duikers, are often solitary or move in small groups. Thus, there is a chain of consequence: global cooling leads to grass and open space, and grass and open space favor the sociable grazers which, in practice, are the bovids, deer, and the big cursorial perissodactyls, the horses. That is how the world is at the moment.

So here we have established the rules of the game: the nature of the world stage; and the ecomorphic roles that the animals play upon that stage. It is time to look briefly at the players themselves.

A LIGHTNING SKETCH OF OUR FELLOW MAMMALS

In the box on page 131 is a list of the main mammalian groups from the time of their origins in the Triassic, but in particular as they have unfolded since the Cretaceous. Earlier versions of this book now moldering in my files contain accounts of all of them; but even the broad-brushed versions, taken order by order, were almost as long as this entire book is supposed to be. There are just so many of them.

So the following account is highly selective. It omits all creatures who live in the sea, from seals to whales; and gives extremely short shrift to the rodents and bats even though they are of enormous ecological importance and between them include about half of all species of living mammals. In fact, it focuses upon the large terrestrial mammals that have shared our history and have helped to shape our evolution and our ecology. They are, in the main, the marsupials and xenarthrans (the edentates), the ungulates, the carnivores, and our own group, the primates.

But peruse the box if you would and note the broad relationships between the groups. It is not really surprising that the various ungulate orders should be distantly related to each other: that is, that the extinct meridiungulates from South America should have shared a common ancestor both with the perissodactyls (such as horses and rhinos) and with the artiodactyls (such as cattle, deer, and camels), or indeed with the proboscideans who nowadays manifest as elephants. The probable relationship between proboscideans and sirenians, who nowadays include dugongs and manatees, is somewhat more unexpected. It is positively intriguing that somewhere back in the Cretaceous all these various ungulate or quasi-ungulate groups apparently shared an ancestor with the cetaceans—the whales and porpoises. It is a real surprise to find, as anatomists and palaeontologists have done, that the ungulates and cetaceans in turn show many similarities with the carnivores. In the same way, our own group, the primates, show obvious similarities both with the insectivores and with the bats. The point, though, is the one discussed in chapter 3: that evolution is opportunist, and must begin each new radically different group—species, family, order—with some probably eccentric representative of whatever animals are there already. Thus, herbivore can give rise to carnivore and vice versa, and unless the organs (or even the underlying genes) are irretrievably lost, then hoof may become claw and claw may become hoof. Evolution ducks and weaves.

In the sketch of our fellow mammals that follows, I would ask you to note six points in particular. First, that there are broadly different ways of being a mammal. Cats, horses, and human beings are variations on only a single theme; but the platypuses, marsupials, and creatures like the armadillo demonstrate that creatures very different from ourselves also belong in the class of mammaldom.

Second, and most intriguing, see how every different lineage has striven in its time to do many different things, and explored many different ecomorphs, and that most ecomorphs have in their time been enacted by many different lineages. Many different creatures have been tanks, like the modern rhinos. Many have been giraffelike. Several quite unrelated groups have adopted the mode of the proboscidean. This radiation of lineages and swapping of roles is an important feature of what Elisabeth Vrba called the "dance through time."

Third, note the way that some conservative groups sometimes give rise to groups that become highly adventurous, and vice versa—that some groups that begin by being highly varied finally produce just one or a few lines that really take off. Thus, the conservative tapirs produced the adventurous rhinos, and our own prehominid ancestor produced both the conservative chimps and the highly variable hominids. But then, out of the adventurous rhinos only one group—the tanklike rhinoceratids—came through, and the only hominid to emerge from the array is ourselves.

The fourth striking theme is the change of fortunes—how some families or even entire orders have disappeared altogether, like the uintatheres and the chalicotheres, the oreodonts and the protoceratids, the creodonts and the nimravids, the diprotodonts and the borhyaenids, and all the vast array of South American meridiungulates.

Note, correspondingly, how many modern groups are sadly depleted from former days, such as the perissodactyls and the proboscideans. And note finally that you cannot tell at any one time who will succeed at later times. Those piggy relatives of the camels, the oreodonts, were the dominant herbivores in North America for tens of millions of years, and now they are gone completely. The bovids, on the other hand—antelope, cattle, sheep, and goats—have triumphed in the past few million years as, of course, has the genus *Homo*. But fortunes change. The point is salutary.

Then note how mobile the various creatures have been—that many groups have bestridden the entire globe (though usually missing Australia), and that where an animal lives today often bears no relationship at all to its place of origin or to its previous excursions. This restless globetrotting is the second aspect of the dance through time.

Finally, we may perceive the host of fine creatures who are now long gone yet survived until the emergence of *Homo,* and indeed of our own species, *Homo sapiens.* In later chapters I discuss whether this is simply coincidence or whether, in fact, our own ancestors pushed those animals into oblivion.

Let us begin our survey, then.

A SIMPLIFIED CLASSIFICATION OF THE MAMMALS

The following attempts to show the main groups of mammals mentioned in this book and how they are related to each other, while keeping polysyllabic names to a minimum. An asterisk indicates that the group is extinct. The animals in brackets after some of the formal names are merely *aides mémoires:* examples that are mentioned in the text.

Continued on next page

CLASS MAMMALIA

SUBCLASS PROTOTHERIA

infraorder	cohort	superorder	order	suborder	family
			Monotremata (duck-billed platypus)		

SUBCLASS THERIA

infraorder	cohort	superorder	order	suborder	family
Trituberculata*					
	Pantotheria* (ancestors both of metatherians and of eutherians)				
Metatheria					
		Marsupialia (the marsupials)			
			Didelphoidea		
				Didelphidae (American opossums)	
				Borhyaenidae* (borhyaenids)	
				Thylacosmilidae* (sabertoothed South American marsupials)	
			Diprotodonta		
				Thylacoleonidae* (sabertoothed "marsupial lion")	

infraorder	cohort	superorder	order	suborder	family (continued)
					Macropodidae (kangaroos)
					Diprotodontidae* (diprotodonts)
					Palorchestidae* (marsupial equivalent of giant sloths)
	Eutheria (commonly referred to as placental mammals)				
		Edentata (edentates)			
				Xenarthra (armadillos, sloths, anteaters, glyptodonts)	
				Pholidota (pangolin)	
		Epitheria (includes most living mammals)			
				Insectivora (insectivores: hedgehogs, shrews, moles)	
		Glires			
				Lagomorpha (rabbits, hares)	
				Rodentia (mice, rats, squirrels, porcupines, etc.)	
		Archonta			
			Dermoptera (flying lemurs, or colugos)		
			Chiroptera (bats)		
			Primates		
				Prosimii (primitive primates including lemurs and tarsiers)	
				Anthropoidea	

infraorder	cohort	superorder	order	suborder	family (continued)
Eutheria	Epitheria	Archonta	Primates	Anthropoidea	Cebidae (New World monkeys: woolly, spider, etc.)
					Callitrachidae (New World monkeys: tamarins and marmosets)
					Cercopithecidae (Old World monkeys: vervets, baboons, macaques, etc.)
					Hylobatidae (gibbons)
					Pongidae (chimps, gorillas, orangutans)
					Hominidae (humans, australopithecines)
		Ferae	Creodonta* (creodonts)		
			Carnivora (true carnivores)		
				Fissipeda: Miacoidea* (the earliest true carnivores)	
				Fissipeda: Feloidea (catlike carnivores)	
					Viverridae (civets)
					Hyaenidae (hyenas)
					Felidae (cats)
				Fissipeda: Canoidea (doglike carnivores)	

infraorder	cohort	superorder	order	suborder	family (continued)
					Mustelidae (weasels, otters, badgers)
					Canidae (dogs)
					Procyonidae (raccoons)
					Amphicyonidae* (bear-dogs)
					Ursidae (bears)
				Pinnipedia (seals, sea lions, walruses)	
		Ungulata (hoofed animals)			
			Arctocyonia* (very primitive ungulates: condylarths)		
			Dinocerata* (uintatheres)		
			Embrithopoda* (arsinoitheres)		
			Artiodactyla (even-toed ungulates)		
				Suina	
					Suidae (pigs)
					Tayassuidae (peccaries)
					Hippopotamidae (hippopotamuses)
				Tylopoda (tylopods)	
					Merycoidodontidae* (oreodonts)
					Protoceratidae* (protoceratids)

infraorder	cohort	superorder	order	suborder	family (continued)
Eutheria	Epitheria	Ungulata	Artiodactyla	Tylopoda	Camelidae (camels, llamas)
				Ruminantia (ruminants)	
					Moschidae (musk deer)
					Cervidae (deer)
					Giraffidae (giraffes)
					Antilocapridae (pronghorns)
					Bovidae (cattle, antelope, sheep, goats, goat antelopes)
				Acreodi* (Andrewsarchus)	
			Cetacea (whales and porpoises)		
			Perissodactyla (odd-toed ungulates)		
				Ceratomorpha	
					Rhinocerotidae (modern rhinos)
				Ancylopoda* (chalicotheres)	
				Hippomorpha	
					Palaeotheriidae* (palaeotheres)
					Equidae (horses)
					Brontotheriidae* (brontotheres)
			Hyracoidea (hyraxes)		

infraorder	cohort	superorder	order	suborder	family (continued)
			Meridiungulata* (South American ungulates)		
				Litopterna* (litopterns)	
				Notoungulata* (notoungulates)	
				Astrapotheria* (astrapotheres)	
				Pyrotheria* (pyrotheres)	
		Tethytheria			
			Proboscidea		
				Moeritherioidea* (Moeritherium)	
				Deinotherioidea* (deinotheres)	
				Elephantoidea	
					Gomphotheriidae* (gomphotheres)
					Mammutidae* (mastodonts)
					Elephantidae (true elephants, including mammoths)
			Sirenia (dugongs and manatees)		
			Desmostylia* (desmostylians)		

The above is mostly based on the classification shown in Dougal Dixon et al., The Macmillan Illustrated Encyclopedia of Dinosaurs and Prehistoric Animals (New York: American Museum of Natural History, 1988).

AN OVERVIEW OF THE MAMMALS

Note, first of all, that the great class of mammals is divided broadly into two—two quite different ways of being, a primary division that must have taken place in the Jurassic, when the dinosaurs were at their height. The groups that have come through, and dominate, are the Theria: the creatures that we think of as mammals, which nourish their unborn offspring in the womb via a placenta. The other group, the Prototheria, lay eggs. The only prototherians that remain today are the monotremes—and there are only six species of them: the duck-billed platypus of Australia and five echidnas from Australia and New Guinea. But the prototherians were ecologically significant in the distant past and more recently there lived a platypus as big as a sheep. Now the monotremes remind us that mammals did indeed evolve from reptiles and show what the halfway stage might have been like, for they suckle their young in true mammalian fashion, but the laying of eggs is genuinely reptilian.

The therians in their turn are divided into three groups of which one (the Trituberculata) is extinct and is simply the ancestor of the other two. The two surviving groups are the Metatheria, alias the marsupials, and the Eutheria, which contain all the rest—whales, bats, horses, dogs, rhinos, cats, us, lemurs, mice, armadillos, and what you will. These two great groups, which between them are what most people mean by "mammals," probably divided from each other in the early Cretaceous, perhaps as long as 120 million years ago. So we see again that the ecological Age of Reptiles was also the age in which the mammals laid out their own ground rules.

ALTERNATIVE MAMMALS: THE MARSUPIALS

Ecologically, these days, the marsupials are the lesser group, confined to Australia and New Guinea, South America, and with a few errant types (the opossums) in North America. They live in New Zealand (possums and wallabies) only as introductions, imported by Europeans. But the marsupials seem to have arisen in North America—or that, at least, is where the oldest fossils are—and spread by routes that are not altogether clear but presumably involved a trip through Eurasia into Gondwana, when Australia was still a part of it. There are fossil marsupials in Europe and Africa, and one bonus of the possible greenhouse effect will be to facilitate paleontology in Antarctica, which should reveal a great many more. It seems, though, that the principal home of present-day marsupials, Australia, is historically an outpost.

Marsupials have all too often been presented as also-rans. Nowadays many are taking a terrible bashing in Australia at the hands of introduced eutherians, like the cats and foxes that kill whole suites of the smaller ones or the water buffalo and sheep that transform entire landscapes. The South American types are supposed to have suffered at eutherian hands after the collision with North America in the Pliocene, which allowed the Great Interchange. But as we have seen, any kind of animal is liable to suffer when faced for the first time with a ready-made predator or competitor from another continent. In many instances marsupials do extremely well in competition with eutherians, as the opossums have done in North America, and as many Australian natives do today. Possums in Sydney and other cities are among the world's great feral animals, like the raccoons of North America and the foxes of urban Britain. I was once frightened half to death by a possum that leapt from a trash can outside the Sydney Opera House.

In fact, marsupials should be seen not as inferior mammals but as alternative ways of being a mammal. Like the eutherian mammals, they do nourish their young briefly in the womb via a placenta and then suckle them with milk after birth. The key strategic difference between the two groups lies in the emphasis that they give to each stage. For marsupials the phase in utero is cursory. The babies are born as virtual embryos, and the burden of infant nutrition is borne by lactation, for which purpose they are generally cosseted in pouches. Eutherians generally split placental and lactational feeding more evenly, and give birth to more advanced babies (although some, like those of bears, are remarkably tiny at birth). But in the harsh and unpredictable climate of Australia, the emphasis on lactation brings great advantages, for if drought sets in, then the joeys in the pouch simply die as the milk dries up and the mother can start again in better times; but eutherian embryos often cling to the mother as good parasites and sometimes, if the bad times roll during pregnancy, both the mother and the fetus perish. The marsupial strategy is harsh, but realistic.

Marsupials may also seem somewhat limited in form—that is, in their range of ecomorphs—compared to eutherians. But that, too, is not quite fair. There are no marsupial whales or even giraffes or elephants, and the gliders are not quite in the same league as the eutherian bats. But when looked at over time, the marsupials are far more various than is now apparent.

For example, present-day marsupials seem low on predators. But until the 1930s, Australia had a doglike marsupial: not the modern dingo, which is a true dog from Asia, but the thylacine, or Tasmanian wolf—*Thylacinus*. Thylacines are placed in the family Thylacinidae. At least five other families of marsupials have produced predators that between them have matched at least some of the eutherian otters, dogs, bears, and cats—so there is convergence both between marsupials and eutherians, and between different families of marsupial. South America was the home of the extremely varied family Borhyaenidae. One of them, *Borhyaena*, was bearlike and had flat, bearish feet, although, oddly, it also had a long, heavy tail. It presumably hunted by ambush. Another borhyaenid, *Cladosictis*, was otterlike: from the late Oligocene to the early Miocene it must have dashed through the undergrowth after small mammals and reptiles and plunged into the rivers after fish. The marsupial yapok, *Chironectes*, of the opossum family Didelphidae, still lives an otterlike life in the Andes.

Two different marsupial families produced creatures the size of leopards or even bigger that for all the world resembled saber-toothed cats, or at least did the same job. The more catlike were from the South American family Thylacosmilidae of which the type genus is *Thylacosmilus*, which lived from the late Miocene to the early Pliocene. *Thylacosmilus* had lost its incisors and, as in saber-toothed cats, its canine teeth formed long, bladelike, slashing teeth. Like the sabertooths, too, *Thylacosmilus* was robust and muscular, built for ambush and clinging. On its lower jaw it carried horny flanges that apparently protected the projecting sabers when not in use, like a sheath, and may also have served as permanent sharpeners. In contrast to the sabertooths, the canine teeth of *Thylacosmilus* grew continuously like the incisors of rodents (a fact that can be inferred from the structure of the roots, which remain open in teeth that continue to grow but is closed in those which, like the teeth of humans, reach their allotted size and then stop).

Thylacoleo of Australia, in the family Thylacoleonidae, shows a quite different approach to the saber-toothed genre. It, too, was leopardsized. But unlike *Thylacosmilus* or the cats, its sabers were formed from its incisor teeth.

Peculiarly, too, it had huge thumbs that bore commensurately enormous claws. Presumably, it used these to grip its prey—kangaroos? diprotodonts?—as it brought its killing incisors into play. The nearest living relatives of *Thylacoleo* are the modest phalangers. *Thylacoleo* itself died out only at the end of the Pleistocene, about 10,000 years ago. Perhaps it was helped to extinction by the first aborigines.

The marsupials have also invented an ecomorph: that of the

Macropodidae, or kangaroos, for although the eutherians include some small bipedal hoppers, there are no big ones. Macropods probably evolved their bounce in Australian tropical forests by negotiating the tangled undergrowth. Such forests are now largely confined to Queensland, but in the past, as the world's climate shifted and Australia changed latitude, they have raced in a body almost over the entire continent. "Raced" is hardly an exaggeration, for the great forests of Queensland have probably not been in the present location for more than 8,000 years (and neither, for that matter, has that ecological miracle the Great Barrier Reef; for it is a post–Ice Age phenomenon). Today the red and the great gray kangaroos of the genus *Macropus* are the biggest macropods, but there were giants in the past including *Procoptodon* of the Pleistocene, which stood three meters tall—an astonishing ten feet. *Procoptodon* had a short face like a rabbit but also extraordinarily large hind feet, each of which, as in ostriches, had only one functional toe (the fourth).

Until recently, too—the latest types may have been contemporaries of the first aborigines—Australia harbored a suite of giant wombatlike creatures (although they are not closely related to wombats) of which at least one, within the genus *Diprotodon,* had the size and form of a modern rhino. Australian children are brought up with the legend of the bunyip; a large but amiable and mournful beast that is supposed to roam the outback—the equivalent of the Tibetan's yeti and the Loch Ness monster of Scotland. But the bunyip may not be a fable at all. It may be a true folk memory of *Diprotodon,* transmitted generation to generation by the aborigines. *Diprotodon* died out only a few thousand years ago and folk-tales can last easily as long as that. Other marsupials have resembled mice and gerbils, many were like raccoons, and several marsupial families produced members that are or were very like European moles. The whole order today includes 282 species, but over time there can hardly have been fewer than 30,000, and probably many more. Among the many extraordinary creatures that remain are the koala, kangaroo, yapok, wombat, quoll, and opossum, but the world has clearly lost some magnificent representatives—including many in human times.

Marvelous though the marsupials are, however, it has to be admitted that the eutherians are even more various, although the present types are only an arbitrary sample of what has gone before. Note, first, from page 133, that the eutherians are further subdivided into two main groups: the epitherians, which include most of the creatures we think of as mammals—horses, bears, whales, and the rest—and the xenarthrans, or edentates. The edentates now are sadly reduced to tree sloths, armadillos, and anteaters. But, like the marsupials, they have had a much more glorious past than is apparent now, and, like the largest marsupials, the later ones encountered our own ancestors and suffered at their hands.

THE EDENTATES

The edentates are tremendously diverse. The tree sloths, armadillos, and anteaters are the living types, dumped together in the single order Xenarthra; while the pangolins have their own order, the Pholidota. "Xenarthra" means "foreign joint." All other mammals, from giraffes to pipistrelles, have a statutory seven vertebrae in the neck but the xenarthrans have one or more extra ones. No one knows whether the edentates are really a single lineage, a true clade, or whether they represent several different major subdivisions of mammals that have nothing much in common except their unlikeness to epitherians.

But it is clear that they are sadly faded compared to former glories. The xenarthrans evolved in South America, apparently in the Paleocene, and several of them successfully invaded the north when the Panama bridge was established in the Pliocene. The living edentates are all fairly small—none larger than an anteater—but there are two major extinct groups that included some giants. The first of these are the giant sloths, which first appear in South America in the Oligocene.

Giant sloths are also known as ground sloths, and artists commonly depict them standing on their hind feet like kangaroos and slashing at the vegetation. However, their hind legs are stumpy and bent and equipped with huge claws, and seem most unsuited to standing or walking. I suggest, therefore, that giant sloths were in fact arboreal, just like their extant relatives; and that they used their hind legs and claws as grappling hooks. The only reason for doubting this is that giant sloths were so large. But so are grizzly bears, which are excellent climbers, and there are plenty of enormous trees in the Americas. In fact, the traditional scenario that places giant sloths on the ground seems to me to make no sense at all.

Oddly, the diminutive tree sloths that are still with us today were apparently the ancestors of the giant sloths, and have outlived their more adventurous scions. Yet for 30 million years or so giant sloths were highly successful. They did well in North America after the north met the south. They abounded in Florida, their fossils are found well up into California, and they died out on both continents only after human beings arrived. They too might have been victims of human beings. Nearly fifty different genera are known, which means that over time there must have been scores of species, and as the epochs passed some types grew bigger and bigger. Biggest of all was *Megatherium* from the Pleistocene of Patagonia, Bolivia, and Peru. It was six meters (20 feet) long including the tail, which it presumably used as a prop while it fed. *Megatherium* must have weighed 3 tons, as much as a fair-sized Asian elephant.

Another giant, *Neomylodon*, survived almost until the present day, and samples of its skin can still be found. The skin, too, is uniquely edentate, for beneath the shaggy hair it is studded with plaques of bone. So as if they were not formidable enough, with their massive arms and slashing claws, giant sloths also wore a kind of chain mail.

Indeed, edentates more than any other mammal have had a taste for armor. The armadillos still have, of course, though none today can compare with some of their giant forebears from the Pleistocene. Most armored of all, however, were armadillo relatives known as the glyptodonts, which flourished in South America from the early Miocene, about 20 million years ago. Like the giant sloths, the glyptodonts are known to have produced some fifty genera.

The glyptodonts were the mammals' answer to the giant tortoise. Like giant tortoises, they probably lived in semidesert and so were obliged to graze in extremely open territory. Thus, they needed protection against predators for whom, otherwise, they were sitting ducks. After all, the only other strategies for desert herbivores are to run away like antelopes, or duck beneath the surface like ground squirrels, or hide and hope the danger will pass, like baby plovers; but for an armadillo relative, these are not options. Presumably—like modern edentates—glyptodonts would have allowed their body temperatures to fluctuate more than somewhat, which would have helped them in a poor, desert economy.

The biggest glyptodont of all was the type genus *Glyptodon*, which grazed on the plains of Argentina in the Pleistocene and was 1.5 meters (5 feet) tall and 3.3 meters (11 feet) long. Its huge globular "shell" was fashioned from a mosaic of bony plaques that suggest the surface of a golf ball. Glyptodonts also had a powerful armored tail, which at least in *Doedicurus* from Pleistocene Patagonia was whimsically tipped with a ball of spikes, as on a medieval fighting mace. The last glyptodonts died out only 15,000 years ago, and they feature in the legends of Patagonian Indians just as diprotodonts are apparently remembered in the folk history of Australian aborigines. So here are yet more creatures (to add to the diprotodonts, giant kangaroos, and ground sloths) that could be victims of our own ancestors.

But the bulk of modern mammals are epitherians. These are hugely diverse and the following account is egregiously curtailed. In fact, I can deal with only three main groups: the ungulates, the carnivores, and the primates.

THE UNGULATES

Three main groups of ungulates prevail today: the perissodactyls, which carry their weight on their middle toes like horses, rhinos, and tapirs; the artiodactyls—including the pigs, hippos, camels, and ruminants—which bear their weight equally on their third and fourth toes (and hence are "cloven-hoofed"); and the proboscideans, represented by the elephants. But over time there have been at least sixteen separate orders of hoofed mammals that thereby qualified as ungulates. All of them are vaguely related, yet they are distinct enough to be hived into three huge superorders (as shown starting on page 135). Those superorders also include some creatures that have lost their hooves or even their feet altogether—like the sirenians and the whales, and the long-gone, bottle-bodied, maritime desmostylians. The ungulates are also related, distantly but unmistakably, to the carnivores. Some of the earliest ancestral types, like the huge and irredeemably hideous *Andrewsarchus*, with its vast head and crocodile jaws, were probably meateaters—scavengers—and yet had hooves. *Andrewsarchus* has from time to time been shuffled from order to order.

Between them the ungulates have explored an enormous range of ecomorphs, most of which still have their representatives: rhinos show us what it is to be a tank; tapirs, pigs, cattle, and sheep are standard sideboards; many quite different living ungulates are cursorial; while giraffes, camels, and elephants demonstrate a further range of big, specialist forms. For good measure, the tiny hooved hyraxes, which in the Bible are called "cavies" and indeed resemble big guinea pigs, are ungulates, too, perhaps related to perissodactyls, as Richard Owen suggested in the nineteenth century, but perhaps more closely akin to elephants, as some have suggested since. Hyraxes were not always as small as they are now, for they had pig-sized relatives in the past that were far more obviously ungulate. The ungulates have also lost one complete morph: that of the chalicotheres, which were big perissodactyls with appropriately horselike heads, but they apparently stood on their hind legs like giant kangaroos and hacked at the vegetation with huge claws that were uniquely bifurcated, each one resembling a toasting fork. One of the South American ungulates, *Homalodotherium*, was also giant-sloth shaped. It seems odd that a form that was so clearly favored—it was adopted by at least three unrelated groups and was highly successful for millions of years—is no longer represented at all.

Overall, the living and extinct ungulates demonstrate beautifully the principal themes of this chapter: that any one lineage may adopt many morphs; that any one ecomorph may be explored by many different lineages, which arrive at the same conclusions via quite different genetic routes; that there are spectacular changes of fortune, so the triumphs of today may be the disasters of tomorrow; and that any one lineage may, over time, race over the surface of the globe so that where it lives today may bear no relation to its origins or previous peregrinations. Ungulates indeed may be more mobile than most because, after all, mobility is one of their specialties. Finally, we may note that many extinct ungulates were clearly flourishing until they confronted our own ancestors.

There are too many groups to treat one by one. Instead, I shall just pick a few ecomorphs and lineages that illustrate the themes. First—a common ecomorph.

A PLETHORA OF TANKS

The tanks we know today, par excellence, are the remaining five species of rhino; but in their time at least half a dozen orders of ungulates, plus some marsupials, have essayed the tankish form.

The earliest ungulate tanks of all were the uintatheres, such as the late Eocene *Eobasileus* of North America, which stood 1.5 meters (5 feet) at the shoulder and was 3 meters (10 feet) long, and massive with it. On its head *Eobasileus* carried six bony protrusions like the knobbly ossicones that nowadays adorn the heads of giraffes—two on the nose, two between the eyes, and two high on the forehead. In its upper jaw it bore daggerlike canine teeth, proving (as do gorillas, pigs, Indian rhinoceroses, and primitive deer) that carnivores have no monopoly on big canines. But the arsinoitheres were even bigger—like *Arsinoitherium* of early Oligocene Africa. They, too, were very rhinolike, carrying massive twinned horns side by side on their noses. But rhino horn is made from keratin, the hard protein produced in the skin that is also the stuff of nails and hooves. Arsinoithere horns were of hollow bone. Like uintatheres, arsinoitheres also bore a pair of ossicones high on their heads.

Then there were two groups of perissodactyls that were more closely related to horses than to rhinos, and yet were more rhinolike in form. First, there were the brontotheres, like *Brontops* of the Lower Oligocene, which stood 2.5 meters high and again had protective and aggressive horns on its head. And there were various members of the paleotheres that achieved extraordinary variety in their briefish run in the Eocene and early Oligocene, and included some forms that were horselike, some tapirlike, and some that were tankish. Thus, there have been close relatives of the horse that were very like modern rhinos, and some ancient rhinos that were like horses, while (less surprisingly) both horse relatives and rhinos have produced more generalized, tapirlike forms. There have also been several tanks among the South American ungulates, including the stocky *Toxodon*, which was nearly 3 meters long. *Toxodon* survived until the Pleistocene, and may have fallen foul of the first human invaders.

Finally, the marsupial diprotodonts of Australia were 3 meters long and also robust. They survived for 10,000 to 15,000 years after the first people arrived in Australia. Whether those aborigines finally saw them off remains debatable, but the rapid decline of the modern rhinos shows how easily such big animals can be pushed aside when the pressure is on.

Taller creatures with big noses to compensate may be seen as variations on the tank.

ELEPHANTS AND ELEPHANT LOOK-ALIKES

Elephants belong to the order Proboscidea, which of course takes its name from the nose. Now it is reduced to just two closely related species in two genera—the Asian elephant *Elephas* and the African *Loxodonta*—but since the order's first appearance in the Eocene, it has embraced at least 160 species in 44 genera and perhaps many more. Furthermore, the elephants represent only one proboscidean suborder out of three.

The oldest proboscideans are the moeritheres, which arose in the Eocene as piglike creatures in northern India. They are probably ancestral to the other proboscideans, although the best-known genus, the tapirlike *Moeritherium*, which occupied Egypt in the Eocene and early

Oligocene, lived alongside modern-looking elephants in Africa for several million years and so it cannot be their ancestor.

Then there are the curious and wonderful deinotheres (such as *Deinotherium*), which appeared in the early Miocene and died out toward the end of the Pliocene after a successful run in which they remained almost unchanged for 20 million years. In general form they resembled modern elephants, except for their extraordinary tusks, which were only in the lower jaw and pointed downward rather than forward or upward. Perhaps the deinotheres used them for digging or for stripping bark.

But the ultimately successful group, which in their time have spanned all continents except Australia and had a huge ecological impact from deep into the Arctic to the tropics, are the suborder known as Elephantoidea. Both the living genera belong to just one elephantoid family, the Elephantidae; but there are two extinct families that in their time were of huge ecological significance. The oldest and most curious are the Gomphotheriidae. Some of the gomphotheres resemble modern elephants in general body form and even in their heads, but some of the earlier types had long lower jaws that contained tusks, just as the upper jaws did, so their heads were long like an anteater's, but bristled with protruding teeth. In yet other types, known as the ambelodonts, the lower tusks became broad and flattened, thus forming the whole lower jaw into a giant shovel, presumably for rooting out aquatic vegetation. Gomphotheres reached North America via Beringia and then pressed on into South America when the opportunity arose in the Pliocene.

The second elephantoid family are confusingly called the Mammutidae—confusingly, because the name sounds like "mammoth" but in fact refers to the mastodons. Again, they are very elephantlike in general form, but their grinding teeth are quite different in structure. Mastodonts also reached South America via North America and some biologists have claimed that the genus *Cuvieronius* survived there until several centuries after Christ. The claim is highly contentious, however.

The last of the three elephantoid families are, of course, the Elephantidae, whose huge grinding teeth are unique among mammals. There have been scores of species and subspecies of all shapes and sizes, some very hairy like the mammoths and some almost naked like the modern types, some with curly tusks (like mammoths) and others ramrod straight (the straight-tusked elephants), some that were even bigger than the modern Africans.

But in addition, elephants have shown a remarkable propensity for miniaturization. The problem is the one we have met both in this chapter and the last: big homoiotherms need an enormous amount of food—enough not simply for individuals but to sustain entire populations. Big mammals cannot survive on small islands in viable numbers, and if they are to survive at all, they must produce diminutive races which, with time, may become distinct species. In the previous chapter we met the dwarf red deer of Jersey. Cyprus and other islands of the Mediterranean harbored dwarf hippos until well into human times. But dwarf elephants are known from islands worldwide. They have been found on islands off California. Some of the most extreme types came from the Mediterranean: *Elephas falconeri* from Crete, Malta, Sicily, and Cyprus was less than a meter high. On Wrangel Island, north of Siberia, dwarf mammoths apparently survived until the time of the pharaohs. Indeed, a painting in one of the pharaoh's tombs shows a brown bear and a mysterious elephant that is no bigger than a sheep but has well-developed tusks and so is clearly adult. Its tusks are curled, its head is domed, and the beast is clearly hairy. In a letter to *Nature* (June 2, 1994, p. 364), Israeli archaeologist Baruch Rosen suggests that this astonishing beast could well be a dwarf mammoth, taken as a gift to the Egyptian court. There may have been no direct trading links between Egypt and northern Siberia, but we can well imagine such an exotic beast being shuffled from one end of Asia to the other like some precious jewel over the course of its long elephantine lifetime.

We also saw in the last chapter that giant poikilotherms can survive on islands in viable numbers, and cited the dragon of Komodo, which is the world's largest living lizard. In *Nature* (April 30, 1987, p. 832), Jared Diamond of the University of California Medical School, Los Angeles, has suggested that Komodo dragons evolved their huge size specifically to prey upon the dwarf elephants of the small Indonesian islands. Most modern lizards are small enough to bask on an elephant's toenail. That fact that lizards once preyed upon elephants shows what astonishing possibilities lie just beneath the surface.

Proboscideans have produced the most striking elephant morphs and we are fortunate that the living types are among the finest and biggest; yet the proboscideans have not had a monopoly on the elephant ecomorph. At least two orders of the peculiar South American ungulates, the astrapotheres and the pyrotheres, produced big-bodied, short-necked forms that also had trunks, as is indicated by the position of the nasal opening on their fossilized skulls. In short, the ecomorph of the elephant has been shaped in large part as all creatures have by the principles of engineering. The elephant morph is a logical form to adopt: a huge and almost invulnerable body with a limblike nose with which to manipulate and communicate with the outside world.

Proboscideans have always tended to be elephantlike; the more tapirlike moeritheres were simply early types. But there are three ungulate lineages in particular that have been extraordinarily diverse: the rhinos; the tylopods, which is the group that includes the camels; and the collection of long-gone and little known South American creatures collectively known as meridiungulates.

THE SOUTH AMERICANS: MERIDIUNGULATES

When South America began its long separation as an island, it must have had on board a group of those ancient Cenozoic creatures whose relatives in other continents became, for example, the perissodactyls and the artiodactyls. On South America they evolved into at least four quite different orders, which are loosely grouped together as meridiungulates. All are now gone. Some suffered badly after the Great Interchange, while several survived until the coming of human beings, about 12,000 years ago.

In their day, however—a day that lasted at least 40 million years—they radiated to emulate most if not all of the ecomorphs that other ungulates, and indeed some nonungulates, explored in the rest of the world. The meridiungulates are not widely known and do not feature in most standard texts. But they provide some of the most extraordinary examples of convergent evolution the world has seen.

Two of the four main meridiungulate orders—the litopterns and the notoungulates—were particularly diverse. Some litopterns resembled hipparions, but *Thoatherium* was as one-toed as a modern *Equus*, although it never had long jaws and high-crowned teeth like a modern horse—showing that individual features of different animals may converge, while others remain different. But then, high-crowned teeth and long jaws evolve to cope with grass and *Thoatherium* died out in the Miocene, before South American grassland truly came into its own. Contrariwise, *Macrauchenia* was a litoptern with the general form of a camel—a long-necked, long-legged browser, big but not hugely heavy. But the form of its skull shows it also had a proboscis, perhaps like an elephant's. *Macrauchenia* was the last survivor of all the litopterns and died out only in the Pleistocene. There seems no reason why such a well-equipped creature should have died out.

The final meridiungulate order, the wonderfully various notoungulates, reached their peak in the Oligocene, when there were nine entire families. We have already met two of them: *Homalodotherium*, shaped like a giant sloth, or chalicothere; and the rhinolike *Toxodon*, which again, like *Macrauchenia*, lasted until the Pleistocene.

THE ASTONISHINGLY VERSATILE RHINOCEROSES

The Rhinoceratoidea have been the most various superfamily of large mammals of all times. The primitive *Hyrachyus* was a runner the size of a collie dog, while the later *Juxia,* also a runner, was as big as a horse. Many rhinos have been hippolike: *Teleoceras* was one. Most were hornless, and many were for all the world like their ancestral tapirs. Indeed, only the members of the existing rhinocerotids ever had horns, and these have often been bizarre. The single horn of *Elasmotherium* was enormous: its base extended over both nose and forehead. Several species, like *Diaceratherium,* had nose horns side by side, like rifle sights. Some members of the rhino superfamily took the form of a modern giraffe, and achieved this form long before the giraffids did. One of them, as we have seen, was by far the biggest land mammal that ever lived. Between them, early rhinoceratoids dominated ecosystems on every continent except Antarctica, South America, and Australia. Though they are often supposed to be primarily African, they reached Africa only 25 million years ago, which was quite late in their history. In fact, the rhino superfamily arose in North America, and for a long time the North American rhinos were extremely important, but they became extinct during the last great period of late Miocene cooling, around 5 million years ago. The Panama land bridge that might have enabled them to live on in South America opened too late for them to take advantage of it.

The rhinos descended from the tapirs, which have remained resolutely tapirlike since the Eocene. We might in fact suggest that the rhinos have been the adventurous arm of the tapirs; just as we, human beings, may be seen as the adventurous arm of the anthropoid apes.

WHY HORSES ARE HORSES

The perissodactyls as a group are extraordinarily varied—witness the rhinos and the paleotheres—but the horses, although there have been hundreds of species that have lived all over the globe (except Australia), seem simply to have become steadily more horselike as the Cenozoic has worn on. The first acknowledged equid appeared in North America in the early Eocene as a woodland creature the size of a fox and is officially known as *Hyracotherium* (because it was at first thought to be a hyrax) but is popularly and more aptly called eohippus, or dawn-horse. It had four toes on each of its front feet and three on each hind foot. Then over the following epochs *Hyracotherium's* many descendants in general grew bigger and bigger, steadily lost their surplus toes, and tended to develop high-crowned teeth (a condition known as hypsodonty) and the long, deep faces that such big teeth require to accommodate them. In practice, this evolutionary course followed two distinct lines. One group, known colloquially as hipparions—a general name commandeered from just one of the many genera—retained three toes on each foot. In the other line the loss of toes continued until, as in the extant genus *Equus,* only the middle toe remains. The last hipparions died out in Africa in the Pliocene, only about 3 million years ago, and their footprints are preserved in fossil mud alongside those of *Australopithecus afarensis*—one of the earliest-known members of the family Hominidae and possibly, or probably, our own direct ancestor. The one-toed equids are with us still though sadly reduced to just seven species, including the zebras and asses, which are all placed in the single genus *Equus.*

This apparently single-minded evolutionary progression enormously impressed Victorian and early-twentieth-century zoologists. They saw it as a prime example of orthogenesis—straight-line evolution toward some apparently presighted goal; and although many post-Darwinian biologists ostensibly stripped theology from their scientific thinking, many seemed to cling to the notion that horses were somehow destined to become horses, and had stuck rigidly to their script. Such an interpretation, with or without overt theological connotations, can be found in every standard text of evolution. Children are brought up to believe that orthogenesis is real, and indeed is the norm.

Modern zoologists have condemned this orthogenetic interpretation, for they reject the idea of destiny and find the notion that God directed things superfluous. The old-style paleontologists, say the moderns, suffered from wishful thinking. They misread the horse's fossil record just as they misread that of human beings. They liked to believe that evolution did indeed pursue paths that were inevitable. So they interpreted hominid fossils to show human beings becoming steadily brainier and more like us; and interpreted the fossils of man's chosen companion as if it were the ambition of ancient equids to provide vehicles for us to ride upon. In truth, say the modern biologists, ancient horses were perfectly capable of evolving in a whole host of different directions at any one time, just like every other animal. What about the tiny *Nannippus* of late-Miocene North America, which reversed the trend to hugeness and hypsodonty, and went back to the woods? In fact, the idea of destiny in its theological form should be rejected. Horses did not grow bigger and toothier because they were destined to do so. Yet *Nannippus* remains very much an exception. The general trend, seen in genus after genus among horses, and in hipparions as well as one-toed nags, is to become bigger and toothier. Those facts will not be denied. But we do need a new interpretation.

So here it is. As the Cenozoic wore on, the world grew cooler and drier, and by the time of the Miocene, grassland was taking over from forest. Like many other ungulates, horses over time became better and better adapted to eating grass. This explains their hypsodonty—and one North American hipparion even developed open-rooted teeth: teeth that grow throughout life to compensate for wear. At the same time, horses became more and more cursorial, the better to cover the ground. They also became bigger—because grass is poor fodder and suitable only for big animals, which have a lower metabolic rate than small ones.

Yet many other grass-eaters—like the deer and the antelope—did not generally evolve the huge bulk of the horse (although a few have done so). But then deer and antelope are ruminants; and as we have seen, ruminants are more efficient herbivores than horses. When the body size is large, the difference in efficiency makes little difference because food requirements are low relative to body weight. But when the body is smaller and the metabolic rate is higher, the difference in efficiency does make a difference. Hence, big horses could compete with big ruminants, but small horses could not compete with small ruminants. Hence, the option of becoming small was not really open to horses, because they could not out-compete the ruminants once the latter got into their stride. *Nannippus* slipped through the net briefly, but only briefly. The equids may well have wanted to radiate in all directions, as some modern biologists would have us believe they did. But in practice their survival depended on bulk. So their evolution was not orthogenic in the Victorian sense. They were not trying to become anything in particular; they were just outcompeted when they contrived to be anything other than big. But the appearance of orthogenesis is there nonetheless. It is idle to deny the actual narrowness of horse evolution.

The big-bodied cursorial option has served the horses well, however. Both the three-toed and the one-toed lineages spread out of North America into Eurasia and then into Africa, and eventually into South America when the opportunity arose in the Pliocene. Wherever they finished up, they often radiated again, and some of the emergent forms migrated back to where they had come from. Thus, there has been much toing and froing between Eurasia and North America whenever Beringia has emerged above the waves. Overall, the pattern of horse distribution has mirrored one of those superior fireworks, which burst to form an array of stars, each of which then bursts again to give another generation of stars, and so on. *Equus* itself evolved in North America about 4 million years ago, and migrated to Eurasia and so into Africa.

After the last separation of North America and Eurasia, however, around 8,000 years ago, horses finally disappeared from both Americas. They, too, it seems, fell to the first human colonists.

VARIATIONS ON A THEME OF CAMEL: THE TYLOPODS

Tylopods practice foregut fermentation just as the ruminants do, but differently: their stomachs have only three chambers instead of the ruminant four. The only tylopods left to us are the camelids— the camels and llamas and their ilk—but there is a great deal more to the tylopods than that. In fact, the group as a whole is thought to have evolved around 40 million years ago (late Eocene) from piglike ancestors ("Suina," on page 135); and one extremely successful group of tylopods, the merycoidodontids, or oreodonts, retained the piglike mien. Throughout the Oligocene, until the end of the Miocene, the oreodonts were the dominant herbivores of North America. The oreodonts, like the passenger pigeon, show how the mighty may fall; although the oreodonts' demise, unlike the pigeon's, cannot be laid at the human door, for they died out long before the genus *Homo* even evolved.

Other tylopods, including some oreodonts, took many different forms. Some resembled rabbits. But the finest and most significant apart from the oreodonts were the beautiful, strange, deerlike Protoceratidae, which diversified in the warm south of North America. They had simple horns on their heads like cattle, but the males also had extra bony hornlike outgrowths on the tops of their muzzles, well in front of their eyes, again like the foresights of a rifle, which in some species were long and forked like antlers. Perhaps they were used for sparring and perhaps merely for display. Protoceratids arose in the late Eocene and evidently survived the great North American decline of the late Miocene, only to die out in the early Pliocene.

Camels themselves are known for their resistance to drought, but their real skill is their ability to starve. After all, camels do need to drink every few days (up to a hundred liters at a time), while many desert antelope have no need to drink at all except for a lick of dew. No, the crucial advantage of camels over ruminants is an ability they share with pigs: to convert surplus energy quickly into fat, and to mobilize that fat again if the food runs out. Ruminants cannot do this. They need a constant input of carbohydrates. Overall, then, camelids can endure greater uncertainty and more prolonged privation than ruminants can. Today their last strongholds are in the deserts of the Middle East and central Asia and—as llamas and their ilk—in the high Andes of South America.

Ecomorphically, camels are the tylopod answer to the giraffe: the long-necked *Aepycamelus* from the middle and late Miocene of North America was particularly giraffelike. But they first arose in the late Eocene of North America as short-legged creatures less than a meter high. They peaked in the mid-Miocene around 10 million years ago, and did not migrate to Eurasia or Africa until about 5 million years ago. They did not reach South America until the Pliocene interchange. Thus, they were most widespread in the early Pleistocene, when they ranged throughout the Americas and much of the Old World—but only a little after that they collapsed almost everywhere.

They finally disappeared from their North American land of origin only about 12,000 years ago, not long after the first people arrived. Now, like the horses that disappeared at the same time, camels are again doing well in North America as feral animals, showing that as far as they are concerned there is nothing wrong with the terrain.

So far, the story of the ungulates has been somewhat gloomy. Many wonderful groups have gone, including some that might have been swept aside by our own ancestors. The remaining group, the ruminants, have over time lost at least as many fine representatives as, say, the perissodactyls have done. But in terms of species and range they are flourishing still.

THE TRIUMPH OF THE RUMINANTS

Ever since the Miocene the genre of large herbivory has belonged increasingly to the ruminants. Presumably—no other explanation seems necessary—their supreme ability to digest bulk fodder efficiently has enabled them to dominate both the forest and the ever-spreading grassland over a greater area of the world than any other surviving group. It seems odd, though, that they have never produced any megagiants comparable with the rhinos, elephants, uintatheres, titanotheres or brontotheres of other orders. The giraffes are the biggest ruminants, and the cold Pleistocene produced some huge deer and cattle, but nothing in the proboscidean or rhinoceratoid class. Perhaps this is preemptive competition again. Even the ruminants may have been unable to invade niches that were already so competently occupied.

There are six families of ruminants. Two are small and perhaps ancestral: the chevrotains (Tragulidae), which seem to retain the features of the very earliest ruminants, and the musk deer (Moschidae), which perhaps demonstrate the erstwhile link between the giraffes and the deer. Of the other four families, two are much depleted—the pronghorns (Antilocapridae) and the giraffes (Giraffidae); while two, the deer (Cervidae) and the cattle, antelopes, and sheep (Bovidae) are now the most speciose of all large families of mammals.

The antilocaprids are now down to one species—the pronghorn—but they have had a fine past. They apparently arose in North America during the Miocene, and together with the tylopod protoceratids they showed how to live on the plain in cursorial herds without being either a deer or an antelope. They differ from antelope most obviously in their headgear. As in antelopes (and other bovids), the horns of antilocaprids have a body core and a horny sheath (which is made of the hardened protein keratin). In contrast to bovids, however (although with a superficial resemblance to deer), the sheath, but not the bony core, is shed and regrown each year. Also, antilocaprid headgear can take extraordinary forms. The horns of the modern pronghorn look fairly unremarkable—like those of a chamois (a bovid) though with a single fork (like the antlers of a young roe deer). But some extinct types had several pairs of horns arrayed along their faces like protoceratids, and the horns of others were elaborately branched. Antilocaprids also have the most highly evolved feet of all the artiodactyls, for they retain only the functional third and fourth toes, and all trace of side toes has disappeared.

All but one of this fine family disappeared during or before the Pleistocene, though it is not easy to see why. They were and are extremely swift, probably capable of speeds up to 80 km per hour. Presumably they once were chased over the plains by the cheetahlike cats of the North American Pliocene, which may or may not have been close relatives of the modern *Acinonyx*. Thus were enacted scenes precisely like those of modern Africa—but on a different continent and with a different cast. Nature is endlessly inventive, yet endlessly reinvents.

The deer, Cervidae, first arose in the Oligocene, presumably in Eurasia, and, in stark contrast to the antilocaprids, they remain the only large herbivores to compete in variety and range with the bovids; there are thirty-six living species, ranging almost from pole to pole. As a group they are remarkable for their branched and often enormous antlers, which are made of bone but are shed and regrown each year, with each year's contribution being bigger than the previous year's. Antlers grow allometrically, both within a single animal as it ages and between different species, which means that big animals not only have bigger antlers than smaller ones but also have antlers that are bigger in proportion to their bodies. Thus, moose, which are the biggest of living deer, have huge, flattened antlers, and some extinct types had even bigger ones. Biggest of all was a relative of the modern fallow deer called *Megaloceros*, which is often called the Irish elk, since its antlers have been found in such abundance in the Irish bogs. Its body was about the size of a big moose, but its antlers could span 3.7 meters (12 feet) and weigh 40 kg (100 lb.). It has sometimes been argued that the Irish elk "outgrew its strength" or caught its wonderful antlers in overhanging trees. This must be nonsense. Irish elk were highly successful—not exclusively Irish at all, but flourishing in the late Pleistocene throughout the Old World as far as China and Siberia. If they were really so damagingly constructed, they could hardly have evolved at all. But this was yet another creature to come up against the formidable genus *Homo*—from *H. erectus* to *H. sapiens*. This, rather than any overhanging tree, was probably their undoing.

The Giraffidae probably evolved from deer sometime in the Miocene. The only two remaining species live in Africa (the giraffe and the okapi), but the family almost certainly arose in Eurasia; zoogeography is mocked again. Most extinct giraffids were very unlike the modern giraffe. *Sivatherium*, for example, was a huge bull-like creature from the Pliocene and Pleistocene of India whose ossicones were flattened and branched to resemble the antlers of a moose, although they were not made to be shed. Various unrelated lineages, including camels, litopterns, and rhinos, have just as much claim to the giraffelike eco-morph as the giraffids do.

The Miocene also gave rise to the family that has become the most speciose of all large animals, the Bovidae—which include the cattle, antelope, sheep, goats, and a mixed bag of goat antelope such as the chamois, takin, saiga, and musk ox. Now there are 120 bovid species in 46 genera, but in the Pleistocene there were at least twice as many. They are now at their most various in Africa, and the oldest-known types date from the Sahara of about 20 million years ago—yet the bovids may not owe their origins to Africa. They could just as soon have arisen in Europe or Asia. They did not reach North America until the Bering Strait was bridged in the Pleistocene, when bison and the mod-ern forebears of bighorn sheep and mountain goats flooded across into a continent that had already lost many of its native animals to a change in climate.

The plethora of antelopes in Africa accounts for much of the pres-ent variety of bovids: they contrive to carve up the vegetation and espe-cially the grass between them, so there is scope for many different specialists. Elizabeth Vrba contends that their variety evolved in waves, with entire suites of species appearing at intervals in Africa since the Miocene, each in response to some global shift in climate. The most re-cent to unfold have been the gazelle genus *Gazella,* which arose about 2.5 million years ago, apparently prompted into being by the same cli-matic shift that encouraged *Homo* to evolve. In modern gazelles we are privileged to see the process of species formation still in action, and for this reason it is often hard to tell one from another: Arabian and Dor-cas gazelles are only now being sorted into species, and there may be more types than we are now aware of. *Gazella* and its almost exact con-temporary *Homo* make an intriguing and instructive ecological and evolutionary contrast. *Gazella,* as herbivorous specialists, have formed at least fifteen species but stayed in much the same area, while the hy-pergeneralist *Homo* is now reduced to one species that is spread all around the world.

Cattle reached some of their greatest heights in the Pleistocene. The cold favored large body size: *Pelorovis* of Africa had horns that spanned four meters. In Eurasia and North America, bison achieved consider-able variety as they shifted from continent to continent in vast herds, with several long-horned types in addition to the presently more com-pact forms. The ones that feature in European cave paintings are long-horned. I am sure that in bison we see an evolutionary principle at work that is well recognized among plants but seems to have been little recognized in animals—that of introgression. That is, creatures separate into subpopulations that then evolve independently to form what for all the world look like new species, but then the apparently different types come together again to form what is effectively a hybrid. If the separation has not proceeded too long and the two parent species click then the hybrids fare very well. The fossil record suggests that this has occurred in bison, and their ecology—their ability to occupy entire continents and to circumnavigate the globe when the opportunity is present—makes them admirable candidates for such a process. Most paleoanthropologists concede that a similar process may well have taken place between subgroups of the equally peripatetic *Homo.*

This concludes our lightning tour of the ungulates: creatures which, above all, shaped the ecology of the open plain that our own genus, *Homo,* evolved to inhabit in such sharp contrast to its forest ancestors. We have preyed hard upon the ungulates—and still do. Since long before we came on the scene, however, they have been the prey of specialist mammalian carnivores.

CARNIVORES AND CREODONTS

At the very start of the Cenozoic 65 million years ago there lived a creature now called *Cimolestes* which, at the time, would have been hard to distinguish from dozens of others of its tree-shrew-like contemporaries. But within a few million years—by the mid-Paleocene—*Cimolestes* had given rise to two entire mammalian orders that were dedicated to carnivory. First into their stride were the Creodonta, which appeared in the Paleocene about 60 million years ago, radiated spectacularly until the Oligocene around 30 million years ago, but disappeared entirely by the late Miocene, 8 million years ago (the last lived in Pakistan). The second group, the Carnivora, appeared early but put up only a modest showing until the Oligocene, around 35 million years ago, after which they have flourished. Nearly 240 wonderful and often beautiful species are with us still, in 98 genera, from the tiniest weasels the size of a gerbil to walruses and elephant seals, Siberian tigers (the largest of all cats) and Kodiak bears (the largest of all living land carnivores). As always, however, we find that the living cast of creatures is but a fragment of what has gone before, that at many times in the past there have been entire roll calls of major predators where now there are one or two or none at all, and that the extinct forms include entire ways of being and ways of life. And of course in both creodonts and true Carnivora, we find the familiar pattern: each order and each family within each order often adopting many different styles of life, and each style emulated independently by many unrelated creatures.

Meat-eating is a high-risk strategy. Game theory shows that predators fare best when they catch prey that is roughly their own size, but any creature as big as themselves puts up a fight. Scavenging may look easier, but corpses occur patchily and there is fierce rivalry for what there is. But the risk is worth the game. Meat is very high in nutrients compared to plants, and carnivorous mammals get by with occasional meals—just a few meals a day for a small carnivore or a few a week for a large one—while herbivores must spend almost their entire lives in foraging. A wild elephant feeds for seventeen hours a day.

The secret of successful mammalian carnivory lies, it seems, with the carnassial teeth: cheek teeth modified into scissorslike blades, top and bottom, that shear through skin and flesh. Many Carnivora have in fact abandoned specialist flesh-eating and those that have—including omnivorous but often fruit-eating raccoons and fish-eating seals—have also often abandoned carnassial teeth. But for the committed meat-eaters, both creodont and carnivore, the carnassials are the tool of the trade. Yet it may be that the carnivores in the end surpassed the creodonts because their teeth were less committed to meat-eating. The carnassials of carnivores are no less fierce than those of the creodonts, but they occur further forward in the mouth; that is, they are fashioned from the rear premolar and the first molar, leaving molars behind the carnassials for crunching and grinding. The carnassials of creodonts were formed from the rear-most teeth. This meant that the carnivores could more easily eat plants if the meat supply failed, as modern dogs (and especially foxes), bears, and badgers are happy to demonstrate. Creodonts would have found this more difficult. Perhaps this gave the carnivores the edge when the going got really tough. As can hardly be overstated—and as successful creatures, from kangaroos to human beings, demonstrate—the art of survival is to scrape by in the worst of times, not merely to flourish in the best.

Over time there were forty-five or so genera of creodonts in two families, the short-lived Oxyaenidae, which died out in the Eocene around 40 million years ago, and the much more various Hyaenodontidae, which evolved in the Eocene and ranged through North America and Eurasia. Between them they adopted the forms and presumably the lifestyles of bears, dogs, hyenas, raccoons, cats, weasels, and otters. The bigger ones preyed upon brontotheres, chalicotheres, and rhinos. The hyenalike *Hyaenodon* of the Miocene, 20 million years ago, was the creodont equivalent of the modern lion, both hunting and scavenging. *Apataelurus* reinvented the sabertooth. The largest of all creodonts was *Megistotherium* of the early-Miocene Sahara, which was heavier than a modern bison.

Thus, the creodonts did more or less everything that the land carnivores did. But the true carnivores, we must assume, did it better.

THE DOG GROUP AND THE CAT GROUP: THE "TRUE" CARNIVORES

The very earliest of the true carnivores were pine-marten-like creatures that lived 65 to 60 million years ago and were known as miacids. They then gave rise to two separate groups. In the New World there arose the Canoidea (doglike animals), which came to include the Canidae (dogs), the Ursidae (bears), the Mustelidae (weasels, otters, badgers), and the Procyonidae (raccoons and coatis). The modern pinnipeds (marine carnivores) are believed to have arisen from mustelids, and so form a parallel group to the otters; but since they play little part in our own history, I will not dwell on them. In the Old World the miacids gave rise to the Feloidea (catlike animals), which now include the Viverridae (civets), the Herpestidae (mongooses), the Felidae (cats), and the Hyaenidae (hyenas). But in the Oligocene, 30 million years ago, the Bering land bridge came into being for the first time and there was some interchange of carnivores between the two land masses; although the canids did not enter the Old World at this time, and in fact made their first appearance in Eurasia only in the late Miocene, around 6 million years ago. A few North American carnivores also managed to raft to South America before the Pliocene convergence, including the raccoon known as *Cyonasua*, which evolved into the huge *Chapalmalania*. *Chapalmalania* was like a giant panda—another fine piece of convergence.

Hundreds of species have disappeared from these carnivore families, and some entire ecomorphs have gone with them, including all the sabertooths. Three entire Carnivora families have disappeared, too: the father of them all, the Miacidae; the Arctocyonidae, or "beardogs," which had bodies like bears and heads like dogs (convergence here of bits of animals) and thrived from the Eocene (50 million years ago) to the late Miocene (5 million years ago); and their almost exact contemporaries, the Nimravidae, which were for all the world like cats, although they are now generally allotted a separate family of their own.

DOGS, BEARS, AND RELATIVES: THE CANOIDEA

The Canidae themselves are an ancient group, dating from the Eocene around 40 million years ago; and although the very earliest were short-legged and mongooselike, they quickly adopted the dog mode and have not changed much since, except for a considerable enlargement of the brain (and presumably of corresponding intelligence). In the early Oligocene there were only five genera; by the late Miocene, 10 to 6 million years ago they peaked at forty-two; and nowadays there are between ten and twelve, depending on who is counting. Although the canids have become especially significant in Africa and Eurasia—wild dogs, bush dogs, wolves, jackals, and foxes—they remained exclusively American until the late Miocene. They have taken just 6 million years to conquer the Old World.

Most dogs have been doglike—hunting by running and chasing, often in packs; some scavenging, and with a penchant for omnivory—but some have broken the mold. Some of the earliest kinds were cat-like. But the borophagines, such as *Osteoborus* had big bearlike bodies, short faces, and huge crushing teeth; that is, they filled the bone-crushing role of the modern hyena. In fact, hyenas, which as cat relatives evolved in the Old World, made a very poor fist at migrating into America when the Bering land bridge opened, even though the modern spotted hyena *Crocuta* is one of the most successful of all modern carnivores. *Crocuta* apparently never reached America, but if it had, it might have found the bone-crushing dog *Osteoborus* to be a formidable rival. For its part, *Osteoborus* remained exclusively American, from the Miocene to the Pleistocene.

One of the most spectacular of all dogs, and resolutely doglike, was the dire wolf *Canis dirus,* which lived in Pleistocene California. It was like a modern wolf only much bigger, and was evidently a scavenger. The fossil remains of several thousand individuals have been found in the La Brea Tar Pits in Los Angeles. Dire wolves evidently had some fierce scraps with the massive sabertooth of the day, *Smilodon.* Many bones of each have been found deep-scarred by the teeth of the other.

Although modern studies of body proteins and DNA suggest that bears (Ursidae) diverged from raccoons (Procyonidae) in the Oligocene, 30 million years ago, with the giant panda diverting from the rest of the ursids in the early Miocene around 20 million years ago, the fossil record reveals no bears until well into the Miocene. In short, they are a comparatively recent group. Most bears have been bearlike—big and versatile omnivores—although the modern polar bear in practice is almost carnivorous, the sloth bear of Asia is a termite-eater, and the spectacled bear and giant panda veer toward vegetarianism.

But again we have lost some glories. The biggest bear these days (and the biggest of all land carnivores) is the Kodiak, a subspecies of the grizzly. But the cave bear, which braved the worst of the Pleistocene ice ages in northern Europe and Russia and survived into recent times, was even bigger. For tens and perhaps hundreds of thousands of years, it irritated our predecessors: *Homo erectus, H. heidelbergensis, H. neanderthalensis,* and finally *H. sapiens.* Late paleolithic *H. sapiens* may well have hastened the cave bear's extinction. But although cave bears are believed to have been vegetarian, they must have been formidable opponents. Whether they lived in caves as a matter of course is unknown; but they certainly hibernated in them, apparently in large groups, as the weather became particularly impossible.

Yet other bears have assumed the habit of dogs and become runners. The giant short-faced bear *Arctodus simus* was the biggest true land carnivore that ever lived—almost half as tall again and at least twice as heavy as a modern grizzly. It lived in North America from the early Pleistocene, around 2 million years ago, until recent times—a mere 10,000 years ago. But it had long legs for running (presumably) and short jaws for gripping; in fact, it was built and perhaps hunted like a giant Rottweiler. It must at least have terrified the first human beings in North America, and Dr. Valerius Geist of the University of Calgary has speculated that *Arctodus simus* and other bears may even have delayed the human invasion of that continent. Geist points out that even in recent centuries the less formidable grizzlies of North America have often made life very difficult even for people with modern weapons. Even more of a runner was *Hemicyon* (its name means "half dog") of Miocene Eurasia and North America, which was as big as a modern brown bear but more lightly built, and may have hunted in packs like a wolf.

CATS, HYENAS, AND OTHERS: THE FELOIDEA

Also dating from the Miocene—around 15 million years ago—are the Hyaenidae. Now they are specialist bone-crushers; carnivore paleontologist Alan Turner of Liverpool University comments that the spotted hyena, *Crocuta*, which is the most successful of the modern types, "can crush and digest anything it can get into its mouth." *Crocuta* can make a meal from the skeleton of a zebra after lions have taken all that they possibly can, munching the steeliest of bones like celery, and digesting not only the marrow but also the protein within the bone itself. They excrete stools rich in calcium that fossilize to form coproliths.

But as Dr. Turner has shown, the hyenas did not adopt this extreme and formidable specialization until the dogs arrived from the New World in the late Miocene. Before that, they had themselves been more doglike. *Crocuta* itself in recent times succeeded not only in Africa, where it still thrives, but all through Asia and Europe as well (there are some in my local Yorkshire caves), yet it failed to trek through Siberia and into Alaska when the Beringian land bridge opened. The only hyena that did make the crossing was *Chasmoporthetes*, which was obviously a long-legged sprinting hunter rather than a scavenger—the hyena's answer to the cheetah. But *Chasmoporthetes* fared badly in North America, though it did well throughout Eurasia and Africa. Meanwhile in North America, as we have noticed, bones were being crushed by the hyenalike dog, *Osteoborus*. So hyenas have played at being dogs and speedy cats during their comparatively brief sojourn on Earth, while dogs have at times emulated hyenas. The biggest and most formidable bone-crushing hyena of all was *Percrocuta* of middle- and late-Miocene Africa and Eurasia. *Percrocuta gigantea* of China was as big as a modern lion.

Two families can claim to be cats. The true cats are the Felidae, which of course are with us still (with around thirty-six species); but the felids in their early days were accompanied by catlike creatures that now are placed in a separate family, the Nimravidae. Both groups arose in the Eocene and got into their stride in the Oligocene, but the nimravids faded and disappeared by the late Miocene, about 8 million years ago. Both the nimravids and the felids have demonstrated two ways of being a cat: either to be a biting cat, like all modern cats, or to be a sabertooth, as many extinct cats were. People mistake sabertooths just as they mistake Irish elks and dinosaurs. They did not die out because their body form was too extravagant. They were extremely successful animals for tens of millions of years. As already outlined above, marsupials from at least two different families adopted the sabertooth form—*Thylacosmilus* and *Thylacoleo*. At least three different lineages within the Felidae have independently adopted the sabertooth mode. And the nimravids produced *Nimravus* in the Oligocene and early Miocene of Europe and North America with its long, low, leopard-sized body—and its sharp saber teeth. Thus, the mammals as a whole have produced convincing, catlike sabertooths at least six times.

In actuality, the felid sabertooths had teeth of various sizes. Those of *Meganteron,* which apparently spread from India into Eurasia, Africa, and North America from the late Miocene to the early Pleistocene were relatively short (though longer than a biting cat) and cats with this design are called dirktooths. *Homotherium* from the Pleistocene of Africa, Eurasia, and North America had longer teeth than the dirktooths but flatter and more recurved than the biggest sabertooths, a design known as "scimitar-toothed." *Homotherium* apparently walked on flat feet like a bear and died out only 14,000 years ago.

But everyone's idea of a sabertooth was the Pleistocene *Smilodon,* with species in each of the Americas. Its teeth were hugely long, and its jaws opened to 120° or more to give them full purchase. In cross section they were flattened from side to side like the blades of a saber, not round in cross section as often wrongly depicted. The backs of each saber were serrated, like steak knives. Exactly how the sabertooths, scimitartooths, or dirktooths killed their prey is uncertain (how could it be known?), but features that all sabertooths have in common, whether felid, nimravid, or marsupial, offer a clue. For one, they all tend to be muscular, particularly in the chest and forelimbs.

Smilodon is exaggeratedly so, with chest and arms like a wrestler, and comparatively short hind legs. In other instances we have seen convergence of individual features—the litoptern with a camel-like body and an elephantine trunk; the arctocyonids with bearish bodies and doglike heads—but in sabertooths there is no such mosaicism. No sabertooth ever developed the long running legs of a cheetah, for example.

In short, the constancy of form in sabertooths suggests that whatever lineage they belonged to, they were built for ambush rather than a chase. Presumably, they leapt at their prey—a brontothere, a rhino, a chalicothere, a mammoth—from beneath and held on tight around the neck, "getting inside," or "closing" with the opponent, as a boxer would say, away from the victims' slashing horns or claws or tusks. Then apparently they somewhat carefully sliced their sabers into the throat, perhaps severing the carotid or jugular and producing a faint in seconds. The teeth, with their flattened cross section, could not have simply been thrust through the skull as sometimes depicted, for they would have broken.

Neither, probably, did they simply disembowel their victim and wait for it to die. Severe damage to the unprotected throat, and particularly to the windpipe and the blood vessels to and from the brain, seems the most likely method. Killing a large animal is not easy. Here we see again that huge ecological success attended those creatures—a wide variety of them—that evolved an effective method. Just one trick is all it takes.

The biting cats evidently separated from the sabertooths around 15 million years ago—mid-Miocene—and have by now divided into four commonly recognized genera: the big cats, *Panthera*, and the small cats, *Felis*, are the big groups, while the cheetah is conventionally placed in its own genus, *Acinonyx*, and the clouded leopard is classed as *Neofelis*. Among the modern cats *Neofelis* has by far the biggest canine teeth relative to its size—as if it was yet again reinventing the sabertooth line. Molecular evidence now suggests that *Acinonyx* is not as different from the other cats as its anatomy suggests; it may indeed be a close relative of the North American puma. The genus *Felis*, however, may be a mixed bag, with the South American small cats such as the ocelot and the margay having divided from the other *Felis* in the Miocene, around 12 million years ago.

Whatever the truth of their relationships, all three genera have in their time spread from Eurasia into North America and Africa. North America harbored extremely cheetahlike animals in the Pleistocene—which may indeed have been true cheetahs—and had bona fide lions, which entered via Beringia around 35,000 to 20,000 years ago. It now seems, too, that some of the North American lions may in fact have been tigers, so tigers may not be so exclusively Asian as generally supposed. Jaguars, now exclusively American, clearly originated in Eurasia. The cave lion of Pleistocene and recent Europe was a subspecies of the modern lion—*Panthera leo atrox*—but 25 percent bigger, the biggest felid that ever lived, greater even than a modern Siberian tiger. Archaeological evidence suggests that the last ones died out in the Balkans only at about the time of Christ. Once more the apparently ancient world meets the present day.

Finally, we may note the richness of the carnivore fauna in any one time and place. Even nowadays in some parts of the world several species each of big cats, dogs, bears, or hyenas may divide the prey between them; but in times past the cast was often bigger, and included in particular the sabertooth predators of the megaherbivores and perhaps the fearsome Rottweiler running bears. Carnivores are pertinent to us too because, like us, they are intelligent and sociable, and of course are hunters; in a different way they faced and solved the same problems as our own ancestors.

Clearly, human beings have had a tremendous impact upon the carnivores, and still do, for we have brought many to the point of extinction. Equally clearly, the carnivores have had a tremendous impact upon us. Perhaps they affected the speed and direction of our migrations around the globe. Beyond doubt, they helped to sharpen our wits in our long history on the plains of Africa and Eurasia.

At this point it is time to introduce our own group.

THE PRIMATES

From the 1730s to the 1750s the Swedish biologist Carolus Linnaeus initiated the methods of taxonomy and systematics by which animals are named and placed into groups. It was he who first formally established the binomial system of naming species that still applies: generic name first, and then specific name, as in *Homo sapiens* and *Panthera leo*.

But Linnaeus cannot be classed as a modern. Most modern biologists seek a system of classification that truly reflects relationships, as discussed earlier, in which different members of the genus *Homo* or the genus *Panthera* are supposed to be closely related to each other literally—that is, to share a recent common ancestor. But the idea that different species may indeed have shared a common ancestor implies evolution. Lions and tigers, though both placed in *Panthera,* are clearly different; and if they did have a common ancestor, then each has altered since that time in its own particular way. Such alteration over time is evolution. But serious consideration of evolution did not begin until the late eighteenth century, several decades after Linnaeus, and he did not entertain it. He supposed merely that God had a tidy mind and had created different creatures in orderly groups. If God chose to provide several variations on a theme of *Panthera,* why not? With such a godly background Linnaeus was happy to impose value judgments of a theological nature. Thus, he gave the world the term "primates," to include human beings, apes, and monkeys and their relatives. The term of course means "first." Other mammals and birds he called "secundates," and the rest of the animal kingdom were "tertiates." Only the term "primate" has stuck, and most people have forgotten its implication of hierarchy. In theory (though not always in practice), modern biologists prefer to eschew the notion that any group of animals is "higher" than any other.

Indeed, the notion of hierarchy should be avoided—unless, as I argue in later chapters, we are talking only of technical ability in carefully defined areas. Yet Linnaeus did have a point. The primates are undoubtedly different. Oddly, however, what makes them special is their extreme generalness.

ANOTHER ALTERNATIVE: THE PRIMATES

Look at any mid-Mesozoic mammal of the kind from which all subsequent mammals have evolved. Biologists commonly compare the general shape of such creatures to that of a modern tree shrew. I have said that a tree shrew is shaped roughly like a mouse; but without too much exaggeration, we might compare it to a newt. In other words, the typical Mesozoic mammal has the general form of the tetrapod—that is, of the very first animals that ever came on land. It has four legs to which it gives more or less equal emphasis—no particular bias toward the hind limb, as in many a dinosaur or kangaroo—and it has five toes on each limb, giving a foot that in a general way provides grip; just enough, for example, to cling to bark or perhaps to slap down on an insect. It stands bent-legged on the permanent qui vive, and because it is small, it has an enormously high ratio of strength to weight. To cap the design, it has a long tail for balance.

This standard tetrapod ecomorph, as we have seen, serves the bulk of mammal species even today—notably among rodents and small marsupials and carnivores—and, stretching the point only slightly, we can say it has also served a host of reptiles (including the modern lizards) and the tailed amphibians. It is indeed a good all-purpose form that answers the problems of life on land with the simplest possible principles of engineering: a leg at each corner, everything kept small to avoid problems of weight, a tail to even things up, and away you go.

Most mammals that have aspired to be bigger than stoats and rats have modified this basic form, as we have seen. The springlike legs became stilts and then pillars. The pentadactyl foot has often been reduced to a peg. Tails have often been abandoned. Even limbs have at times been sacrificed, or, in the case of the bats or indeed the birds, fantastically re-created as wings. The carnivores kept their five toes, but shortened them, adding claws for weaponry, as in cats, or reducing the foot to a pad, as in dogs (which might just as well be hooved for all the use they make of their feet). In short, most mammals that aspired not to be rats or polecats broke out of the mold by severe modification of the basic tetrapod ecomorph.

The outstanding exception has been the primates. From the beginning, and emphatically, they retained their four-limbedness. Some types, like the lemurs and bush babies, developed exceedingly powerful hind limbs and can leap as prodigiously as any kangaroo. Yet their forelimbs remain important to them, both for locomotion and for other purposes; they have not been downgraded like the forelimbs of an ostrich or a tyrannosaur. Indeed, the apes, from the Oligocene on, were able to reexploit the forelimbs as prime agents of locomotion, and invented the hand-over-hand mode of swinging known as brachiation. Then again, some of the committed brachiators have reduced the thumb and turned the remaining fingers into hooks, but on the whole the primates, in sharp contrast to the dogs or horses or whales or bats, have hung on to the basic five fingers of the tetrapod and turned them to good purposes. They use them as pokers and prodders and in particular have swung the thumb around, and often the big toe as well, so that they oppose the rest and convert a perfunctory grip into a firm grasp.

In short, while other mammals have made a virtue of specialization—transforming the tetrapod endowment of limbs, toes, and tails, and sometimes spurning it altogether—primates have preferred simply to improve on what they were given. Modern primates take many various and extraordinary forms, from human beings to galagos, but the basic general form is always clear to see. This is why, as Linnaeus and all taxonomists since have found, primates are so hard to define. They possess very few features, if any, that are not possessed by other mammals, and particularly by the most primitive; indeed, their differentness lies in their extraordinary conservatism. But generality leads readily to versatility, and the species that has finally prevailed, and in the end has transformed the ecology of the whole world, is the one which, more than any other, has exploited this generality most emphatically. Human beings have had such extraordinary success precisely because we have not converted ourselves into horses or bats or lions. As generalists, we are able to emulate any and all of them. Thus, as discussed in greater depth in chapter 6, we are the animal equivalent of the Turing machine: the universal device that can be turned to any task.

Natural selection has favored physical conservatism and hence generality primarily because the primates elected to stay in the trees. As we will see, they are a remarkably ancient group, who got well into their stride during the Cretaceous when the dinosaurs were still supreme. I love the notion of the earliest primates, pointy-nosed like a modern tree shrew and vocal as a lemur or a squirrel, scolding the dim heads of *Diplodocus* as they probed among the canopy for shoots. Thus, the rise and radiation of the primates coincides with that of the angiosperms, and indeed they are protégés of the angiosperms, exploiting the novel fruits and succulent leaves but above all the complex and variable geometry of the new-styled trees.

For trees—and particularly complex and variable trees—are hard taskmasters. Of course, they encourage the ability to grip, first in a general way and then with the increasingly opposable thumb for good measure; and although many specialist birds do well in the trees with only two legs, and some snakes climb remarkably with none at all, common sense suggests that a versatile limb at each corner is an advantage.

Of outstanding significance is the mobility of the joints. Primates (or some of them at least) have uniquely versatile shoulders. In this respect, contrast human beings with cats. Cats are lissome, elegant, agile; every fine epithet of physicality applies to them. We compare the finest athletes to them, and dancers. Yet cats cannot move as we do. Little old ladies roll their shoulders in the warm-up for aerobics—creakily, perhaps, but they do it nonetheless. If we tried to make a cat do the same, it would cry with pain, for we would dislocate its joints. We can spin our arms through almost all the points not simply of the compass but of a sphere. No other creatures can do this. Only the boneless trunk of an elephant or the tentacle of an octopus or the multiboned neck of an ostrich has such universal mobility. The significance of this hypermobility often goes unremarked, yet it can hardly be overstated. For without the mobility of the shoulder—and of the wrist, with the convenient standard elbow in between—we could not make proper use of our hands. If a horse had hands and all the nerves and brains to go with them, it still could not thread a needle, because its to-and-fro forelimbs would not allow it to bring those hands into appropriate juxtaposition.

Here again, then, we see the opportunism of evolution. For our wonderful shoulders are the gift of our apish ancestors who invented brachiation. We have abandoned the way of life that gave rise to brachiation and therefore have abandoned the skill itself—and yet have made supreme use of that skill in a new context. Thus, too, in ourselves we see the supreme conservatism of the primates, yet with each individual feature developed: the primitive pentadactyl limb converted first to the gripping hand and then to an organ of unparalleled dexterity; the forelimb first made strong and mobile for the purposes of locomotion and then freed to become the perfect and necessary foil for the extraordinary hand.

Yet physique is only the raw material of action. What the animal does is what counts. Here again the trees played a vital part because, for reasons that are commonsensical, they demanded a commensurate development of the nervous system. Life in the three-dimensional world of trees requires excellent vision. Hearing is particularly useful because the animals are so often hidden from each other, while their enemies hide from them. Scent is at least as useful in the trees as on the ground, and many modern primates make extensive use of pheromones and other olfactory signals. More than this, though, life in the trees demands balance, and hence coordination, in particular of hand and eye. Where the eye prescribes, the body must follow precisely, for any failure is severely punished.

In practice—as the modern lemurs demonstrate—marvelous senses and extraordinary coordination do not necessarily lead to all-around intellectual excellence.

Lemurs are lovable, but those who love them best tend also to acknowledge that they are dim. Nonetheless, a powerful principle in evolution is that of preadaptation: a muscle, an appendage, a gland, a set of nerves that have evolved for one purpose (or as we should say, in response to one set of selective pressures) is often at some later time opportunistically pressed into service for another. Human evolution seems to have relied heavily upon the coercion of basic primate qualities that initially evolved for quite different purposes, including the mobile arms. Similarly, it is not fanciful to suggest that the extreme development of the human nervous system builds upon the complexity that the primates in general have evolved to ease their lives in the trees.

So much for generalities. What of the creatures themselves?

THE VARIETY OF PRIMATES

Biologists recognize about 200 species of living primates. They are a speciose group, as might be expected in forest species of modest size. Another 250 species are known from the fossil record. But the oldest known primate fossils are about 65 million years old, and the same kind of logic that tells us that there must have been at least 200,000 species of mammals in times past also tells us that there must in reality have been at least 6,500 primate species throughout the Cenozoic. So we know only about 3.8 percent of the primates that lived throughout that time.

From what we know of it, though, the evolutionary tree of the primates seems to be basically Y-shaped. That is, in the beginning there was an ancestral primate stock known as the plesiadapiformes, or protoprimates: the creatures that in a general way resembled modern tree shrews. These protoprimates then divided to form two main prosimian lines. One of those lines is now represented by the lemurs, lorises, and bush babies; the other line has given rise to the modern tarsiers, tiny big-eyed creatures that superficially resemble bush babies and now live in Indonesia (although their fossil remains have been found in many places, including London's Thames River). Whether the simians arose from the lemurlike line or the tarsierlike line remains uncertain, though most people favor the tarsier line. The prosimians and the simians are collectively known as euprimates to distinguish them from the protoprimates.

The oldest known primate fossils—that is, protoprimates—date from the Paleocene, 65 to 56 million years ago, but the protoprimates began to decline in the Eocene and by 35 million years ago they had more or less disappeared except for one species in North America. The first known euprimates appeared in the early Eocene—around 55 million years ago—in Belgium, France, England, and Wyoming. They were mouse-sized or rat-sized prosimians, and included species from both arms of the Y—some lemurlike and some tarsierlike. Finally, the earliest known simians—effectively, Old World monkeys—are from Africa, dating from the boundary of the Eocene and Oligocene, around 36 million years ago. Thus, the modern groups all arose some time before the last protoprimates disappeared, which means that the modern types arose from only some of the protoprimates, leaving others to carry on as before. It is assumed that the apes arose from Old World monkeys, the New World monkeys having found their way to South America at some early but unknown date by an unknown route and henceforward enjoyed (or not) the company of the meridiungulates, edentates, and marsupials that we met earlier. Both the Old World monkeys and the lemurs produced ground-living types—baboons, gorillas, hominids, and a giant gorillalike lemur in Madagascar, which, as we will see in chapter 8, survived until the time of Charlemagne and the Venerable Bede—but the New World monkeys seemed to have remained stubbornly in the forests. They have therefore lived only where the forest has been, within hailing distance of the Equator. In the form of spider monkeys and some others, however, the New World monkeys have produced some of the most fabulous acrobats of the entire animal kingdom.

But if the oldest known primate fossils are only 65 million years old, why do biologists suggest that the primates as a whole are much older than this? Well, it is clear first of all that the protoprimates were already diverse by the time the first fossils appeared, which suggests a long pe-

riod of evolution before that. The same point—plus their wide distribution—applies to the first-known euprimates.

Logic dictates, too, that the oldest known fossils cannot possibly be
the oldest representatives of their kind. Fossilization is a rare event,
after all; and when animals first appear, they are rare. The earliest fossil
bones are therefore likely to date from a time when their erstwhile
owners were already common. Logic similarly dictates that if an animal
is particularly unlikely to form fossils—as primates seem to be—then
paleontologists are particularly unlikely to find the very earliest types.
In fact, this logic can be translated into a mathematical formula (see
Robert D. Martin, "Primate Origins: Plugging the Gaps," *Nature*, May
20, 1993, pp. 223–234). The fewer fossils there are (relative to the calculated number of extinct species), the older the group is liable to be,
relative to the number of fossils found.

In practice, as we have already noted, only 250 or so extinct primates
are known out of a calculated possible total of 6,500; in other words, we
know only about 3.8 percent of the extinct types. The relevant math
tells us that the oldest known fossils are therefore likely to be about 30
percent younger than the true age of the group. Given that the oldest
known primate fossil is around 65 million years old, the probable age
of the actual oldest primate is probably around 85 million years. At that
time—way back in the Cretaceous—the dinosaurs still had 20 million
years to run and several evolutionary flourishes to undergo. The same
kinds of calculations should also be applied to the euprimates, and although the oldest of these date only from the Eocene, they, too, are
then pushed back into the Cretaceous.

In fact, we should not stop with this first round of calculation. For
if the oldest primate really does date back to 85 million years ago
then there must have been even more species over time than the 6,500
that are calculated now. So the percentage known is even smaller
than expected, and the the true age has to be pushed back even further. In practice, paleontologists seem reluctant to push protoprimate
origins much before 85 million years ago, but some suggest that the
first euprimates, lemurlike and tarsierlike, may be as old as 77 million
years. So they, too, may have overlapped the dinosaurs by an astonishing 12 million years. In short, creatures very like the lemurs that
now can be seen to such advantage in, say, the Jersey Zoo may well
have batted the snouts of dinosaurs as the importunate reptiles thrust
their heads into the canopies of the newly flourishing angiospermous
trees.

Similarly, the math suggests that the first known simians probably emerged not 36 million years ago, which is the date of the earliest fossils, but as long ago as 52 million years ago—at the boundary of the Paleocene/Eocene. Such an early division does, however, help to explain how it is that the New World monkeys have remained so similar to the Old World monkeys in most essentials, even though the two groups separated so early. It also explains how monkeys got into South America in the first place; they probably arose before South America split from Africa to became an island. In practice, however, the earliest known fossils of South American monkeys date only from the late Oligocene, a little over 20 million years ago.

We should ask, of course, whether such calculations should also be applied to all the other groups we have discussed so far. The answer in principle must be yes, but the results would be far less dramatic. For example, living on the open plain and in such numbers, horses have provided us with many thousands of fossils, almost an embarrassment of riches. We probably know a high proportion of the extinct types, and the adjustment needed to extrapolate from the earliest known fossil to the origin of the group would be minor compared to the correction needed for primates. After all, some periods, such as the Oligocene, have yielded very few primates indeed, though we know there must have been plenty. There have long been good reasons, too, for supposing that the primates are ancient. The latest revisions of antiquity are dramatic, but not surprising in principle. It seems clear, too, that there were no large mammalian herbivores of any kind in the Cretaceous, so the true ungulates must all be assumed to date from the Paleocene on. Primates and a few other groups of small mammals got under way as early as they did precisely because they did not interfere with the dinosaurs. The emergence of big mammalian herbivores and carnivores was quite simply forestalled so long as the dinosaurs occupied the large animal niches.

That, then, is an outline of the mammals. Yet we should acknowledge that many other animal groups have also done wonderful things throughout the Cenozoic, including of course the fish, which have diversified more than any other vertebrates; the insects, which have exploited the angiosperms more thoroughly than any other group and today provide most of the species on Earth and a huge proportion of the animal biomass; and the reptiles, which prevail in many an environment that is too meager to support viable groups of mammals—including deserts and islands, and even desert island continents like Australia. This passing comment is all I can give them, however. But it would be foolish not to give half a dozen paragraphs to the birds.

A WORD ON BIRDS

The more you look at birds, the more you realize that the tits and finches that bring such charm to the garden give a misleading impression. At present, there are about 8,600 species of birds—roughly twice as many as there are mammals—and they are divided into about twenty-five orders. About 60 percent of them—more than 5,000 species—belong to just one of those orders, the Passeriformes, or perching birds. Most passerines are small, like tits and finches. The biggest are the crows, and the biggest of all is the raven. Clearly, the passerines flourish because they fill the niche of the small daytime flier, feeding mainly on insects and seeds, which is not occupied by the mammals. Indeed, the birds as a whole seem to play second fiddle to the mammals just as the mammals, for such a vast period of time, played second fiddle to the dinosaurs.

Only when we look at the fossil record as a whole, and turn our eyes away from the ubiquitous passerines, do we see the true potential of birds. We see indeed that the passerines were comparative latecomers to the global party. Like the bears, they did not truly come into their own until the Miocene. Before that, the role of small daytime fliers was filled mainly by the Coraciiformes, now represented by kingfishers and bee-eaters. Taken as a whole, however, the birds may be seen as miniature dinosaurs that could, if only the niches were again vacated, re-create a world ecology very like that of the Mesozoic, an ecology which, as then, would be dominated by great terrestrial herds. Even today on the plains of Kenya the ostriches, bustards, ground hornbills, and secretary birds form a kind of post-Mesozoic subculture. Their chief predator—at least of their young—is another bird: the giant Marshal eagle.

The only flightless member of that Kenyan quartet is the ostrich, which is the largest of all living birds and belongs to the ratites, the only major group of birds that has apparently been flightless from their time of origin in the Paleocene. The other ratite survivors are the rheas of South America, the kiwis of New Zealand, the emus of Australia, and the cassowaries of New Guinea and northern Australia. But the greatest ratites of all were the moas of New Zealand of which there were about fifteen species, some smaller than a modern emu but some far taller even than an ostrich. They survived for just a few centuries after the Polynesians arrived on New Zealand to become the Maoris, about a thousand years ago. The moas could dominate New Zealand because those islands had no native mammals. Yet they were not free of harassment even before the Maoris arrived. They were preyed upon by a giant eagle, one of the biggest that has ever lived, which was short-winged like the modern harpie eagles that prey upon monkeys in South America, so it could duck through the trees. But the heaviest ratites of all, and hence the most massive birds of all time, were not the moas but the elephant birds of Madagascar. The biggest species stood ten feet tall and weighed half a ton.

But many other birds besides ratites have reevolved flightlessness. Flight takes a huge amount of energy, and the wings need constant up-keep, so the birds seem always to be looking for the easier, terrestrial option. On islands, freed from mammalian predators but at risk from crosswinds, species after species in a great variety of groups has re-verted to flightlessness. There are or have been flightless pigeons, like the dodo of Mauritius and the solitaire of Rodriguez, flightless ibises from Réunion and Hawaii, a flightless parrot (the kakapo) from New Zealand, flightless cormorants, auks, and of course penguins, and many flightless geese and ducks. The Hawaiian goose (which in fact flies well) is now known to be only one of an entire suite of geese and ducks, some of them flightless, that lived on that island until recent centuries. Some flightless types became truly formidable, like the carnivorous diatrymids, which flourished briefly from the Paleocene to the Eocene in Eurasia and North America, and the terrifying phorus-rhacids of Patagonia in South America, which had heads as big a horse and huge eagle beaks, and flourished from the Eocene right through to the Pleistocene. They must have hunted in packs like wolves—or indeed like some dinosaurs—to bring down a galloping litoptern in full dash. Diatrymids and phorusrhacids were relatives of the modern cranes and rails. The moorhen is a rail. Within its tiny breast is the spirit of *Tyrannosaurus rex;* and within its gene pool is at least some of the potential to release that spirit. It lacks only for opportunity. Perhaps its time will come again.

These, then, are the principal creatures with whom we and our ancestors have shared this planet. They have helped to shape our evolution and our history, and we increasingly shape theirs. It is time to look more closely at ourselves.

CHAPTER 5

HOW WE BECAME: THE FIFTH COLUMN

The "Fifth Column" of this chapter's title are the primates. I see them over a period of 80 million years or more creeping up on other creatures, always a little different from the rest, doing things their own way, rarely being dominant but usually having a presence at least in the tropics, and finally producing a lineage that evolved into human beings who have altered everything. This is a literary metaphor, in the manner described by Misia Landau, but I think it is a helpful one: the heuristic and so legitimate use of myth.

To be sure, primates did not set out to be subversive. It never entered any of their heads that one day one of their number would evolve into us, and that we would then transform the world. I will argue in chapter 9 that there is more to the notion of progress than most scientists now acknowledge, and we have already observed that evolution has an innate tendency to pursue some lines rather than others. But I certainly do not suggest that our own evolution has been driven inexorably from protoprimate to us along some metalled route. If things had gone differently at any stage—if the climate had not changed as it did; if particular gene pools in particular places had not been exposed to particular pressures—then our own particular species or even our entire lineage would not have appeared.

Yet if we look back at our own history, we do see a wonderful chain of circumstances. To a degree unique among mammals, the creatures that we can reasonably acknowledge as our direct ancestors retained virtually the whole array of basic tetrapod characters, and then over time developed them.

Primordial fingers became powerful and dexterous hands and forelimbs expanded into hypermobile arms, while the senses and nervous system grew in complexity to provide outstanding balance and coordination. Our ancestors evolved these characters in response to the pressures of their own times—as adaptations to ways of life that were quite different from ours. Yet they provided us nonetheless with the raw material to create an ape that could abandon the trees, but then employ its arboreal inheritance to exploit its new environment in ways that no other animal had ever before attempted. If it was logically and practically possible for an animal to design its own ancestors, then we could hardly have chosen better.

In this chapter I want to ask how the final transformation took place—how one of the forest apes became ourselves.

PONGIDS AND HOMINIDS

Linnaeus found no difficulty in establishing discriminating characters of the different tribes of apes, but very great in scientific contra-distinctions between the genera man and ape; but it is to be hoped that he had not met with many individuals of either kind that had produced any practical hesitation in determining his judgment.

—Samuel Taylor Coleridge, "Allegory"

Virtually all details of human evolution have been fiercely fought over at some time and some still are,* but one point at least is universally agreed: that human beings are indeed descended from apes, and that our nearest living relatives are the apes of Africa, the two species of chimpanzees in the genus *Pan,* and the gorillas, *Gorilla.* These ideas were first formally enunciated in 1871 by Charles Darwin in *The Descent of Man and Selection in Relation to Sex* (John Murray & Sons, London). Chimpanzees and gorillas are conventionally classed together with the orangutans, *Pongo,* in the family Pongidae. Humans and their extinct relatives are placed in a separate family, the Hominidae.

At first sight, it seems easy to enough to distinguish hominids (members of the Hominidae) from pongids (Pongidae). At least, as Coleridge commented, it is hard to confuse individuals. We have much bigger brains, both in absolute terms and relative to body size. We walk upright, while the apes use their knuckles as forefeet and are essen-

---

*The principal arguments over the past hundred years or so are excellently described by Roger Lewin in *Bones of Contention* (London: Penguin Books, 1987).

tially quadripedal. Smaller but sometimes sharper anatomical distinctions also help the diagnosis—notably, that pongids have big, prominent canine teeth that they use for display, while the hominids' canines became more and more like incisors as evolution proceeded and are used for eating. Even so, many biologists have complained, as Linnaeus did, that the distinctions are matters of degree and the line is hard to draw. Indeed, modern studies of the kind described in the box below reveal that the DNA of modern humans is extraordinarily similar to that of modern chimps ("99 percent overlap," as the biochemists tend to say) and only slightly less similar to that of gorillas. This is less than the difference between many pairs of species in some other groups that are virtually considered to be the same animal. Orangutans, on the other hand, the Asian wing of the Pongidae, are far more different from chimps and gorillas than the two African genera are from hominids.

MOLECULAR PHYLOGENY

Nowadays molecular biologists—or, less grandly, biochemists—can throw light on various issues of animal relationships: How closely are different animals related to each other? Who might be the ancestor of whom? When did two related species first share a common ancestor?

The logic of their approach is simple. Over evolutionary time, body proteins alter, and so does the DNA (deoxyribonucleic acid) that provides the codes for those proteins, just as gross anatomical features alter. However, there are two distinct differences between proteins and DNA on the one hand, and gross anatomical features on the other. The first is that there are straightforward objective ways of measuring the difference between the proteins or nucleic acids of different animals. For example—and this is a technique first reported as long ago as 1902—the difference between body proteins from different animals can be judged by the relative strength of their reaction when exposed to particular antibodies. The differences between two samples of DNA can be judged by seeing how tightly they cleave together when given the opportunity to do so, which is a very simple measurement to make. Or, if there is time and opportunity, different protein or DNA samples can be analyzed and then compared. Of course, gross anatomical features are also measured, but it is much harder to quantify the differences and similarities of two femurs or molars from different creatures than it is to mix two samples of DNA and measure the subsequent change in viscosity.

Second, the gross anatomical features of an animal are exposed to

continued next page

continued from previous page

the outside world. They matter. They determine the survival of the animal. They are, in short, subject to natural selection and can change rapidly (or be pressured by natural selection not to change at all). Thus, two related animals that live different kinds of lives can quickly evolve different features, while unrelated animals living the same lives come to resemble each other through convergence. Often it can be hard to distinguish the homologous features—those that look the same because they evolved from a common ancestor—from those that are merely analogous, where resemblance is due to convergence.

Of course, proteins and nucleic acids also matter, and they, too, are subject to natural selection. But their molecules also contain large sections that seem to do very little, and so they are not subject to natural selection because small differences in their structure make little difference to their owner's survival. Yet these apparently nonfunctional regions continue to change as the generations pass, because as time passes, more and more random genetic mutations take place. Mutations in features that affect the animal's function are likely to be eliminated (or very occasionally favored) by natural selection. But mutations in parts of molecules that do not affect the animal's life simply accumulate. Hence, the degree of difference between two animals in the nonfunctional regions of their proteins or nucleic acids is related directly to the time that has elapsed since those animals last shared a common ancestor. Thus, mutations within proteins or nucleic acids can provide a molecular clock.

Note, finally, that in practice, DNA occurs both in the nucleus and in organelles in the cytoplasm of the cells known as mitochondria. Both nuclear and mitochondrial DNA are used for comparative purposes. Mitochondrial DNA tends in general to mutate more quickly than most nuclear DNA and is good for working out relationships between closely related species or of populations within species.

Of course, in addition to DNA and proteins, all living cells also contain RNA (ribonucleic acid), which acts as an intermediary between the DNA and the proteins. RNA changes only very slowly over time, but this means that it can provide a molecular clock that operates on a very long time scale. Hence, it is useful for working out relationships between large groups of creatures—classes, phyla, or even kingdoms—whose most recent common ancestors may have lived many hundreds of millions or even billions of years ago.

As noted in the text, studies of both DNA and various proteins show that chimpanzees and humans are extremely closely related and shared a common ancestor within the past few million years.

For this kind of reason, then, many biologists argue that the present classification makes no sense. A family boundary is drawn between two species that are obviously closely related—chimps and humans—while two that are known to be more distant—chimp and orang—are grouped together. It would surely make more sense to put the humans and all the great apes together in one family, which in practice would be the Hominidae, since this name was coined first (by Linnaeus himself in 1758) and, by the rules of nomenclature, takes precedence. Taxonomy is supposed to represent real relationships, but the pongid-hominid division seems to cut right across the biological realities.

I have great sympathy with this latter view—I like the idea that human beings and animals form a continuum—and yet feel, on other grounds, that the present taxonomy contains another quite different but at least equally important biological truth—not evolutionary, but ecological.

In practice, the classification of living creatures is guided by two quite different criteria. The dominant criterion is indeed that of phylogenetic relationship: who is literally related to whom. Thus, as described in the boxed material on pages 131–137, biologists seek to place living creatures in clades, the members of which are presumed to share a common ancestor and to be more closely related to each other than to anything else. But biologists also, like it or not, classify animals in a less purist fashion, according to grade. Animals placed in the same grade are, broadly speaking, related, but for various reasons they may not form a true clade. The reptiles—class Reptilia—are a prime example. The Reptilia could be recognized as a clade only if we allowed it to include the birds and the mammals, which both descended from reptiles. Contrariwise, the living reptiles that are all placed in the same class are not all closely related to each other. Indeed, it is clear on various grounds that birds are much closer to crocodiles than crocs are to lizards. The similarities can be seen in detailed anatomy and in the fossils—and even in behavior. Baby crocs have been seen sitting in the lower branches of trees, chirping for their mother (crocs are excellent mothers), for all the world like sparrows in a suburban hedge. Ecologically, however, crocs are clearly more lizardlike than birdlike. In particular, crocs and lizards are both poikilotherms. So in practice, while acknowledging the phylogenetic relationships of birds and crocs, it is sensible to stick with tradition and leave birds as Aves while keeping crocs with lizards in the pseudoclass Reptilia.

So it is, I suggest, with pongids and hominids. Genetically, the pongid chimp and the hominid *Homo* are extremely close; so close, indeed, that it may be possible to form hybrids between the two. We cannot take this for granted, because in practice the barriers between species do not depend only upon the overall genetic distance between them but on the possession or nonpossession of genes that create specific barriers to successful mating. We can see this principle in action among the tree snails (in the family Partulidae) of the high volcanic islands of the Pacific. Some pairs that are obviously extremely similar, both in appearance and in their genes, are nonetheless unable to mate successfully, whereas other pairs that are obviously much less closely related can mate to produce hybrids. However, I simply do not know whether hybridization has ever been attempted between chimps and humans; it would be very easy technically to inseminate a female chimp with human semen, but if anyone carried out such an experiment in some far flung laboratory, they would probably keep quiet about it.

But in practice, ecologically and morphologically, humans and chimps are quite different creatures. Morphologically, human beings have invented an entirely new ecomorph. Indeed, as discussed further in the following chapters, they have developed a whole new way of life; ecologically, a chimp is hardly more similar to a human than it is to a ground hornbill, which is also sociable (like chimps) and goes hunting in family groups. Indeed, it would not be entirely foolish, if classification is to reflect the broad reality of life and not simply the niceties of genes, to place human beings in a new kingdom. In some significant ways they are as different from all other creatures as plants are from bacteria.

But does this not imply disrespect for the chimp? Is it not better to emphasize the chimp's similarity? I would argue precisely the opposite. It is proper for us—indeed, it is the only tolerable stance—to respect all other living creatures, and for that respect to be as nearly absolute as possible. We are powerful, but our power can properly be seen as a gift. It was not of our making; and, I suggest, power should always acknowledge the principle of noblesse oblige. In short, our respect for chimps should not be dependent on their genetic similarity to us. That is merely another form of anthropocentrism. That they are our fellow creatures, and obviously wonderful, is enough. Is it suggested, after all, that we should accord more respect to chimps than to the beautiful, soulful, but more distant orang? Anyway, we have seen that the pongids—at least the existing ones—comprise Pan, Gorilla, and Pongo. As for the hominids, paleoanthropologists over the years have classified the copious but still frustrating scattering of fossils in many different ways and have spawned a host of generic names. Some of these have been abandoned: like Pithecanthropus, and Eanthropus, which was the name once attached to the spurious "Piltdown man." Most specialists today recognize either two genera of hominids, or three. I will follow the latter course.*

THE THREE GENERA OF HOMINID

The three known and recognized hominid genera all arose in Africa. The first to arise was *Australopithecus,* which sounds Australian but in fact simply means "southern ape." (The scientist who coined this somewhat infelicitous cognomen was himself Australian, though it would be mischievous to suggest that this biased his nomenclature.) These early australopithecines had thinnish crania and smallish jaws and are said to be "gracile." In general, they were between a meter and a meter-and-a-half in height, with crania (and hence with brains) that were roughly the same size as a chimp's.

Then, it seems, *Australopithecus* produced two branches, which evolved in very different directions. One branch developed thick-boned skulls and jaws, and are said to be "robust," and these robusts are commonly placed in their own genus, *Paranthropus.* The other branch remained gracile but developed bigger brains; and this branch became the genus *Homo.* Thus, the evolutionary tree of the Hominidae is generally seen to be essentially Y-shaped: gracile australopithecines in the stem giving rise to robust paranthropines on the one hand and gracile hominines on the other. Incidentally, some paleoanthropologists include the genus *Paranthropus* within the genus *Australopithecus,* and hence recognize only two genera: *Australopithecus* and *Homo.* They argue that the different robust species may have evolved independently of each other, from different australopithecine ancestors, and therefore are not truly related. In this case, *Paranthropus* would not be a

---

*In *Nature* on May 4, 1995, p. 88, Tim White and his colleagues (see text) proposed that the oldest known hominid, which they had previously called *Australopithecus ramidus,* should be re-named *Ardipithecus ramidus.* At the time of writing this footnote— late July 1995—paleoanthropologists are not yet universally agreed that *ramidus* is different enough from other members of the *Australopithecus* genus to warrant a new generic name. For present purposes, therefore, I am sticking to the original name of *Australopithecus ramidus.* But if the name *Ardipithecus* does become accepted then all statements in this text concerning the number of accepted hominid genera must of course be increased by one.

"natural" grouping. I have no views on this. I am following the three-genus grouping, which also has strong supporters, mainly, it seems to me, because it is tidier.

The question now—as it has been ever since Darwin—is, which of the apes did the hominids derive from? Who was the immediate ancestor of *Australopithecus*? The answer is that nobody knows, because although we have good fossils of ancient hominids, we have no fossils at all from apes that could be candidate ancestors. The chimps—the common chimp and bonobo (formerly known as the pygmy chimp)—are our nearest living relatives, but neither is likely to be our ancestor. Both are too specialized: both have clearly developed characteristics of their own since the split with the hominids. It is more realistic to suppose that the chimps and the hominids simply shared a common ancestor—and this creature, "the common ancestor," is the one that everyone now would dearly like to find. The common ancestor is, in short, the modern equivalent of the old-style "missing link." But on various grounds it is very reasonable to assume that the common ancestor was much more like an ape than a modern human; and since chimps, for all their specializations, are very obvious apes, it is reasonable to think of the common ancestor as an essentially chimplike animal—which might indeed, when it is found, be placed in the genus *Pan*. But for the rest of this account I will simply call the common ancestor "the common ancestor" or "the ape ancestor."

The next essential question is, when? When did the hominids arise from the pongids? Or, more specifically, when did *Australopithecus* diverge from *Pan*?

PONGIDS BECOME HOMINIDS

The issue of when the hominids arose from apes has exercised biologists ever since Darwin proposed in 1871 that such a transition had in fact taken place. One powerful tradition has maintained that the split took place a very long time ago—in fact, that it happened well before the emergence of the modern pongids had appeared themselves, which means that the pongids in general and the chimp in particular could not be our ancestors. Beyond doubt, the study of human evolution has been bedeviled by prejudice, and we might detect here the desire of some eminent professors to place as much distance between themselves and chimps as possible. To be fair, however, protagonists of the hominids-are-ancient-and-separate tradition tended rather to suggest that an organ as remarkable as the human brain must have needed many millions of years to evolve. More specifically, the excellent paleontologists Elwyn Simons and David Pilbeam, both formerly at Yale, suggested in the 1960s that the early Miocene ape *Ramapithecus* showed hominid features in its jaw; and extrapolations from this suggested that a discrete line leading to hominids might have begun as early as the Oligocene, more than 25 million years ago. Similarities between us and chimps would then be a matter of mere convergence. By no means everyone agreed with such a view, but it was certainly influential. The diagram of human evolution on page 642 of the second edition of J. Z. Young's *Life of Vertebrates* (London: Oxford University Press, 1962)* shows just this: the hominids going their own sweet way even before the Miocene began.

That idea is now dead. Indeed, a stake has been driven through its heart. Some paleontologists, like Elwyn Simons, argue that paleontology itself killed it off. That is, more Miocene fossils turned up that on the one hand were virtually the same as *Ramapithecus* but on the other hand were obviously related to orangs. So *Ramapithecus,* too, must be in the orangutan rather than the gorilla-chimp-hominid lineage. In practice, however, the *coup de grâce* to the hominids-are-ancient-and-separate idea has been delivered since the 1960s by the newly emerging techniques of molecular phylogeny, by which the relationships and possibly ancestry of animals is determined not by differences in their gross anatomy but by differences in body chemistry.

Specifically, Vincent Sarich and Allan Wilson of the University of California at Berkeley first examined the albumen protein in the blood, and then looked directly at the DNA of humans, chimps, and other primates. (See the Molecular Phylogeny box on page 178.) The albumens and DNA of humans and chimps proved remarkably similar; so similar, said Sarich and Wilson, that the two must have shared a common ancestor only 4 to 6 million years ago, not in the Oligocene, but in the very late Miocene or even the Pliocene.

At the time, that conclusion was stunning. At first, many paleontologists simply felt that Sarich and Wilson were wrong. But now the two are vindicated: a 4 to 6-million-year divergence—call it 5—has become

---

*I cite *Life of Vertebrates* not as a primary source, which in this context it obviously is not, but as a highly influential one, whose author, one of the outstanding zoologists of the midtwentieth century, tried hard to reflect the orthodoxy of his day. Young's espousal of the hominids-are-ancient theory thus reveals the zeitgeist.

the orthodoxy. The ancient *Ramapithecus* is no longer plausible as a human ancestor, and the realization has grown—thanks not least to Stephen Jay Gould and Niles Eldredge—that huge evolutionary changes, including the growth of the human brain, can take place quickly when the conditions are right. The time since the late Miocene embraces all of our prehistory that is specifically hominid. For us, the Pliocene-Pleistocene is truly the day before yesterday.

So the next obvious question is, how? How did the chimplike common ancestor evolve into us?

HOW THE FIRST HOMINIDS BECAME

The three principal species of australopithecines have been discovered in reverse order of antiquity. The youngest is *Australopithecus africanus*, found in a lime pit in South Africa and named by Raymond Dart in the 1920s. Older than *africanus*, and apparently the species that lasted the longest is *A. afarensis*, discovered by Don Johanson and his colleagues on a hillside at Afar, Ethiopia, in 1972. The oldest of all is *A. ramidus*, found in 1992, again at Afar (in a place called Aramis), and reported in *Nature* by Tim White and his colleagues (T. D. White et al., "*Australopithecus ramidus*, A New Species of Early Hominid from Aramis, Ethiopia," *Nature*, September 22, 1994, pp. 306–312).

Datewise, *Australopithecus ramidus* is tantalizingly close to the supposed age of the common ancestor. Fragments of seventeen individuals were found lying on fossilized volcanic ash (tuff) that is dated at 4.4 million years. Furthermore, *ramidus* has precisely the combination of characters that paleontologists were hoping to find and indeed expecting: more apelike than *A. afarensis*, less apish than a chimp, but with qualities of both. *Ramidus* comes from the Afar word *ramid*, meaning "root," a pleasing piece of nomenclature.

Between them, the different *ramidus* fragments provided a complete set of teeth—including one deciduous, or milk, tooth—plus the bases of two skulls and, most unusually, all three bones from a left arm. There is plenty here to go on even though there are some frustrating omissions.

Thus, the teeth reveal an animal that really does seem to be halfway between an ape and the later, more humanlike *afarensis*. The canines are still prominent, but compared to a chimp's, say White and his colleagues, they appear "low and blunt." The bottom canines are also worn, as if they were used for eating rather than display. The enamel of the canine and other teeth is thin like a chimp's, not thick like a human's. It is not as thick even as that of *afarensis*. The milk tooth was a first, lower, milk molar—a tooth that has always proved particularly instructive in deciding who among the apes and hominids is related to whom. In *ramidus* this tooth is narrow like a chimp's, not wide like a human's: "far closer to that of a chimpanzee than to any known hominid."

The arm bones are broken, so it is impossible to say whether in life they would have been long like a chimp's or shorter like the known hominids; but the bits that do remain show a mosaic of chimpish and hominid details.

But the key quality of the hominids is their bipedalism. The crucial issue, therefore, is whether *Australopithecus ramidus* walked upright and, if so, whether it did so confidently and with all the appropriate anatomy and reflexes. But the bones that could really tell us—in particular the pelvis and the feet—are missing. There is a clue, though, in the base of the cranium. For in animals that walk on four legs, the place where the neck bones join to the skull, the foramen magnum, is far back, so when the animal walks, its horizontal head is thrust out in front of the horizontal backbone. But in humans and the other australopithecines the foramen magnum is farther forward, so that the head balances on top of a vertebral column that is held vertically. The pieces of *A. ramidus* skull show a foramen magnum that is significantly farther forward than in a chimp. The inference is that this indeed was an upright animal.

The next species in order of age, *Australopithecus afarensis,* leaves us with far fewer doubts. The first *afarensis* that was found by Don Johanson and his colleagues included almost half the skeleton. This was the famous Lucy, named after the Beatles song *Lucy in the Sky with Diamonds,* which at that time resounded through the Johanson camp. *Afarensis* have now been found throughout East Africa, in Ethiopia, Tanzania, and Kenya, and they range in age from more than 4 million years, to less than 2.5 million. Lucy herself is around 3 million years old.

If any creature can properly be called a typical australopithecine, it is *afarensis*. Its brain, at 400 to 500 ml, is not conspicuously larger than a chimp's, but *afarensis* does have a somewhat less protruding face and canine teeth that are more prominent than ours but far less so than those of any ape or of *ramidus*. The arms are relatively longer than a human's, but far shorter than a chimp's. Most excitingly, though, we can answer the crucial question that for *ramidus* remains uncertain. We know how *afarensis* walked.

For one thing, Lucy's skeleton does include much of the pelvis and some bones of the legs, which are proportionally longer than a chimp's and apparently more stiltlike, like a human's. Uprightness is suggested. What clinches the matter, though, is the wonderfully felicitous find by Mary Leakey in the late 1970s, some six years after Johanson's initial discovery. She came across footprints that were made by three *A. afarensis* as they strode across volcanic ash at Laetoli, Tanzania—ash that fell 3.6 million years ago and is now fossilized. The prints show very clearly what the anatomists hitherto had been able only to guess: that *afarensis* could walk happily on two legs, effectively in the heel-and-toe fashion of true human beings. Incidentally, those fossil prints were crisscrossed by the tiny hoofprints of a three-toed hipparion. There are scenes from history I would love to have witnessed. The Sermon on the Mount would be one. The little band of *afarensis* and the three-toed ponies on the Pliocene mud of Tanzania would be another.

The discovery of Lucy and the footprints of her fellows finally answered a question that had concerned all paleoanthropologists ever since Darwin: whether the first hominids became bipedal before they became big-brained, or whether they acquired their big brains first. For reasons that in retrospect seem largely mythological, leading scientists of the early twentieth century favored the latter view: that human evolution had been brain-led. It was largely for this reason that they fell so easily for the hoax of Piltdown, in which, in 1912, parts of a modern human skull and the slightly doctored jaw of an orangutan were stained to make them look old and buried side by side in a gravel pit in Sussex.* This was what they were looking for: a creature with apish features in which the brain had clearly developed first. That Piltdown was

---

*Although some paleoanthropologists were never taken in by Piltdown, the ones who were included some of the most distinguished and expert of their time—Sir Arthur Keith, Sir Grafton Elliot Smith, and Sir Arthur Smith Woodward in Britain, and Henry Fairfield Osborn in the United States. Who planted the spurious bones is one of the great whodunnits of the twentieth century, although the leading candidate is their discoverer, the lawyer and amateur bone-seeker Charles Dawson. Eminent professionals should not have been fooled by the mock-up, but it met their preconceptions precisely: Arthur Keith's belief that the hominid line was ancient, and separate from the pongid; and Elliot Smith's firmly held belief that hominid evolution had been brain-led.

a hoax was finally revealed by chemical tests at the British Museum in 1953. As we have seen, the actual facts of the case did not become clear until Lucy et al. appeared in the 1970s: that our australopithecine ancestors were slim and upright, broadly similar to small humans though with relatively longer but far from apelike arms; and with a face that was somewhat flatter than an ape's but topped with a small apish cranium. The big brain was imposed later onto this quasi-human form.

Finally, the youngest of the gracile australopithecines, *A. africanus*, is southern African rather than East African, as *afarensis* and *ramidus* were. In general, *africanus* looks more human than *afarensis*, with a less prominent face and more incisorlike canines. *Africanus* seems to have appeared about 3 million years ago, and died out some time after 2.5 million years ago—not as long-lived as *A. afarensis*.

Then there were the robusts, who some prefer to call *Paranthropus*, which is the convention I follow here. They were not in fact conspicuous bruisers: at less than a meter and a half they were no taller than the graciles. But they had heavier skulls that in some were crested for muscle attachment, and huge jaws with millstone teeth, presumably for grinding coarse vegetation. As Elisabeth Vrba has commented, "They tried to chew their way out of trouble." Conservative classification recognizes two species from East Africa: the earlier *P. aethiopicus* from about 2.6 million years ago and the later *P. boisei* who presumably descended from *aethiopicus* and lived from about 2.3-to-1.2 million years ago. In southern Africa lived *P. robustus*, from approximately 2 million years ago to roughly 1 million years ago—the most recent australopithecine (defined broadly) to survive.

Finally, the hominines (members of the genus *Homo*) first appeared around 2.2 million years ago in the form of *Homo habilis*. Events after that occupy most of the rest of this book. First, it is pertinent to ask how

---

These preconceptions sadly affected the anthropologists' response to the first report of *A. africanus* in 1924, which was the first australopithecine to be found and in some ways was therefore the paleoanthropological discovery of the century. Their failure to acknowledge the importance of *africanus* in turn confused the entire field of paleoanthropology for the following three decades. In fact, *africanus* was reported and described in February 1925 by Raymond Dart, an Australian anatomist, who also gave the australopithecines their somewhat cumbersome appellation. Arthur Keith, Elliot Smith, and Arthur Smith Woodward denied that *A. africanus* was any kind of human ancestor partly because they were already wedded to Piltdown, and partly because both *A. africanus* and Raymond Dart were colonials, and therefore not pukka, whereas they (and Piltdown) were true Englishmen (or in Keith's case a Scotsman, which is the next best thing). This sounds bizarre, but is generally considered to be the case.

these various hominids were related to each other—who really gave rise to whom.

RELATIONSHIPS

Exactly how the three gracile australopithecines are related to each other, and to the later paranthropines and hominines, is unknown. *Ramidus* may be the first true hominid, or there may be other distinct species between *ramidus* and the missing common ancestor. It seems highly likely, however, that *ramidus* gave rise to *afarensis,* and that *afarensis* gave rise to *africanus.*

Beyond that, there are several theoretical possibilities. Perhaps *africanus* gave rise to *Homo,* leaving *afarensis* to produce the paranthropines. Perhaps *africanus* gave rise to the paranthropines, while the hominines derived directly from *afarensis.* Perhaps *afarensis* generated *africanus* and the paranthropines and the hominines, in which case *africanus,* like the paranthropines, was simply a dead end. Perhaps *afarensis* themselves, despite their variability and ubiquity, are off the main line. We just do not know. But it does at least seem reasonable to assume that *afarensis* is the ancestor both of the paranthropines and of the hominines; although for the time being *africanus* will have to be left floating, perhaps as a human ancestor, perhaps as a paranthropine ancestor, and perhaps as an evolutionary twig that blossomed for a time in the south and then faded.

So now we should ask the biggest question—why? I suggested in chapter 3 that big evolutionary changes generally take place only when the ecosystem is given a kick, and the various lineages—those that have the genetic wherewithal to do so—are forced to change. So what did the kicking in this case? Why did the australopithecines become? The broad answer to the question "Why?" surely lies in the global changes discussed in chapter 2: the cooling of the Earth that has taken place throughout the Cenozoic as the Himalayas rose and leached the insulating carbon dioxide from the atmosphere. The cooling has not occurred smoothly, for that is not the way of things, but in jerks, and in the late Miocene there came a sudden fall in temperature and, with it, increasing aridity, for coolness and dryness go together. The habitat changes when this happens, and animals that cannot simply move out, die; and so there was a huge die-off of native mammals in the late Miocene of North America. In particular, forests retreat when the world cools and dries, for trees above all need water. In the Miocene almost worldwide it was the grasses that stepped opportunistically into the breach. They had played only minor roles since the warm, wet Eocene, but now they came truly into their own.

At first in late Miocene Africa the margins of the great central forest turned into more modest woodland. These woods were closed at first—the canopy meeting overhead—but they became increasingly open as the cooling continued. Then, at around 2.5 million years ago, as the Pliocene neared its end, came another burst of cooling and drying and much of the woodland gave way finally to the kind of savannah that we see today—wooded in parts, but much of it merely dotted with scattered acacias.

The common ancestral ape that lived in the late Miocene forest of Africa evidently produced scions for all occasions. One lineage became the chimps, which adapted more and more to the dwindling but still-vast forest. The other lineage became the hominids, which adapted first to the more open woodland, and then, when the time came, to the newly expanding savannah. In general, the gracile australopithecines were creatures of the woods, although *afarensis* finally adapted to a wide range of territory, both wooded and more open, while the genus *Homo* truly rose to the challenge of open country and became a creature of grassland. Indeed, *Homo* appeared right on cue; soon after the global cooling of 2.5 million years ago.

We have direct evidence that *A. ramidus*, the creature closest to the common ancestor, was a creature of woodland. A marvelous assemblage of plant and animal remains from the immediate surroundings shows that this was the case (these accompanying fossils are recorded in a separate paper: Giday Woldegabriel et al., "Ecological and Temporal Placement of Early Pliocene Hominids at Aramis, Ethiopia," *Nature*, September 22, 1994, pp. 330–333). Thus, there were thousands of seeds of *Canthium*, a plant commonly found in African woods and forests. Primates were the commonest large vertebrates, including baboons and colobus monkeys—the latter accounting for 30 percent of all the fauna. The most frequent bovid by far was a medium-sized kudu, another woodland creature.

Then there was a pygmy hippo, *Hexaprotodon*, an extinct genus—another woodland creature, like the modern pygmy hippo of West Africa. It seems that the pygmy hippos declined as the woodland bovids rose to prominence, while the common hippo, *Hippopotamus*, escaped bovid competition by staying in the water. There were pigs alongside *ramidus*, too; plus the three-toed horse, *Hipparion;* and *Ceratotherium praecox*, the early form of the modern African white rhino. There were two giraffe species, both very much like the modern giraffes (and perhaps they should be included in the modern species). Then, too, there was a marvelous assemblage of proboscideans—not a lot of them, but they were there. There were members of the modern family, Elephantidae. There was a straight-tusked gomphothere, *Anancus*. And there were some of the amazing deinotheres, *Deinotherium*, with their lower-jawed, downward-pointing tusks. At the other end of the size scale, there was a small squirrel, plus two species of bush rats and a mole rat.

The carnivores were equally impressive, and clearly geared to the complete range of herbivores. At the small end was the dwarf mongoose *Helogale;* and there were otters, civets, canids, cats, and hyaenids. Then at the top of the predatory size range was the puma-sized saber-tooth *Meganteron*, ancestor of *Smilodon*. And finally, to cap them all, there was the first bear that has ever been found in East Africa, of the genus *Agriotherium*. The African woods and forests have been sadly deprived of bears ever since. We tend from our northern vantage point to think of grizzlies and polar bears, but bears are just as at home in tropical forests and may still be found in the forests of India (sloth bear), South America (spectacled bear), Asia (Asiatic black bear), and Southeast Asia (sun bear and giant panda). Why Africa has had no bears in living memory south of the Sahara is a mystery.

Finally, for good measure, the *ramidus* fossils were accompanied by those of bats, birds, tortoises, and crocodiles. For *Australopithecus ramidus,* some of these creatures would have been food, some competitors, and some mortal enemies. But if *ramidus* had had the wit or the leisure to be naturalists, they would surely have been in heaven. The fauna they lived among was truly fabulous.

That the first australopithecines lived in such woodland was crucial to their evolution, and hence has been crucial to ours. If they had remained in dense forest, it is hard to see how they could have evolved at all, for dense forest seems to demand the ecomorph of the chimp. If they had been forced to transpose in one swoop from forest to grassland, then surely they would have failed, not least because of the jaws of lions and hyenas.

But woodland is perfect country to promote the kind of evolutionary changes that we can actually see when the common ancestor, a forest ape, became a hominid. For such woodland is of the kind that in medieval England was known as forest: clumps of closed woodland with bushy clearings (which in Africa grew as the Pliocene wore on). England's existing New Forest shows the form exactly. Sherwood Forest, home of the legendary Robin Hood, was also of that type. Indeed, such forest covered vast swathes of medieval England, and huge numbers of people made their living from it: cutting and coppicing trees (cutting them to the ground, and harvesting the long, thin branches that spring from the stumps), burning charcoal, hunting game, grazing cattle, and herding pigs. These people were collectively known as foresters: a very different métier from the present-day graduates in forestry who culti-vate trees for wood pulp and timber. So important were such forests to the medieval economy that half the laws of England were concerned with them: who could graze what, where, and when, and the dire pun-ishments that would befall transgressors. The Magna Carta is largely concerned with forestry, in that traditional sense. In short, the foresters of the English Middle Ages demonstrated by their number, variety, and overall economic importance that woodland is rich and varied territory offering many opportunities to the imaginative incumbent. The imagi-nation of the australopithecine was closer to that of an ape than to a human being, but the stimulus was there nonetheless.

For example, and most significantly, the woodland existence of the australopithecines explains how they evolved their bipedalism. For if they had gone straight into grassland, we would have to ask how they could have survived before they learned to walk and run so well on two legs. But if they had stayed in the forest, then it is hard to see how they would become bipedal at all. Chimps are not bipedal, after all, and nei-ther are monkeys. Why should an ape that is still living in dense forest become bipedal? But in their African woodland, their "medieval for-est," we can envisage them living a double life, half in the trees and half out of them. They may well have continued to sleep in the trees, on platforms, as chimps and gorillas do. They probably took to the trees at the first sign of danger—which in large part explains their survival in the face of lions and hyenas. They would also have used the trees as vantage points. In similar fashion cheetahs sit at the tops of termite mounds to watch for prey and meerkats climb to the tops of bushes to scan the sky for hawks. But increasingly the early australopithecines fed on the ground. Increasingly, they dashed across the clearings both to escape the many dangers and in pursuit of prey. This way of life—re-treating to a safe place and dashing out to feed with a weather eye on predators—is common in nature. We see it in virtually all desert ro-dents, which live in holes, and the small carnivores that prey upon them but keep their eyes open for bigger carnivores, or in rock hyraxes that sally forth to forage but must dodge the desert eagles as they go. Thus, we may envisage Lucy and her tribe, standing upright in the boughs at the edge of the copse, holding a branch above for balance—

a good, safe, all-seeing position highly reminiscent of that of a watchful meerkat. Lucy was found on a barren hillside, right enough. But when she and her kind lived there in the mid-Pliocene, it was a much friendlier place.

But did the early australopithecines really dash after prey? Did they eat meat, and how much of it? This is a crucial issue and has been a highly contentious one. I am quite sure that they did, and that meat-eating has played a key part in human evolution.

AUSTRALOPITHECINES AS PREDATORS

The way that animals live, how they spend their days and the extent to which they socialize, is largely determined by what they eat. Committed herbivores need to eat more or less all day (or all night) if they are to find enough, while big predators eat only occasionally and spend most of their time dozing or frankly comatose. Some creatures need to be sociable in order to feed efficiently, like most vultures, who rely upon each other to seek out cadavers. Other animals, like Sumatran rhinos, need to be more or less solitary if they are to find enough to eat. Clearly, too, diet determines ecological impact: predators stand higher on the food chain than herbivores do, but in general are less numerous than herbivores. What australopithecines ate, then, and how they found their food, is a crucial matter. It largely determines what kind of creatures they actually were and how many there were and what effect they had on all about them.

To a large extent, diet can be inferred from anatomy, but there are drawbacks to this method. We know for example that the fearsome predatory jaw power of the giant panda is expended (even squandered) on bamboo. More to the point, primates tend to be generalists, generally lacking the dental extremes of the most specialist feeders. Patterns of dental wear offer clues to diet, but the deep scars left by the coarsest foods, for example the skins of wild fruits, tend to blot out the milder abrasions of softer fare. But the softer fare—including meat—may have a far greater influence on the animal's welfare and way of life.

As we can see, there is scope for honest doubts about the diets of early australopithecines. But a lot of argument, too, has been based merely on preconception. Thus, in *African Genesis*, published in 1961, journalist Robert Ardrey contended that the australopithecines were ravening carnivores. He also argued that carnivory is a prime source of aggression in animals, and further suggested that modern humans are innately aggressive because they are the scions of carnivores. Others, particularly in the gentler seventies, depicted australopithecines as the world's first flower children, whiling away the African millennia in vegetarian communes. You pays your money, it seems, and you takes your choice.

If we apply common sense, ecological principles, and knowledge of other animals and people that face some of the same problems as the australopithecines, then we must conclude that Ardrey was more right than wrong, although not in the way that he argued. The unsurprising reality, probably, is that the first australopithecines were omnivores, eating everything. In fact, versatility was the key to their immediate success and to their later evolution, including the evolution of the large brain. They may indeed have eaten only small amounts of meat, perhaps not much more than 10 percent of total calories. On the other hand, even such a small commitment to meat can have a profound effect on a creature's ecological success, on its effect upon other animals, and as Ardrey suggested, though not quite accurately, on its personality.

First, I suggest that the notion that the first australopithecines did not eat significant amounts of meat is belied in part by their teeth. Admittedly, although their front teeth were small, the first known australopithecines had big wide molars and premolars, relatively bigger than a modern human's or indeed a chimp's, and clearly designed for grinding vegetation. But were they big and millstonelike enough to enable their owners to subsist exclusively on the coarser vegetation they would find outside the lush tropical forest? Wouldn't active creatures with such teeth and guts have needed meat to carry them through? On evolutionary grounds, too, it would be surprising if they were not meat-eaters. After all, we know that we eat meat, and that our immediate Stone Age ancestors did, and that some of our number today—such as the Eskimos and South American gauchos—eat meat almost exclusively. We also know, thanks to Jane Goodall, that chimpanzees not only eat meat but are formidable hunters. It would be odd indeed if the lineage that runs from the chimplike ancestor to australopithecine to *Homo* had switched away from meat, and then switched back again.

Furthermore, I am sure that the way in which chimps obtain meat is highly instructive in a way that is reminiscent of Ardrey's ideas, though not quite the same. Ardrey argued that because australopithecines were meat-eaters, then this per se made them particularly aggressive. Such a broad generalization surely has to be rejected because in practice the modern committed predators that we know about simply are not more aggressive than many specialist herbivores. I would feel happier in the presence of a wolf than of a zebra stallion, while a wild bull Asian elephant effectively means instant death.

What is true, however, is that specialist predators employ "aggression" as a tactic or indeed as an essential weapon for procuring food. A lion that contrives to bring down a zebra or a sabertooth that leaps at a mammoth runs a tremendous risk. The flailing hooves and the clubbing trunk can be lethal. The predator must do as the boxer does: slip the opponent's guard, and get inside. But that takes nerve, and nerve requires hype. The necessary heightening of mood is achieved by mental preparation and surges of adrenaline. Without it, the predator would be ineffectual and would also endanger its own life. Every rugby player or boxer knows that the easiest way to get hurt is to be halfhearted. Only the rapid and committed strike can succeed.

In short, aggression is not gratuitous viciousness. It is a tactic: a device for raising the physiological tempo and focusing the mind in times of extreme physical exertion. Of course, it is risky to invoke a state of mind in which caution is thrown to the winds. But of all the options available to the hunter, it is the least risky. It is riskier by far to hold back. To avoid the fight altogether is simply not an option—not, that is, if the predator is ever to eat. Shakespeare's Henry V made the general point as well as it can be made at the siege of Harfleur (act 3, scene 1):

In peace there's nothing so becomes a man
As modest stillness and humility:
But when the blast of war blows in our ears,
Then imitate the action of the tiger;
Stiffen the sinews, summon up the blood,
Disguise fair nature with hard-favour'd rage;
Then lend the eye a terrible aspèct;
Let it pry through the portage of the head
Like the brass cannon; let the brow o'erwhelm it
As fearfully as doth a galled rock
O'erhang and jutty its confounded base . . ."

Heroic rhetoric, shameless hype—but also sound advice. Wolves indulge in similar exhortation as they leap and yip before the hunt, egging each other to action. Domestic dogs respond in this wolfish way to the universal cry of "Walkies!" which for them has little to do with exercise but is a symbolic adventure. Jane Goodall observed the rituals of

impanzees. In fact, the hunting technique of chimpanzees,
ired on film, is quite horrible. For hours a sinister mood
overtake them—a conspiratorial silence. Then the mood becomes ...enzied, and together, cooperatively, they encircle and close in
upon some hapless monkey—yelling, screaming, and shaking
branches, inducing stark terror which, in turn, causes the prey to make
mistakes. When the monkey is finally caught, it is torn, literally, limb
from limb in an orgy of excitement. Horrendous. But perhaps from a
survival point of view such apparent excess is necessary. The hype increases the vigor of the hunters and confuses the prey. Chimpanzees
do not have the appearance of more conventional predators, and until
Jane Goodall began her studies many biologists assumed they were vegetarians. But their massive strength, their cooperativeness, their intelligence, and their controlled use of "aggression" makes them very
proficient indeed. Chimpanzee hunts seem usually to succeed, while
lions are content to make one kill in every four attempts, and a tiger
one in ten.

I find it easy to envisage the first australopithecines in their semi-
tropical Mio-Pliocene Sherwood Forest employing very chimpish techniques: hurling themselves en masse from the trees in an orchestrated
frenzy, armed with sticks, and beating a monkey, a pangolin, a small antelope, or a pig insensible before it had time to gather its wits. And
such creatures, noisy, cooperative and frenetic, would in practice have
had little to fear even from the megapredators like leopards and hyenas. Such predators tend to be specialists. They do what they do. These
hyperactive hordes would have been outside their ken: too unpredictable, too weird. The early australopithecines, one might suppose,
would have been creatures to leave alone. Better stick to antelope.

Thus, Ardrey's original thesis is not quite right. The early australopithecines were not in a general way "aggressive," any more than a lion
or a domestic pussy cat is especially aggressive. But like lions and pussy
cats they were able to deploy aggression as a necessary technique for
obtaining food. In practice, however, unlike lions and other cats that
go about their work in appropriately professional silence, the method
of the australopithecines was that of the chimpanzee: the creation of
mayhem and confusion; the method of gang warfare. Professor
Bernard Wood of Liverpool University has pointed out that the names
given to early species of *Homo* are often designed to suggest that our
ancestors were serious-minded people, perhaps with their eye on their
glorious future. Thus, they are called *H. habilis*—"handy man"; or *H. ergaster*—"working man." But, we agreed, *Australopithecus hooliganensis*

might have been nearer the mark. Our first hominid ancestors succeeded, probably, by being at times extremely unpleasant—to a degree which, among other animals, only modern chimpanzees achieve. Controlled bursts of temper are one of the keys to our success. This is a chilling thought. It shows how naive it is to suppose that what comes naturally to us is ipso facto morally right.

To return to our thread: How much meat would the early australopithecines have obtained in this way? Would it have made a significant contribution to their diet? Robert Foley of Cambridge University argues on ecological grounds that the first hominids to make a truly significant switch to meat were members of the genus *Homo,* and notably *H. erectus.* They might have obtained 20 percent or more of their calories from meat, while the australopithecines—like modern chimps—perhaps obtained only 10 percent.

But for three reasons I feel that this putative 10 percent could have been highly significant. First, modern nutritionists generally agree that human beings do not need to eat large quantities of meat. We are perfectly able (and indeed are advised) to obtain the bulk both of calories and of protein from plants. But very few modern nutritionists advocate a vegan diet, with no meat at all. For the protein in plants, though often sufficient in gross amount, tends to be of dubious quality. It often lacks particular essential amino acids. Meat protein, by contrast, is first-class. It contains all the essential amino acids in appropriate ratios. Meat (defined broadly to include offals) also provides minerals such as zinc and vitamins such as B_{12}, which can be difficult to obtain adequately from plants. So meat, for humans, is valuable not as a bulk food but as a guarantor of quality. A little goes a little way—but that little is very advisable. It will convert a marginal or frankly poor diet into one that is more than adequate. Thus, I suggest that australopithecines who ate modest amounts of meat would thereby have increased their ecological range and their survivability out of all proportion to the apparent quantity of meat that they consumed.

There is a second aspect to this, too, which Rob Foley points out. In the tropical rainy season there are plenty of plants to eat, but the animals are hard to catch because they spread themselves far and wide to reap the bounty. In the dry season, vegetation is hard to come by, but animals are easier to catch because they gather in the least awful places—for example, around the shrinking water holes. As marsupials and camels demonstrate in their different ways, and as will become clear later in the context of game theory, one of the principal arts of survival is to get through the bad times. Meat-eating would have carried the australopithecines through the dry seasons. Again, this would have increased their range and survivability enormously.

Third, diet influences social life. Some hunters are solitary killers, like tigers and leopards, but some decidedly are not—like dogs of all kinds, lions, and ground hornbills. Primate hunters—chimps and ho-minids—are definitely of the social kind. As loners, they would proba-bly be fairly hopeless unless they were well armed. Cooperative hunting also requires mutual trust, and in general the fruits of hunting are exploited most efficiently when they are shared, because a good kill provides far more food than any one individual can consume, and a lit-tle for everyone goes a long way. Thus, hunting, even on a modest scale, would have consolidated social bonds, which in any case are strong and intricate in many primates. It would also, probably, have en-couraged division of labor among the sexes. We cannot be sure that australopithecine males would have done most of the hunting, but that is certainly the case among modern hunter-gatherers, and as a tactic it makes sense. Hunting is always dangerous, and men, who are not needed to bear or suckle children but can supply sperm in more than adequate amounts, are more expendable.

However, there is an odd sense in which the apparently small depen-dency on meat is worse for prey species than is total commitment. For the population of a specialist predator must dwindle if it overhunts its prey. The classic example is the northern lynx, whose population rises and falls with that of its principal prey, the snowshoe hare. But a non-specialist predator—one that can eat many different kinds of animals, or eats both meat and plants omnivorously—is not dependent on any one prey species. When the favored prey is reduced, it simply changes diet. But if the favored prey is truly favored, then the predator may pick it off in passing even after it has become rare. Thus, do sea otters make very short work of sea urchins, which they love, subsisting on clams when the sea urchins decline but continuing to zap them in passing nonetheless. Thus, are feral pussy cats in Australia picking off the small marsupials one by one. Thus, could modern whaling ships drive blue whales to extinction if they were not constrained by law, because the steady supply of permitted minke whales provides enough to keep them at sea, and they can simply take the blues en passant. In general, animals like us—which eat meat opportunistically, and simply shift their diet when the favored prey becomes rare—are able to break the ecological law which says that predators are bound to be rare relative to their prey.

The overwhelming point, though, is not that australopithecines were hunters, significant though it is. The key point is the one I stressed in chapter 3: that natural selection will produce significantly new creatures only under the right conditions. Here in the benign Miocene version of Sherwood Forest was everything that was needed. The woodland setting provided endless variety—and endless reward for creatures that could fully exploit that variety. Versatile animals could make their living in a virtual infinity of ways and therefore could hang on even in the bad years, while the more specialist species were subject to the vagaries of weather. The australopithecines responded to that variety. The genes were right: they could partake of fragments of the central gene pool of apes, the common ancestors, which may well have been vast and must have been extremely various just as the modern chimpanzees are various. The population sizes were right (or some of them were)—not too big and not too small—and they were isolated or semi-isolated at the forest edge in precisely the way that was needed to encourage speciation.

And perhaps most significantly of all, the ecomorph was right. Here were creatures with hands that had been honed to grip and arms that were designed to whirl through all the points of a sphere and hence provided the perfect foil for those hands. These features had been evolved for arboreal purposes, but now, more and more, they were freed from the chore of locomotion as the australopithecines increasingly cultivated the art of bipedalism. In all these faculties the hominines picked up where the australopithecines left off.

Before we move on to the hominines, however, there is one theoretical matter that deserves attention if we are really to understand what happened. So far, I have mentioned six different species of australopithecines—three graciles, and three robust paranthropines. But is that all there were? How many australopithecines were there really? The australopithecines we know about seem to include about half a dozen species, spanning the period between 4.4 million years ago *(ramidus)* and something under 1 million years ago (the age of the latest paranthropines). For various reasons, however, we cannot be confident that we know all of them. For a start, there is space, in time and in body form, to fit in at least one more species between *ramidus* and the common ancestor. More generally, we know that the primate fossil record is extremely spotty, and indeed that we probably know less than 4 percent of all extinct primate species. In particular, there are virtually no anthropoid fossils from late Miocene Africa, when the australopithecines apparently split from the apes; and the later fossils, from the Pliocene and Pleistocene, are geographically patchy. In short, a great deal probably went on of which we know nothing.

In fact, broad, arm-waving generalizations suggest that entire suites of species may be missing. For the retreat of the African Miocene forest would have provided exactly the conditions needed for the creation of new species. The wooded margins of the central Miocene forest would have covered a huge area with hundreds of isolated or semi-isolated pockets able to hold many different, self-contained viable populations of middle-sized animals such as apes, all of which could be constantly and repeatedly seeded from the vast central gene pool of ancestral apes.

In short, many different subpopulations of ancestral apes around the wooded margins of the central forest might have taken to the ground in the same way the australopithecines did. Those different types would not be true australopithecines, although they would have been related through their common ancestry, and perhaps resembled the australopithecines through convergent or parallel evolution. Other populations of ancestral apes might have become ground-living quadrupeds—the ape equivalent of baboons. If they did, they have disappeared without a trace, although the gorilla is essentially a ground-living quadruped that has stayed in the forest (perhaps driven back in by the emerging hominids). The australopithecines themselves may have spread to new areas and speciated, perhaps to form a suite of ground-living apes comparable with the array of forest-living monkeys of South America. Semi-isolated around the forest edge, they would have had a much greater opportunity to speciate than the ancestral apes, who would have been more in touch with each other in their more or less continuous central forest; although in truth the modern common chimps are now clearly divided into several (some would say many) different subspecies.

Finally, we may note that if there was such speciation, then the australopithecines we know about may not, in fact, be our direct ancestors. They may merely be aunts and uncles. Our true ancestors—and a good many more besides—may have disappeared without trace. But we can draw comfort: if the australopithecines we know about were not our true ancestors, they must surely have been very similar to the creatures who were.

There are two main objections to such speculations. First, there is no direct evidence for it—although this is less important than it may seem, because we only began the speculation in the first place because we needed to fill in the obvious gaps in the fossil evidence. The second and larger objection is theoretical. We know that evolution is chancy, so although the central pool of ancestral apes may have had the opportunity to produce many different ground-living descendants, in practice only a small proportion of those opportunities would have been taken. In fact, it is very possible that the many theoretical opportunities were seized only once. Perhaps, after all, *ramidus* and *afarensis* represent the only lineage there has been.

Rob Foley has imposed some order on these speculations by applying ecological principles based on direct knowledge of other animals. He has found—perhaps unsurprisingly, but you cannot take anything for granted—that the number of species in a genus of animals of any given body size is related to the total area that that genus is known to cover. The area covered by the australopithecines is not known for sure, but it can be worked out. They do not seem to have left Africa—at least, all claims to have found them in Eurasia are generally considered mistaken. They probably lived at the forest edge, and because the central African forest was huge, its margins were commensurately extensive.

In theory, such an area might have supported twenty or thirty species of australopithecine. But in practice Rob Foley has found that such areas tend to support only about five species of mammals of the kind that are related to each other and are roughly the same size as australopithecines.

In practice, *ramidus* and *afarensis* represent the only australopithecine lineage that we know about between the late Miocene and about 3 million years ago. But after that various others start to come on line (like *africanus* and *aethiopicus*), and from then on, as we will soon see, the fossil record does sometimes show about five hominid species coexisting at any one time—australopithecines alone or australopithecines plus hominines. Thus, around 2.5 million years ago *africanus* coexisted with the last of *afarensis,* plus *P. aethiopicus.* So perhaps, from the mid-Pliocene on, we do know most of the australopithecine types there are to know. But early-Pliocene and late-Miocene types remain enigmatic. It still seems somewhat unlikely that australopithecines were the only chimpish apes to take to the ground, and that *A. afarensis* was the only one of its kind to exploit the benign Pliocene woodland.

One further point, of course, is that each of the known australopithecines may represent more than one species. Some say that the widespread and highly variable *afarensis* fossils come from more than one species. Some suggest that there are two kinds of *africanus*. Others divide *Paranthropus robustus* into two: *P. robustus* from the site at Kromdraai, and *P. crassidens* from Swartkrans.

Here, I am following the ideas of Bernard Wood. As he points out, it is not at all easy to decide how many species are represented by the various fossils. To begin with, hominid fossils are usually fragmented, they come from widely different areas, and those from the same area often represent several or many individuals. If corresponding fragments seem to come from creatures that are very similar to each other, then it seems reasonable to conclude that they do indeed come from the same species. But if they are different, then they may come from the same species, or they may come from a single species that is highly variable. For we know from observation of living animals that some species are extremely variable. Those that live in different regions may differ in size and to some extent in shape, though they may still be perfectly capable of mating successfully if brought together, and so would generally qualify as the same species. We may also expect that long-lasting species would vary over time.

We know, too, that males may be very different from females—usually, in mammals, larger and bulkier. This is sexual dimorphism. Confusingly, some primates are extremely dimorphic (like gorillas and baboons) and others hardly at all (like gibbons and marmosets). This dichotomy has enormous social significance. When the males are much bigger than the females, as in gorillas or elephant seals, we can generally assume polygyny: each alpha male has many wives. When the sexes are much of a muchness, simple pairs are more likely. But when a pile of fossils seems to fall into two size groups, does this imply two coexistent species, or a marked difference between the sexes? This, in practice, can be an extremely hard nut to crack.

To sort the fragmented and scattered hominid fossils into distinct types—which may or may not in the fullness of time be accorded the status of species—Bernard Wood has applied two lines of thought. In general, like many other modern anatomists, he applies the methods of analysis first formulated by the German biologist Willi Hennig in the 1960s: those of cladistics. The basic notion is to decide rigorously which characters that are shared by the various fossils are truly homologous—that is, were inherited from a common ancestor—and which are merely analogous, and have become similar through convergent evolution. Next, the cladist (the analyst employing cladistic methods) must decide whether the homologous features, or characters, truly reflect a close relationship or not. For example, modern chimps and humans have hands that are very similar in detail, which suggests both that they are indeed homologous and that the common ancestor of modern chimps and humans also had such a hand. But the mere fact that chimps and humans have five fingers on each hand (a pentadactyl limb) is not enough. All four-legged land vertebrates (terrestrial tetrapods) have pentadactyl limbs unless, like horses, they happen to have shed a few. Mere pentadactylism is simply a primitive feature that tells you nothing about special relationships. Primates in general hang on to many primitive features and often seem to reinvent primitive features, in the way that Neanderthals reinvented ridges over the eyes (brow ridges). In general, it is far from easy to tell whether shared features are truly homologous and if they are, whether they are not merely primitive.

Second, however, Bernard Wood has drawn a distinction between features that you might expect to vary, and those you would not expect to vary. You might expect, for example, that mere size might vary quite a lot. After all, modern humans come in all shapes and sizes. But there are certain features in which you would not expect much innate variation—for example, in the width of the nose. So mere variation in general body size would not prompt Wood to suppose that a given pile of fossils represented more than one species. But marked variation in nasal width might well suggest this.

By such means Wood has concluded that the present fossils of *A. afarensis* probably represent only one species, although it is highly varied. The known *africanus,* too, he believes, are one species (though there are specimens that others say are different, which at the time of writing he has not yet been able to study). He suggests that the South African robust *Paranthropus* is really only one species, *P. robustus,* and that the proposed *P. crassidens* from Swartkrans is just a variation. He does acknowledge, however, that the East African robust australopithecines really do represent two species: the more ancient *P. aethiopicus* and the more recent *P. boisei.*

Overall, therefore, Bernard Wood concludes that the known types of australopithecine and paranthropine represent five species, while others prefer at least seven, with *A. africanus* and *P. robustus* each divided into two. More fossils might resolve the issue. We must also, of course, add at least one and probably several that preceded and accompanied *A. afarensis* in earlier days.

But however many australopithecines there were, it does seem that one of them, some time after 2.5 million years ago, gave rise to the genus *Homo.* Again, we should ask how and why.

HOW *HOMO* BECAME, AND WHY

I have suggested that the australopithecines arose from a lineage of chimpish apes in the late Miocene as worldwide cooling, probably exacerbated in East Africa by a tectonic rise in altitude, produced a broad band of woodland around the tropical forest. *Homo* evidently arose from a lineage of australopithecines as a second burst of cooling, around 2.5 million years ago, began to transform those woods into harsher territory; more like the grassy savannah of modern Africa. Scions of the chimplike common ancestor probably could not have adjusted to open grassland all in one go. But australopithecines, already moving easily on two legs, were equipped for the adjustment. We may legitimately doubt whether the forest-bound common ancestors would have made the transition to the ground if the climate and the terrain had not altered; and we may just as legitimately ask whether the australopithecines would have abandoned their semiarboreal mode unless a second, further burst of aridity had finally taken most of their trees away from them.

In practice, the oldest accepted member of the genus *Homo* was described by Louis Leakey, John Napier, and Phillip Tobias in 1964. The first-discovered fossils came from Olduvai Gorge in Tanzania and are dated at 1.8 million years ago. Leakey, Napier, and Tobias called their find *Homo habilis*—"handy man." Others have been found since, and the oldest now known, from Malawi, dates from 2.5 million years ago.

To be sure, some paleoanthropologists doubted at first whether *H. habilis* was different enough from *Australopithecus* to be placed in the *Homo* genus, while others doubted whether he differed sufficiently from the later *H. erectus* to be placed in a different species. But *H. habilis* has a significantly larger brain than the australopithecines—around 600 to 750 ml. So almost everyone now accepts both that *H. habilis* is a true *Homo*, distinct from *Australopithecus;* and that he is indeed significantly different from *H. erectus* (whose brain, at 800 to 900 ml, is significantly bigger still). Furthermore, as discussed later, Bernard Wood now suggests that the various *H. habilis* fossils that have been found in Tanzania, Kenya, Ethiopia, Malawi, and possibly those from southern Africa in fact represent at least two species, *H. habilis* and *H. rudolfensis*. Let us just assume for the time being, however, that the first true hominines (members of the genus *Homo*) did indeed appear about 2.5 million years ago.

Homo habilis, too, rather than any australopithecine, is considered the probable author of the recognizable stone tools that were found in his vicinity. The australopithecines who were there earlier must also have used tools, as chimps do; indeed, they must have made tools, for Jane Goodall showed that modern chimps make tools (for example shaping twigs to fish termites from their nests), and it is inconceivable that the australopithecines actually lost such skill. But australopithecines either stuck to wooden tools as befits a woodland animal or, if they used stones as tools, did not fashion them in a recognizable way. That is, they might have used pebbles for clubbing prey or for pulverizing and hence tenderizing the flesh—the hamburger may be as old as the steak in the hominid diet—but apparently they did not chip them into blades. Without such chipping, it is impossible to see whether an australopithecine stone was a tool or not. But *Homo habilis* was a definite chipper. So he was not a mere maker of tools. He was an artificer of stone; and this was a very significant shift.

The varied lifestyle offered to the australopithecines in their pleasantly open forest encouraged the handiness that they had so felicitously inherited from their arboreal ancestors. In *Homo habilis* we see the final fruition of that handiness; indeed, the name *habilis* means "handy." But then we see something more: the interaction and mutual stimulation of hand and brain to create an evolutionary feedback loop in which both were developed to a new and unprecedented degree. Again, everything had to be right for this loop to become established— the gene pool, the ecomorph, the population size, and the right degree of selective pressure: not so mild as to induce no change, and not so harsh as to obliterate. The late australopithecines that became hominines, faced with the new pressures of expanding grassland, provided precisely the combination needed for brains to expand.

We take for granted that brains and intelligence are a good thing, and hence assume in a general way that they ought to be favored by natural selection. But natural selection, as we have seen, is concerned only with the short term, and under what circumstances could it specifically favor a big thinking brain? After all, the brain is an expensive organ whose maintenance requires more than its fair share of the body's energy, so big-brained animals, compared pound-for-pound with smaller-brained creatures, need more food, and more food means more effort and more risk. Natural selection cannot favor such extravagance unless the brain provides a commensurate payoff. And it seems to me that under most natural circumstances a big, contemplative brain is totally useless. What, for example, would a codfish do with the brain of Jane Austen? How would it express its genius? On what would it inscribe its novels? Who would read them? In truth, the literary fish would quickly fall to one with sharper teeth.

In short, a large brain can be favored by natural selection only if its cerebrations can be translated rapidly into useful action, where "useful" must be interpreted in the crude terms of survival. But a dexterous hand can provide precisely the interpretation needed: the ability to turn stones into tools with which the rest of the environment can be adjusted to the animal's needs. Thus, in a truly handy man in a conducive environment, natural selection could indeed enhance brainpower, and did.

By the same token, a superbly dexterous hand is pointless unless instructed by a sympathetic brain. Fit a duck-billed platypus with the hands of Alfred Brendel, come back in a million years, and you would find the hands reevolved into paddles. What use has a platypus for such refinements? But *Homo habilis* was developing the brain that could tell its dexterous hands what to do.

So the two enhanced each other. The more agile the hands became, the greater the challenge presented to the brain, and the greater the rewards if natural selection could rise to that challenge; and the more imaginative the brain became, the more dexterity was favored. To put the matter formally: each organ provided conditions in which natural selection would favor the further development of the other. We see such positive feedback loops throughout nature, notably when two species are locked together in a burst of coevolution. Thus, a particular orchid may become more and more closely adapted to the whims of some particular pollinating moth, which in turn becomes more and more closely geared to the orchid. Each eggs the other on: each provides circumstances in which natural selection favors particular adaptations in the other. In the development of the hominine brain and hand, we see the coevolution of two organs within the same creature. We also see the most significant natural feedback loop that has occurred in the entire Cenozoic, and perhaps throughout the lifetime of the entire planet.

The fossil record suggests, too, that the resulting loop produced results quite rapidly, for the human brain developed from the size of an ape's to the size of a *habilis* in a few hundred thousand years at most, and probably in far less, and from habilis to us in another 2 million. On the geological scale such a burst could reasonably be seen as a punctuation. Certainly, the human brain has not been expanding uninterruptedly since the Oligocene, as many earlier paleoanthropologists supposed must be the case.

More generally, and perhaps more mundanely, we can see how *Homo* adjusted to increasingly open and grassy country, just as we envisaged the adaptation of *Australopithecus* to its woodland. In that open ground *Homo* would have been more exposed and vulnerable to predators than *Australopithecus* in the woods, with less cover and no instant, vertical escape route. The prey, too, would have been more agile: more cursorial antelopes and fewer bush rats. On both counts we can assume a premium on increased cooperativeness, both for defense and for hunting, and with cooperativeness, a premium on social intricacy and intelligence. The need for tools—meaning weapons—also increases in these more vulnerable creatures, again for both defense and offense.

In creatures of open country, too, natural selection must surely favor increased mobility. First, animals of the plain need speed over the ground, again both for defense and offense. But in addition, animals that succeed in a grassy environment need to cover the space. Good grazing occurs patchily, and you need to move from area to area whether you are a herbivore seeking fodder or a carnivore following the herbivores. Woodland is in general more concentrated. The first *Homo* inherited efficient bipedality from their australopithecine forebears. We may envisage that on the open plain they perfected this skill both to become sprinters and, perhaps, long-distance travelers. Indeed, they may have become the first primate nomads (if "nomad" is defined broadly).

The original uprightness of the australopithecines was, I have suggested, an adaptation to life in low trees. But on the plain it served some quite new purposes, so here is yet another example of nature's fortuitousness. On the open plain there is no leafy canopy to shield against the Sun. But recent studies have shown that upright creatures suffer far less from the heat than animals that carry their bodies horizontally, in the manner of most mammals. The reason is that of common sense: that an upright body exposes a smaller area to the Sun when it is overhead and at its hottest. Modern studies have also shown that the breeze is significantly faster and more cooling at a meter and a half or more above the ground, than at ground level. Thus, a five-foot creature standing upright keeps a cool head; and of all the organs of the body, the ever-demanding brain is among the most distressed by overheating. (Heat-seeking reptiles, by contrast, make use of the warm microclimate closer to the ground.)

In short, because of their inherited uprightness, the first hominines were well protected against overheating. This in turn would have given them a unique advantage. For most predators far prefer to hunt at night or in the cool of the morning or the evening. We may reasonably doubt whether any hominids, who rely so heavily on sight, would ever have relished nighttime hunting. But they may well have had the wherewithal to be daytime hunters, even on the burning plains. This would have given them a dual advantage: to attack prey animals when they, too, prefer to stay still and seek what shade there is, and to avoid the direct rivalry of lions and hyenas. Presumably, in those early days on the open plain the first true hominines acquired those additional features that are so peculiarly human. First, the pattern of hair: sparse over most of the body (particularly in people from tropical countries) but retained on the head, where it clearly provided extra protection from the noonday Sun. Second, our extreme sweatiness, a highly efficient device for cooling, is a device that makes prodigal use of water and raises the constant specter of death by dehydration. (Incidentally, I have never favored the occasionally popular notion that human beings acquired their pattern of hair and sweating during a hypothetical phase as semiaquatic animals. The idea is ingenious but gratuitous. The scenario suggested by the fossils—creatures adapting to open country many miles from the sea—seems to say all that is needed.)

Most significant of all though, perhaps, is that the transition from woodland to grassy plain finally released the arms and hands from all commitment to locomotion. They were no longer needed even for balancing in the low boughs. They were freed forever for the thousand and one tasks to which human beings have subsequently put them. They enabled quasi-nomadic *Homo habilis*, if such they were, to travel *en famille*, carrying their infants as they marched. This in turn allowed the youngsters to prolong their childhood and the period of learning and brain maturation that childhood allows. Hand and arm together became the agents of toolmaking, which in turn has two further implications of tremendous significance: first in providing the feedback loop with the brain; and, second, enabling the hominines to throw with increasing force and accuracy. Thus, we became the first specialist daytime tropical hunters and also—crucially—the first big terrestrial predators that were able to kill at a distance. Before we developed our skill as throwers of spears and rocks, only a few minor creatures, like archer fish and spitting cobras, had even begun to exploit the ballistic arts. As will be discussed further in chapter 6, the significance of the spear and bolas, and later the arrow, can hardly be overestimated.

Overall, by perfecting bipedality and developing the arms and hands, the hominines created a new ecomorph, and one of immense versatility and power. For other big cursorial, bipedal animals have generally abandoned the forelimb altogether. The ostrich does not use its wings for propulsion, yet it does precious little else with them either, employing them merely for balance during its headlong sprints, or for sexual display. The forelimbs of *Tyrannosaurus rex* were shriveled to a vestige. In both creatures and virtually all other true bipeds, the mouth and neck serve as a third limb; but hominids can keep their heads and senses out of the action while they fight and even when they feed. A few animals habitually use their forelimbs for purposes other than locomotion, including bears, cats, squirrels, chalicotheres, pangolins, giant sloths, aardvarks, and chimpanzees. But all of them move quadrupedally. Their "hands" must also be feet.

The human being, able to move as well on two legs as most other animals move on four, and employing the liberated forelimbs for all manner of other purposes, is effectively two creatures in one: a kind of centaur, though without the superfluous bodily bulk of the horse. As if that were not enough, the forelimbs are immensely mobile and tipped with dexterous hands, thanks to the many years of development in the trees. Those early arboreal primates were not rehearsing to be us—emphatically not—but they might as well have been. Anatomically, in short, the hominids and particularly *Homo* represent a new form of creature, distinct from all other animals, including the apes. Even without alluding to the brain, it is not speciesist to remark as Hamlet did, "What a piece of work is a man!"

This, then, was the first *Homo*. As already mentioned, the first widely acknowledged species was *H. habilis*, although the known early fossils may represent two distinct species, *H. habilis* and *H. rudolfensis*. Leaving that aside for the time being, one of those two—almost certainly *Homo habilis*—evolved further to produce *Homo erectus*. For the human lineage, as Rob Foley says, *Homo erectus* was the "watershed."

HOMO ERECTUS

Some people doubted the provenance of at least some *H. habilis,* for they had an australopithecine mien. But although the initial discoverer of *Homo erectus* called him *Pithecanthropus*—"ape-man"—there can be

no doubts at all about his humanity. Here was a tall and upright creature fully justifying the adjectival *erectus*. The brain was not as big as a
modern human's—only about 900 ml, whereas ours are either side of
1,500 ml—but it was twice as big as a chimp's of comparable body size.
An *erectus* male would not quite have passed muster in a modern bus
queue, for his eyes peered and doubtless sometimes glared beneath
lowering ridges of bone across the brows—truly, as Henry V put the
matter, "like the brass cannon." And since his cranium lacked the capacity of the modern skull, it also wanted the modern globosity. The
head sloped away behind the eyes, without the onionlike bulge above
each ear. But here in *erectus* was a human nonetheless.

However, the reasons that Rob Foley regards *H. erectus* as a watershed
are ecological rather than anatomical. In particular, he infers on ecological grounds that *erectus* must have eaten a great deal more meat
than the australopithecines, perhaps 20 percent of total calories rather
than the calculated 10 percent of the australopithecines. As I have suggested, even 10 percent had far greater significance for the australopithecines and for their fellow creatures than the figure immediately
suggests. Twenty percent or more is very significant indeed. As a
hunter, *erectus* must surely have been abetted both by his growing intelligence and by his increasing skill with missiles.

More immediately obvious than this, however, and at least as signifi-
cant, is that *H. erectus* was the first hominid traveler, the first to leave
Africa. Thus, the oldest known *erectus* are from Kenya, around Lake
Turkana, and date from around 1.8 million years ago. But the first *erectus* to be discovered was by the Solo River in Java, Indonesia, by the
Dutch anatomist Eugène Dubois in 1891, and is less than 1 million
years old. For decades it has been supposed that the first *H. erectus* left
Africa just over 1 million years ago, but recent *H. erectus* finds in Java
seem to date from as long ago as 1.6 million years ago, which suggests a
much earlier departure. The earlier date would not be surprising. As
we saw abundantly in chapter 4, animals can spread from one continent's end to another in a few hundred years simply by expanding by a
few kilometers per year. However, *erectus* did not spread beyond Southeast Asia, to Australia and the Pacific. He was a landlubber. Sailors
came later.

However, we need not assume that *erectus* simply spread passively beyond Africa merely through the growth of population. I am sure he was
a grassland specialist, and like other grassland specialists, he must have
employed strategies for quartering the ground, and seeking new pastures. Arabian oryx, for example, discover new grassland simply by

sniffing the wind. But they also remember where they have been before. So when they are placed in new territory, they build mental maps within a few years and know exactly where they are within apparently featureless areas of thousands of square kilometers. This has been closely observed in recent years as captive-bred Arabian oryx have been reintroduced to the wild in Oman. Modern-day Australian aborigines apparently have much the same ability. Indeed, such mental mapmaking seems to be usual among animals, a common skill in which most modern humans seem to have become particularly feeble.

In addition, there may also have been true adventurers among the ranks of *erectus* who risked their own lives in exploration. Perhaps, indeed, there were *erectus* Marco Polos. This is a romantic notion, but not mere whimsy. We know after all that most animals have some spirit of adventure—it is a necessary survival tactic in an ever-changing world—and that some are more adventurous than others. An extreme propensity for exploration is a high-risk strategy, but in extremis (for example, when the immediate habitat is already overpopulated) the only alternative to high-risk emigration may be to sit tight and die. As we have seen in the context of aggression, high-risk strategies are sometimes the best options. We know, too, that many young animals, for example young foxes, are banished from the parent territory to seek their fortunes elsewhere.

Second, and more intriguing, the insights of Professor William Hamilton of Oxford University tell us that straightforward Neo-Darwinian natural selection favors creatures who benefit their own kin as well as those who are merely selfish. After all, the kin of the "altruists" contain copies of their own genes, and since natural selection seems to work most powerfully at the level of the individual gene, genes for altruism are favored (provided the altruist can tell kin from nonkin). Human beings are family animals and at least some *erectus*—probably, as ever, the young dispensable males—might have ventured forth as a matter of policy. Worker bees do this, after all, in search of fresh flowers. Then they return to tell the rest what they have found. Perhaps the adventurous *H. erectus* were more like worker bees than Marco Polo, for I doubt if the young adventurers would have put an entire continent between themselves and their kin; and, unlike Polo, they would have traveled more in hope than expectation, for there was no established trade route for them to follow. But the underlying biological principle would have been the same. The adventurer benefits himself by aiding the society—the extended family—of which he is a part.

Indeed, many paleoanthropologists have asked why *Homo* took so long to reach beyond Africa. Successful species do spread, after all. If *erectus* really did not leave Africa until nearly a million years ago, then he must have hung about in Africa for more than half a million years—and *habilis* before him for several hundreds of thousands. Contrast this stay-at-home policy with that of say, lions, which seem to have raced around the world, tracking the herbivores that followed the shifting grasslands or fled from advancing ice.

The answer surely lies largely in the continuity of habitat. Many modern hunting-gathering people live in rain forests—in Southeast Asia, South America, and central Africa—yet we have seen that human beings initially arose by rejecting the habitat of their pongid ancestors. Modern rain forest people have readopted a modus vivendi that does not come easily, just as, on a much longer time scale, whales have returned to the sea from which all vertebrates emerged. Ecologically speaking, the modern people of the rain forest have been pushed to the margins. The Australian aborigines, who had a vast landmass to themselves for at least 40,000 years, never seriously penetrated the huge and apparently rich rain forest that now is in Queensland but which, during the total span of the aborigines, has shifted over much of the continent. Without pressure to move into the forest, they remained as denizens of open space.

Australian aborigines are of course modern people; much more flexible than *H. erectus*. If they prefer to avoid the forest (given that they have the choice), it seems doubly likely that *erectus* would also have done so. Neither would *erectus* have known how to cross big rivers, or felt at home on mountains or in desert. Thus, to reach from Africa to Java he needed a continuity of grassland. There may never have been an uninterrupted route from Africa to Southeast Asia—not at any one time, that is—so the Marco Polo scenario should not be taken too literally. But if appropriate stretches of land came into place intermittently, each reaching farther east than before, then over time he would get to what then would have seemed the edge of the Earth.

Finally, *H. erectus* was apparently the first hominid, which means the first animal of any kind, to harness fire. He did not have to create it. Fires occur spontaneously in nature and many ecosystems—notably grassland—depend upon it. For a creature that can control fire, its significance is hard to overestimate. Fire applied to grassland or to scrub creates more grassland. By using fire human beings of whatever species make more land suitable for themselves. Many animals do this up to a point. Fruit bats, birds, and orangutans disperse the seeds of the forest trees on which they feed and hence increase their own food supply.

Sheep maintain grassland by grazing it, and spread the grass over bare patches by manuring them and carrying seed. But the only nonhuman animal I can think of that creates its own habitat on a massive scale is the beaver which, by flooding the land, makes more good beaver territory. Such a positive feedback loop rapidly creates huge populations; indeed, the beavers grew so numerous that the Hudson's Bay Company was launched on the bounty of their pelts. We can imagine humans creating grassland that encouraged more humans who then made more grassland, and so on. Indeed, the very act of setting fire is a precursor of farming.

In practice, the hominids known broadly as *Homo erectus* may have embraced more than one distinct species. Bernard Wood recognizes two: *H. erectus,* who extended from Africa to Southeast Asia as described, and *H. ergaster,* which apparently remained in Africa. As we will see, this possible division could be of great significance. But whether or not *H. erectus* is taken to include *H. ergaster, erectus* was a highly significant creature. He survived from nearly 2 million years ago to as recently as 100,000 years ago in outposts in Southeast Asia. Thus, he occupied the world for most of the time that hominines have existed and if any species deserves to be acknowledged as the archetypal human, it is not us, but *erectus.* He should also be acknowledged as a tremendously successful species in his own right. He was not merely a rehearsal for *H. sapiens.*

Nevertheless, *erectus* was the forerunner of *H. sapiens.* And in general, whatever *H. erectus* did, *H. sapiens* did more efficiently, on a larger scale, and with more ecological impact.

THE PATH TO *HOMO SAPIENS*

From about 400,000 years ago the fossil record of Africa, Asia, Europe and—of special importance—the Middle East, reveals creatures that have decidedly bigger brains than *H. erectus* (around 1,100 ml and more) and yet are not as brainy or as melon-skulled as fully modern human beings and in general have somewhat *erectus*-like faces. In particular, they tend to have brow ridges of varying degrees of prominence, and their faces tend to protrude.

When you look more closely, however, you see that these post-*erectus* but premodern people vary greatly from place to place and from period to period. Indeed, many biologists over the years have placed the variants in different species. Others, however, have concluded that they

did not know enough about those ancient peoples or how they were related, and have simply lumped them all together in a rough-and-ready
way as "archaic *Homo sapiens.*" One confusing feature is that some of
the more "primitive"-looking types—for example those with bigger
brow ridges and/or with smaller brains—are sometimes shown by modern dating methods to have lived more recently than more modern-
looking types, with flatter faces and bigger brains.

The big-brained hominines—those with brains about the same size
as ours—are of two main types. The ones who genuinely are like us—
the "fully moderns"—first appeared in the fossil record, again in
Africa, around 120,000 years ago.

But there is also a joker in the hominine pack: the Neanderthals.
They had a very ancient-looking face, with brow ridges, which in males
in particular were huge, but also brains which if anything were slightly
larger on average than those of the moderns. They reached their most
extreme, or classic, form in fairly recent times—after 75,000 years
ago—but died about 35,000 years ago. Thus, between 120,000 and
35,000 years ago the world contained two types of big-brained hominines: although, as we will see, they occupied the same area for only
a part of that time.

Overall, we can conclude in a general way that *erectus* (or *ergaster*)
gave rise to archaic *sapiens,* who then gave rise both to true moderns
and to Neanderthals. But within this broad picture three kinds of question remain. The first is the matter of relationships: which *erectus*-grade
people gave rise to archaics, and which archaics produced Neanderthals or moderns. Second, we should explain how *erectus,* archaic
sapiens, and modern *sapiens* all managed to spread themselves throughout Africa, the Middle East, Europe, and Asia. Third, who were the peculiar Neanderthals? Who were their ancestors? Do they have living
descendants, and if so, who are they? And if in fact they all died out,
then why did they do so? The story seems to work out most tidily if we
deal with the Neanderthals first.

THE NEANDERTHALS

The first remains of a Neanderthal to be discovered—those of a
child—turned up in Belgium in 1829. The next—a woman and child—
were found in Gibraltar in 1848. But the first to attract special attention were the top of a man's skull and pieces of his arm, thigh, and
pelvis, which were dug from the banks of the Neander Valley near Düs-

seldorf, Germany, in 1856. This was the first acknowledged Neanderthal man.

But this extremely significant character did not immediately impress the paleontologists. Indeed, they queued up to insult him. Between them they wrote him off as a diseased, freakish, Cossack soldier who had abandoned his uniform and hidden naked in a cave while deserting from some unpromising battle (as one would, of course). In the end, however, an Irish anatomist named William King identified him correctly as a distinct type of human whom he called *Homo neanderthalensis.* Many Neanderthal remains have since turned up in Germany, Belgium, France, Croatia, Italy, Israel, Iraq, and Uzbekistan, so now we know that the Neanderthals were basically Middle Eastern and European, with the most extreme types, the heaviest and most beetle-browed, coming from northern Europe.

Dating is now more certain, too, and shows that the most distinctive, or classic, Neanderthals lived surprisingly recently. They did not reach their climactic form until about 75,000 years ago and lasted another 40,000 years before their apparently rather sudden demise around 35,000 years ago. Since fully modern people first appeared around 120,000 years ago, the two were contemporaries for many thousands of years and remains of moderns and Neanderthals have been found side by side in the Middle East, dating from around 90,000 years ago. Fully modern people were living in northern Europe around 40,000 years ago, and beginning to work stone and make other tools in the delicate and precise fashion that characterizes the Upper Paleolithic. So these effectively modern people lived alongside the Neanderthals for at least 5,000 years.

Classic Neanderthal features clearly took a long time to evolve. People with vaguely Neanderthal-like characters were living from around 400,000 years ago. Existing remains include skulls from Petralona, Greece, and Arago, France; and by 280,000 to 130,000 years ago many had become more Neanderthal-like, as in Swanscombe, England, and Steinheim, Germany.

So what is a classic Neanderthal? To the untutored eye, he simply seems to reinvent or reindulge the features that we popularly associate with primitive human beings. He is very heavy-boned (robust), so that his femur bones are extremely thick-walled compared to the people of today, and his joints are larger. His limb bones are somewhat bowed and by today's standards the forearms and shins are short, relative to the upper arms and thighs. The joints are big and so are the regions of muscular attachment. Neanderthals were stocky and muscular.

Their heads, though, are the most distinctive. Their skulls were as big as a modern's or even bigger, but they were not globose as ours are. They were low-crowned and flat, with a sloping forehead, and were capacitous because they were long from front to back. The jaws were long, too, leaving plenty of room for three molars at the back, and for a space behind them; not like us, in whom the third and rearmost molar sometimes fails to erupt at all and, when it does, is crowded and forms an often troublesome wisdom tooth. The Neanderthal chin is receding, but the front teeth are strong, suggesting to some that the mouth acted as a third limb, a kind of vise. The nose is protruding and wide. The brow ridges are heavy. Neanderthals left artifacts, too, including well-made stone tools, though not as well-made, delicate, or varied as the ones left by modern people from 40,000 years ago and later. The positions of some of the Neanderthal remains strongly suggest that they buried their dead; and the presence of pollen from many different plants around some of the bodies have suggested to some that their funerals were floral ceremonies. Finally, cartoonists and scientific illustrators alike commonly depict male Neanderthals with three days' growth of designer stubble, presumably to emphasise their receding chins. But in fact their Mousterian tool-kits did not include safety razors and it is reasonable to suppose—indeed unreasonable to suppose anything else—that the males had huge bushy beards, which evolved emphatically as secondary sex characteristics. Neanderthals are also traditionally drawn with black hair, but it seems likely to me that such northern creatures would have been blond or ginger. With their great manes of red or yellow hair and their huge round, red beards their heads would have glowed like sunflowers in the slanting northern sun, their faces as round as an orangutan's, though shaped by hair rather than by flesh. It is a fanciful vision but it has more thinking biology behind it than the standard image of the male Neanderthal.

But what do we really know of their lives? Different commentators argue in very different ways. Some writers have presented Neanderthals as mystics: beneath the lowering and hugely muscled exterior lurked gentle if somewhat bewildered moon-worshipers. Some, too, construe the enormous strength of the Neanderthals as a virtue and argue that because of it they did not need to make so many or such varied tools as the putatively more feeble moderns of the same period. But others are less charitable, even derisory. They point out that big bones and big muscles require more food to maintain—which is a theoretical disadvantage—but suggest that the Neanderthals needed their enormous strength to make up for their lack of dexterity. Most commentators agree that their general stockiness is an adaptation to the late-Pleistocene cold: the same shortening of extremities is seen, for example, in modern Eskimos or in Arctic foxes and hares, and the most extreme Neanderthals come from the more northerly latitudes.

Yet there are disagreements of detail and there are puzzles. Thus, it is suggested that the extremely large nose of the Neanderthal acted as a heat-exchange system to warm the freezing air of the Pleistocene northern Europe before it entered the lungs. Fair enough. Saiga antelope, which still thrive on the steppes of central Asia, seem to have adopted the same mechanism. Yet this broad nose is seen apparently evolving in pre-Neanderthal people—well before the classic phase after 75,000 years ago—who were living in a presumably modest climate in the Middle East. We can overdo the "cold-adapted" scenario, since Neanderthals obviously lived in sunny times as well.

The lowering and conspicuous brow ridges, so reminiscent of primitive *erectus* and his predecessors, turn out to be structurally quite different. *Erectus* ridges are solid; here was a hardhead indeed. Those of Neanderthals are puffed out with air spaces. I feel that the ridges—like the huge if hypothetical beards—were for decoration: that they were secondary sex characteristics, selected because they attracted mates, like the antlers of the stag or the tail of the peacock. Such an explanation fits modern thinking very well. For biologists from Darwin on have pondered the evolution of the peacock's tail or the stag's antlers. To be sure, antlers are useful for tackling rivals, but the peacock's tail seems a disaster: it requires huge inputs of energy to produce, yet it is an obvious encumbrance. The latest thinking, however, says that this is precisely the point. The tail is intended to be a handicap. It says to the female and to rival males, "Look at this enormous extravagance! Yes, it does indeed take enormous energy to grow it. Yes, it is a terrible weight. But despite this, I can survive, and this proves that I am better than all the other peacocks!"

This kind of explanation—"the handicap principle"—sounds absurd at first hearing, but the more that biologists have looked at it, both in the field and with the aid of mathematical models, the more it seems to make sense. Males in particular do indeed advertise their hidden reserves of strength and fitness with anatomical and behavioral extravagances that are pure bravura. Herein, indeed, seems to lie the secret of the pongid's big canine teeth. They are not used for eating and seem to play little or no part in fighting, but they are signals. They just say, "Look at these!" Like Shelley's Ozymandias: "Look on my works, ye mighty, and despair!"

Herein, too, perhaps lies the reason why the hominids over time lost the big pongid canines. Some other signal must have taken over from them. Primates are great communicators by eye: some, like woolly monkeys and gorillas, prefer to avoid each other's gaze except in moments of specific aggression; others, like male baboons, flutter colored eyelids to attract attention. The brow ridge exaggerates the stare; again, the eye is caused to "pry through the portage of the head." The big hollowed brow of the Neanderthal, perhaps, took the principle to extremes. It made the stare impressive indeed. It also said, like the peacock's tail, "See what energy I can afford to waste on this gratuitous architrave of bone!" Perhaps, too, the lack of a brow ridge in the fully modern hominid was the obverse of this particular handicap. Perhaps moderns were saying to their potential mates, "See, I can survive even though I lack the brutish strength of those other fellows!" Perhaps indeed modern men emerged first as campus intellectuals. And perhaps such speculation can be taken too far.

Then again, where some see evidence of ritual in Neanderthal graves—in the arrangement of bones and the varieties of pollen—others see accident. As a flood recedes from a cave, it may leave a circle of flotsam as neat as any banquet. A variety of pollen in a grave may simply reveal that the dead person or his burial party had been wandering in the fields.

So much for Neanderthal culture. Of crucial importance, too, is the impact of Neanderthals on other creatures. They were obviously formidable hunters. Their tools suggest this, and it is difficult to see how they could have survived the northern ice ages except by eating meat, as Eskimos and Lapps do today. The rough vegetation of the tundra is strictly for specialist herbivores, with their huge guts and millstone teeth. Yet the herbivores of the north may have made for easy hunting as they gathered in great herds wherever there was grazing. But because the grazing was rough and the climate was cold, the predominant herbivores were big—mammoths, rhinos, bison, horses—and the predators had to have appropriate technique.

But I am sure that Clive Gamble of Southampton University is right to suggest in *In Search of the Neanderthals* (Christopher Stringer and Clive Gamble, London: Thames & Hudson, 1993) that the Neanderthals did not have what I am inclined to call the logistic skills of the fully modern people who shared the continent with them for so many thousand years. Neanderthals, Gamble suggests, hunted in an opportunistic way; killing what was there to be killed capably enough, but not anticipating very astutely from year to year which animals were likely to be where, or seeking to influence their movements. By contrast, he says, their late-Paleolithic modern contemporaries knew how the animals moved and where they were to be found, almost in the manner of a modern whaling captain; and different modern populations in different places increasingly traded with each other, both goods and information, until in effect they quartered the entire continent in a cooperative network that was quite different logistically from anything that any previous animal had devised. Indeed, I will argue in the next chapter that the paleolithic people whom we loosely classify as hunter-gatherers might rather be seen as wildlife managers—which perhaps is how the modus operandi of modern Australian aborigines should be interpreted.

In short, the crucial difference between Neanderthals and moderns does not lie in the shape of their heads or the muscularity of their bodies. It lies in things that do not directly fossilize, suggestive though the fossils and artifacts may be. It lies in their culture and above all in the logistics of their lives: in what the people actually did, and the effect of their actions on the world about them. Logistics are not easily measured, as a bone or a stone tool can be measured. But they can be quantified, as I will discuss in the next chapter.

Such, then, were the Neanderthals. Whether we decide that they really belong in their own species, *Homo neanderthalensis,* or are simply a subspecies of *Homo sapiens,* namely *H. sapiens neanderthalensis,* they clearly were a discrete group who evolved from some archaic stock and developed along their own distinctive lines. Perhaps we might suggest that the Neanderthals are the hominine equivalent of the australopithecine *Paranthropus.* Perhaps we should see the evolution of *Paranthropus* away from the graciles as the first great bifurcation in the hominid lineage; and note that the ones who muscled their way out of trouble were the ones that died out. Perhaps we might see the offshoot of the Neanderthals as the second great bifurcation of the hominid lineage which, as things have turned out, was similarly doomed. Perhaps the Neanderthals did better than they otherwise might have done, and reached the physical extremes that they did, because of the Pleistocene ice ages. Their general stockiness equipped them for the cold; and their muscularity no doubt helped them to take on the big herbivores of the northern plains. Perhaps the first fully modern humans killed them off, as William Golding argued in *The Inheritors.* Or perhaps our own ancestors simply adapted more quickly and adroitly to putative changes of climate as the last ice age built up, and won out in the end by the drip, drip, drip of competition.

And perhaps there are too many "perhapses." It is time to move to a theoretically more tractable problem, though one still shot with disagreement. Who, among the plethora of ancient hominines, was related to whom? Who gave rise to whom, and when and where?

RELATIONSHIPS

The arm-waving picture is that *erectus* (defined broadly) begat archaic *Homo sapiens,* who begat Neanderthals on the one hand and "fully modern" people on the other. No one disagrees with that broad-brush statement. Disagreement begins when we ask for details and start to look closely at the fossil evidence. In particular, we need to explain how each of the three broad categories of *Homo* became spread through Africa, the Middle East, Europe, and Asia; and how types that are more primitive in appearance sometimes prove to be more recent than those who look more modern.

Basically, there are two broad kinds of explanation, which for reasons that will become apparent are called the Candelabra hypothesis and the Out of Africa hypothesis. Just to anticipate a little: my own feeling is (in common with most practicing paleoanthropologists) that the Candelabra hypothesis is deeply flawed—in general, it is unbiological—but I also feel that it contains an element of truth that should be taken seriously. On the other hand, I feel that the Out of Africa hypothesis is often stated too simply and that it needs elaboration before it can really work. But let us take the two main kinds of idea one by one.

CANDELABRA

The Candelabra hypothesis is championed mainly by Milford Wolpoff of the University of Michigan, who presents a view of human evolution from *erectus* on that seems unimprovably simple. He begins by arguing, as almost all paleoanthropologists now do, that *Homo erectus* arose in Africa; and then became the first ever hominid to migrate out of Africa, through the Middle East, and down to the farthest reaches of the Asian mainland—a mainland that at times in the past million years or so has embraced the present-day islands of Sumatra and Java. No one disagrees thus far.

But Wolpoff then argues that all the different groups of *erectus,* wherever they happened to be all around the world, evolved into the various kinds of archaic; so that by about 400,000 years ago, all the *erectus* groups everywhere in the populated world had turned into, and hence been replaced by, groups of archaic *sapiens*. Then, he says, all these different archaic groups, dotted around the populated world, continued to evolve until by about 100,000 years ago, and after they had each evolved into fully modern *sapiens*. He claims, in fact, that all the modern races of human beings have arisen in situ from the archaics who lived before them in the same region, and that those archaics in their turn had previously evolved in situ from the *erectus* who were there before them.

In this scenario the Neanderthals were just one particular race who happened to develop from *erectus* people in western Eurasia. Modern people arose from a different band of *erectus*. But, says Wolpoff, the moderns did not wipe out the Neanderthals. They simply interbred with them. Thus, he says, modern Europeans and Middle Easterners contain genes that are specifically Neanderthal. After all, there are plenty of stocky people around nowadays with big noses. Wolpoff obviously does not regard Neanderthals as a separate species, *Homo neanderthalensis*. At most, he sees them only as a subset or subspecies of ourselves: *Homo sapiens neanderthalensis*.

Wolpoff supports his idea by reference to the fossils. For, he says, within the archaic fossils now found in Asia he can see characters which resemble those of modern Asian people. The same is true, he says, of African or European archaics. In other words, he sees modern racial features welling up in the archaic lineages hundreds of thousands of years ago, and becoming more like the typical features of modern races as time passes. Other paleoanthropologists say they can see no such protoracial features either in archaic *sapiens* or in *erectus*. I would not presume to comment. This is one of those technical discussions that really has to be left to the people who look at bones for a living.

There are broader objections to the Wolpoff scenario, however. For if it is really the case that different *erectus* populations in different parts of the world each evolved first into archaic *sapiens*, and then into moderns, this would imply a truly extraordinary degree of parallel evolution. Parallel evolution is really a special (and therefore rarer) form of convergence. That is, two or more lineages are said to converge when they start out differently (or at some point become different) but then begin to resemble each other more and more, like the horses and some of the litopterns. But parallelism implies that two or more lineages continue to evolve along similar lines, so that at any one point in their evolution they remain in step with one another. Parallelism certainly happens, for the reasons discussed in chapter 3: that natural selection is far more constrained than many modern biologists have liked to suppose, and different lineages of animals do tend to produce similar solutions in response to comparable selective pressures. So in theory, yes, the *erectus* people of China could have turned into modern Chinese over tens of thousands of years while the *erectus* people of Scandinavia transformed into modern Swedes. But such a degree of parallelism over such a time really does begin to stretch the credulity. Furthermore, we know that although modern Chinese people do look very different from extreme, blond Swedes, we also know that the genetic differences between the two are minute. If the Chinese and the Swedes really had arisen in situ from the archaics and the *erectus* that were in southern Asia or in Scandinavia before them, then we would expect to see a much greater genetic difference between the modern people than in fact is the case.

Wolpoff does answer such objections, however. For he argues that although the different *erectus*, and then archaic, populations evolved separately in different parts of the world, nonetheless they were all in touch with their neighbors. Since they were in touch, they exchanged genes through miscegenetic matings. In fact, he says, there was enough gene flow between neighboring groups to ensure that they all evolved along broadly similar lines—and enough, too, to ensure the near homogeneity of the entire modern human species, but not enough to override the local effects of natural selection in each region, which shaped each racial type and went on shaping them over tens or hundreds of thousands of years.

Thus, Wolpoff and his supporters see the human family tree shaped like a candelabra. The stem is formed by *H. erectus,* which divides to form a series of parallel candleholders, each with an archaic at its base and a modern *sapiens* at its tip. Hence, the nickname of the Wolpoff hypothesis.

In common with many other biologists, I find this scenario broadly implausible. Evolution does not seem to work this way. The refinement—the flow of genes between the lineages that would keep them in step—would not work unless the degree of flow was exactly right—enough to ensure broad similarity, but not enough to obliterate local difference. Yet I do believe that Wolpoff is correct to emphasize the importance of gene flow between populations—again, the phenomenon of introgression. I am sure that introgression plays a far greater part in animal evolution than most zoologists admit (although botanists are generally happy to acknowledge its importance in plants).

More of this later. It is appropriate first to discuss the alternative picture.

THE "OUT OF AFRICA" HYPOTHESIS

The Candelabra hypothesis does have an element of Out of Africa. It acknowledges, after all, that *erectus* probably arose in Africa and then spread outward. But Candelabra envisages only one diaspora—that of *erectus*. The hypothesis that is properly called Out of Africa, championed not least by Dr. Chris Stringer at London's Natural History Museum, requires more than one.

However, different people seem to describe the Out of Africa hypothesis in slightly different ways. Some seem content with only two emigrations, but some prefer three. Let us look at each in turn.

OUT OF AFRICA I: THE TWO-MIGRATION VERSION

The two-migration version of Out of Africa supposes conventionally enough that *H. erectus* was the first to emigrate—at least a million years ago, and perhaps a lot more. Then, effectively as in the Candelabra story, different *erectus* populations in different parts of the world continued to evolve, and in particular most of them developed larger brains. Hence, more or less independently, different *erectus* populations evolved into a variety of archaic *Homo sapiens,* one of which developed into the Neanderthals. Thus, this version of Out of Africa sees the archaics as updated versions of *erectus* who arose in different parts of the world from whichever *erectus* people happened to be there. So far, then, this version of Out of Africa resembles Candelabra.

But unlike Candelabra, this version of Out of Africa supposes that most of these archaic *Homo sapiens,* who had arisen locally from *H. erectus,* were dead ends. They did not progress to become "fully modern *sapiens.*" They simply died out.

For modern *sapiens* himself, this version has it, emerged only once, from a particular group of archaics who had evolved from *erectus*-grade hominids who had stayed in Africa. These unique, fully modern *sapiens* first appeared in Africa around 120,000 years ago, and some time afterward they also migrated out of Africa and into Eurasia. The earliest fully modern *sapiens* bones outside Africa have been found in the Middle East, dating from around 90,000 years ago.

As these new people spread out, they must have encountered populations of archaics who were already ensconced. Some of the archaic populations who had occupied the route in earlier times may well have disappeared by the time the new modern *sapiens* arrived, perhaps killed off by the fluctuating climate of the Pleistocene. We simply do not know how many archaic groups the newcomers encountered. But we do know of one group, in the Middle East and Europe, who were firmly established and waiting to greet the new emigrants from Africa. For we know (since any other suggestion is inconceivable) that the new, modern *sapiens* must have encountered the Neanderthals. Indeed, the 90,000-year-old modern skeletons from the Middle East were found side by side with Neanderthals. We also know that the moderns had reached Europe by 40,000 years ago and at that time the Neanderthals still had another 5,000 years to run. The prolonged coexistence of the two groups is undeniable. Whether they lived amicably together, or in a state of cold war, or on different sides of the hill, or as colonials and subjects, we are all free to speculate. There are certainly many countries nowadays—indeed, this is the rule rather than the exception—where different groups of people who do not have much in common, or like each other, at least manage to occupy the same space over time, effectively by living in parallel psychological universes. Perhaps some similar social interaction took place between the first modern Europeans and the Neanderthals. It is, however, abundantly clear that the moderns came storming through this long relationship and have multiplied to occupy the entire Earth as no other creature has ever done before, while the Neanderthals have faded away. Whether or to what extent there was intermarriage between the moderns and the Neanderthals before the latter disappeared is a matter for speculation. It is almost inconceivable that there was none at all. It does seem likely, then, that specifically Neanderthal genes are with us still, just as Milford Wolpoff suggests.

So this version of Out of Africa admits only two emigrations: first of *erectus,* who first spread themselves around and then evolved further into various kinds of archaics; and then of the fully modern *sapiens* people who evolved in Africa from one particular group of archaic stay-at-homes. However, this version seems to suffer from the same kind of objections that are leveled at Candelabra. It suggests that different *erectus* populations in different places all evolved along similar lines so that they finally produced a roughly similar suite of archaic *sapiens*. Well, again we may simply acknowledge that natural selection is subject to restraints, and that different lineages often do tend to evolve along similar lines. Even so, the objection holds. It is at least worth looking for an alternative hypothesis.

The most obvious alternative is to suggest that more than two groups of hominines emigrated from Africa.

OUT OF AFRICA II: THE MULTIMIGRATION VERSION

Since I feel that the multimigration model is the most convincing so far, I would like to describe it in full, beginning at the beginning with the ancient *Homo habilis* in Africa.

According to Bernard Wood, as you may recall, *H. habilis* is really two species: *H. habilis* proper, and *H. rudolfensis*. The latter came to nothing, but *H. habilis* evolved into *H. erectus*. But, again according to Wood, *H. erectus* is also two species. The one that most people call *H. erectus* is the one that migrated out of Africa, and indeed was discovered by Eugène Dubois in Southeast Asia. The other one was *H. ergaster*, who stayed in Africa.

According to the multimigration model, *H. ergaster*, the stay-at-home in Africa, is the true ancestor of modern *H. sapiens*, not the migratory *H. erectus*. For according to this model, *H. erectus*, having migrated, did not evolve further in any significant or particularly observable way. They simply lingered on—albeit for a very long time indeed in some places, including Southeast Asia, where, as mentioned, they may have survived until 100,000 years ago. It was *H. ergaster*, and only *H. ergaster*, still in Africa, who evolved to produce a creature who should be acknowledged as an archaic *Homo sapiens*.

So then, this multimigration model has it, some of the new archaics emigrated from Africa, again leaving some at home. These archaics then spread through Eurasia, pushing aside the *erectus* populations they happened to find in situ. These archaics then adapted to whatever locality they found themselves in. Those in northern Europe adapted to the extreme cold and became the classic Neanderthals.

But, this scenario has it, the archaics who had moved out of Africa did not evolve further, or at least not in ways that are revealed in the fossil record. But further evolution to a significant and observable degree did take place in the archaics who had remained in Africa. For these African archaics gave rise to people who could pass muster as, and indeed were, the first fully modern *Homo sapiens.*

So then these new, modern sapiens did exactly the same thing as the archaics and the *erectus* people had done before them. Some stayed at home to become modern Africans. Some migrated into Eurasia and pushed aside the archaics who had got there before them—including the Neanderthals—or perhaps mated with them, just as the two-migration version supposes.

Just to cap the point, there is empirical evidence to support the overall notion of Out of Africa in either version; for modern studies of mitochondrial DNA from different races of modern humans suggest that they all shared a common ancestor who lived around 200,000 years ago in Africa. Thus, the molecular studies coincide at least roughly with the fossil picture: that the first modern humans arose from an African archaic, and that they arose as a single entity.

However, some may argue that the three-wave, or multiwave, version of Out of Africa is too ad hoc: that it is a bad thing in science simply to add complexity gratuitously, and that we have added too many diaspora. But I would take the fight to such opposition and argue that the traditional models—Candelabra and the two-wave version of Out of Africa—are far too conservative; and indeed I would even argue that the stripped-down three-wave model that I have just described is also too conservative. In a philosophical vein, I would argue that the scientists' predilection for the simplest possible explanations—which manifests in a desire for conservatism—can lead us all astray. For in practice, whenever we are able to observe nature directly, we see that the reality is almost invariably more complex than it seems, so explanations that are irreducibly simple (like Candelabra or the two-wave Out of Africa model) are, I would say, innately likely to be wrong.

But there is a more subtle point, which is that simplicity should be sought not at the level of observation (what actually happens), but at the level of principle (what causes things to happen). We have seen after all with reference to climate that very simple forces often—indeed, generally—produce extremely complex effects. The mathematics of chaos explains why this is so—how it is that simple effects can have such enormously complex consequences.

Spurred by such thoughts, and acutely aware of my own presumptuousness, I would like to offer a fourth kind of account.

THE RECIPROCAL-CASCADE MODEL

I am supremely unqualified to pass detailed judgment on any particular fossil, hominid or otherwise. If an expert tells me that such-and-such a bone belonged to a Neanderthal or an *A. afarensis*, I take his or her word for it. It is ludicrous for the nonspecialist to do otherwise; science is a collective pursuit in which we have to trust each other (although it will always remain true that outsiders are free to select their own experts). The account that follows, then, is a mind game. For I do feel qualified to play mind games, and suggest that in some fields, at least, they are not played enough. The essence of such games is to define and to lay out possibilities, and it is logically possible to lay out all of them, especially if you include a category entitled "Possibilities We Haven't Thought of Yet."

The fossil record and the archaeological evidence from Stone Age times is poor. Of course, there are a great many fossils and artifacts, and they are studied by some excellent scientists who glean truly astonishing amounts of information from them. But the fossils and artifacts are sparse nonetheless relative to the number and variety of peoples from whom they derive and to the intricacies of their lives. It is entirely possible that pivotal events and populations have disappeared without any trace, and that we will never have direct knowledge of them. We can only pick up scraps. So there is room for speculation—in fact, speculation is necessary if we are to make any headway at all. By the same token, it is absurd to become too wedded to any particular line of speculation. We must always face the question "Who knows?"

I would like modestly to suggest that if the human fossil record is too sparse to tell us what we want to know, then we might reasonably see what we can learn from the records of other creatures. The one that leaps to mind is that of horses. Of course, they are not quite like us, for they are strict herbivores and we are not; and diet affects the way that animals move around the world, and the kinds of social groups they form, and so on. But we and they do have interesting similarities. We are both creatures of open space; we are both big and mobile; we are both able to traverse a continent in a few decades if the conditions are favorable; we are both slow-breeding (although they have a much shorter generation interval than we do); and we have both been extremely successful.

Horses originated in North America, just as hominids and hominines almost certainly originated in Africa. They then spread themselves around the world just as humans did. The question is, what was their pattern of spread? Did they move out of America in one wave, or two, or many? Was their evolution subsequent to *Eohippus* more reminiscent of Candelabra or Out of Africa I or II? Well, it seems to be like none of them, although it has elements of all three. In fact, we see horses radiating in North America, and sending out various waves in various directions—to Eurasia and then to Africa when the opportunities arose, and to South America in and after the Pliocene. Then we see the new groups settling down in new habitats, then radiating in their turn, sending out new offshoots in all directions, including the places from whence they had come in the first place. We cannot directly observe introgression. Yet I maintain—albeit primarily as a matter of faith, backed by other people's studies of modern frogs and ducks— that introgression must have taken place: that new horses heading back from whence they came must at times have encountered others that were still on their way out, and that matings took place at least occasionally. To be sure, we do not see such matings between the different species of zebra that now inhabit Africa. But then the three modern species of zebra have drifted far apart behaviorally, and animals that live very different lives will not come together in the wild. If there were many more types of African horse, forced closer together ecologically and geographically, it is at least reasonable to suggest that we might see genetic interchange more often. Overall, as I commented in chapter 4, the pattern of equid spread was like that of a superior firework: a cascade that produced a further cascade and then a third and so on. In fact, since the sparks often go back from whence they came, we might call this a "reciprocal cascade." It is a highfalutin expression but it does have a kind of ring to it.

Horse evolution has been enacted over 50 million years, while hominine history has lasted a mere 2 million. But I do not see that this makes a great difference. Two million years is plenty of time; thousands of times longer than is needed for a flourishing group of creatures to spread from one end of a continent to the other.

Taking the parallel with horses, then, I do not see why we need suppose that a conservative view of hominine history is innately to be preferred. Instead of asking, "Did human beings migrate from Africa once, twice, or three times?" why not admit that the hominine stewpot in Africa might have thrown out dozens of migratory waves? Why not admit—as with horses—that any one of those waves, once settled in some far-flung haven, might then have thrown out scions of its own, perhaps spreading back into Africa? Or that the ones that had returned could again move out (in a way specifically reminiscent of elephant history)? Why not? It is not good philosophy simply to argue that there is no direct evidence for such a melee. There is no direct evidence for any of the more conservative hypotheses, either. Neither is it good philosophy (although many scientists would probably argue that it is) simply to suggest that conservative explanations are innately preferable because they are more "parsimonious." Parsimony is a good principle in science—meaning that explanations should not be made gratuitously complex, with conditional clauses added ad hoc to explain away the untidy bits. But this reciprocal-cascade explanation of hominine diaspora is not innately complex. In principle, it is extremely simple. It simply says that successful animals are more mobile and more likely to radiate than is generally supposed, and in fact that they could spread over a continent and back again many times in the kinds of periods in which paleontologists customarily trade—especially, one might add, if they were spurred on their way by the advancing and retreating ice fields of Pleistocene Europe. It also posits the reality of introgression.

To be sure, this simple principle produces an extremely complicated scenario: we must envisage bands of creatures racing hither and thither in a quite bewildering fashion, continent-hopping as they go (between Africa and Eurasia and back) and perhaps swapping genes through rape or tryst and all known forms of miscegenation. Horses did this throughout the Cenozoic, and modern human tribes in Europe are known to have milled about in comparable fashion in the long and largely mysterious millennia between the end of the last Ice Age and the rise of the great literate civilizations that began what is conventionally admitted to be history. I really do not see why the hominines, whose history is intermediate in time scale between that of the Equidae and of the pre-Roman tribes of Europe, should not have operated in much the same manner as both of them.

This reciprocal-cascade view of human diaspora would, of course, be totally useless for explaining how a particular archaic skull came to be in a particular place. One might as well try to describe in fine detail how a particular umbrella came to be left at a particular railway station by a person or persons unknown who traveled a hundred years ago from a town that no longer exists on a line whose destination is now obscure. Since the cascade idea is so broad-brush and therefore vague, it could be thrown out on the grounds that it has no predictive value: it suggests no hypotheses that could be directly tested. That criticism might well be just. On the other hand, the cascade idea might well be true.

However, because the reciprocal-cascade notion is novel (I believe) and has not been subjected to learned criticism, I will leave it there. The rest of this chapter discusses only the standard models: Candelabra and the different versions of Out of Africa. All raise an interesting issue. How many species of *Homo* have there been? Indeed, how many species of hominid?

AN INVENTORY OF HOMINIDS

Candelabra supposes that each of the different groups of archaics gave rise to a modern race of modern *Homo sapiens*, and all the different races of *H. sapiens* demonstrably belong to the same species. No one suggests that even the most distinctive modern peoples, the Australian aborigines, are anything but *Homo sapiens*. If all the different modern *sapiens* belong to one species, then, logically, all the different races of archaics who gave rise to them must be one species as well. Indeed, effectively, in the Candelabra scenario, every *Homo* that came after *erectus* and *ergaster* should be placed in the species *Homo sapiens*, although the older types might be acknowledged to be archaic. Even the highly distinctive Neanderthals would qualify only for subspecific status: *H. sapiens neanderthalensis*.

But no such logic restrains followers of the Out of Africa hypothesis, in either version. This idea does not envisage that the different archaic groups reconvened to form the modern *sapiens*. Instead, all the archaics except one are seen as twigs on the hominine family tree, twigs that came to nothing and, more to the point, twigs that remained separate from each other. For people with a penchant for taxonomy, then, it becomes perfectly legitimate to give each of the dead-end archaics a different species name.

Thus, some biologists (including Bernard Wood) give the first archaic that evolved in Africa from *ergaster* the name of *H. heidelbergensis*. (It has a German name because a specimen that is indistinguishable from the basic African type had already turned up in Germany and was christened accordingly.) So it was *heidelbergensis* who spread out of Africa in the second wave of emigration. The various races that *heidelbergensis* have rise to in different locations throughout Africa and Eurasia can also be regarded as separate species: *H. daliensis* in China, *H. javanensis* in Java, *H. rhodesiensis* in what is now Zimbabwe, and *H. neanderthalensis* in Europe. In this arrangement, only fully modern people should be called *H. sapiens*. And overall, in this treatment, we recognize no fewer than six species of post-*erectus Homo*.

But then, before the various archaics and modern *sapiens* came into being, there was *H. erectus* and *H. ergaster,* and before that came *H. habilis* and *H. rudolfensis.* So the total in the genus *Homo* now rises to ten.

Before the genus *Homo* there were *Australopithecus* and *Paranthropus.* Some recognize four paranthropines, but Wood, at least, acknowledges only three: *P. robustus, P. aethiopicus,* and *P. boisei.* The known specimens of *A. africanus* may well represent two species. *A. afarensis* may be only one, but there must have been others before *afarensis.* Now, too, we know *A. ramidus.* Conservatively, then, we might reasonably acknowledge seven australopithecines (including paranthropines), which brings the total known hominid family up to seventeen.

Clearly, through most of the past 5 million years—in fact until the death of the Neanderthals—there have been several species of hominid on this planet at any one time. Around 2 million years ago there was at least one and possibly two hominines, *H. habilis* and *H. rudolfensis,* and probably at least two robust australopithecines, *P. robustus* and *P. boisei.* About 100,000 years ago there may have been five contemporaneous members of the genus *Homo,* including various archaics, the first of the true *sapiens,* and even perhaps the very last *erectus,* still holding out in Southeast Asia. As with cats, horses, and elephants, our family was much more varied in the past.

I like this approach to human classification. At least it expresses the spirit of the thing—the fact that our own family tree has been as bushy as that of any other successful lineage. It was not a cordon, focused on one central stem that led inexorably and orthogenically to us. To suppose that there was such a cordon is merely to reinvoke Genesis in a more expanded form.

Yet the multispecies approach needs qualification. I have suggested that the concept of introgression deserves to be salvaged from the Candelabra idea—the notion that the different emerging lineages of *Homo* swapped genes as they evolved. I have also had much contact of late with attempts to conserve modern endangered animals by captive breeding in zoos. Here there are two contrasting considerations. The first is to keep the different subspecies apart, to conserve their "purity," so at present, for example, Siberian and Sumatran tigers are kept in separate breeding populations. On the other hand, it is essential to ensure that no one population of zoo animals becomes too isolated, or each group will become too inbred. Thus, Siberian tigers from zoos in mainland Europe should from time to time be brought over to mate with Britain's Siberian tigers, and vice versa.

The point is, though, that such breeding plans have given rise to a large body of genetics theory, and this theory shows that the British and European populations of Siberian tigers would effectively be genetically continuous even if the flow of genes between them was remarkably small. In other words, if just one European tiger per generation was brought to Britain—or indeed just one in several generations—this would achieve all the mixing required.

Translate this into the experience of early hominids, from *erectus* on: highly mobile groups, often adventurous, and marauding throughout Africa and Eurasia. It seems inconceivable to me that newly arriving groups would not have mated at least from time to time with the peoples they met along the way. Sometimes the contacts would have been friendly, sometimes forced. But such things must surely have occurred, and if they did, then there would indeed be gene flow between the many groups, precisely in the way that Milford Wolpoff emphasizes. I still think the Candelabra hypothesis is wrong, but I also feel that we can reasonably impose considerable gene flow on to the Out of Africa scenario. So in important details Wolpoff would certainly be right. Modern human beings would indeed contain Neanderthal genes.

This, of course, would make the fossil record even harder to interpret than it is already. Introgression can sometimes be observed directly in modern animals: for example, DNA studies of fish from different rivers in the southwestern desert of the United States reveal that from time to time in the past—during periods of flood—different populations have come together and exchanged genes. Such studies can even show how often such exchanges have taken place. But no such research can be done on ancient fossils that contain no DNA at all, or far too little to allow such interpretation. Introgression among past hominines, if it did occur, would have to be inferred from the fossil evidence, and such inferences would always be uncertain.

But if there was such gene flow between ancient hominids, then many biologists would doubt whether it is legitimate to treat the different populations as different species. After all, a species is commonly defined as a group of organisms that can breed together successfully, but which cannot breed successfully with others. If Neanderthals could indeed mate with modern *sapiens*, then by that strict definition they must both be placed in the same species.

However, the more that modern biologists explore animals (and plants) in the field, the less realistic that traditional definition seems. Clearly, hybridization in the wild is common. Sometimes the hybrids are simply not fertile, in which case the traditional definition of species holds good (because the mating between the two groups is not wholly successful). Sometimes the hybrids are fertile but are clearly at a disadvantage, and fail to make an ecological impact. This is true of the wild hybrids of carrion crows and hooded crows. These hybrids occur in a narrow band in the region of Edinburgh where the carrion and hooded crow populations overlap. But they do not spread out of that band—which suggests that they do not compete successfully either with the purebred hoodeds to the north, or with the pure carrions to the south. But some hybrids are perfectly viable and healthy. Thus, it is that ruddy ducks introduced from the United States are currently interbreeding with the native white-headed ducks of Spain, and the resultant hybrids are threatening to oust the white-headeds altogether. By the traditional definition ruddy ducks and white-headeds should be placed in the same species. But since they look so different and they normally live in separate continents and even behave slightly differently, common sense says that they are separate species. So many modern biologists are content simply to argue that a species is a group of organisms that is evolving as one discrete unit.

Neanderthals and modern *sapiens* probably had very little to do with each other, and were certainly pursuing different evolutionary paths. Thus, they can legitimately be placed in separate species—*H. neanderthalensis* and *H. sapiens*—even though they probably swapped genes from time to time, and even though the modest gene flow between them may have had a significant impact on their respective gene pools.

All this, however, is partly speculative, and partly a matter of opinion—opinion for example on what we ought to mean by "species." What seems to be more or less certain is that by about 35,000 years ago the hominine genus was reduced to just one species, *Homo sapiens*. Certain, too, is that *Homo sapiens* continued the migratory tradition that was begun by *Homo erectus*.

THE TRAVELS OF *HOMO SAPIENS*

Homo erectus stopped at the coast of Southeast Asia, but *Homo sapiens*, arriving much later in those parts, pressed on further. Modern humans reached Australia at least 40,000 years ago—that is, at a time when there were still Neanderthals in Europe—and may indeed have arrived there tens of thousands of years before that, perhaps as long as 80,000 years ago. Those first Australians must have arrived by boat, for although there is not much open sea between Australia and Southeast Asia (since it is loosely straddled by a chain of islands), the two big landmasses have never been in direct contact. Southeast Asia and its archipelago of islands are Laurasian while Australia and New Guinea are Gondwanan, and the sea between them has always been deep. It has certainly been possible from time to time to walk from the Asian mainland to what are now the Indonesian islands, and also from New Guinea to Australia, but not from Indonesia to New Guinea.

We know, too, that within a few more millennia the people of Southeast Asia became truly fabulous sailors, for at least by 30,000 years ago they were pressing on into the islands of the Pacific. These early Pacific people may well have included some of the world's first horticulturalists.

Migration out of Eurasia into the Americas had to wait. The first crossing was made not by boat but on foot, across the huge land bridge of Beringia, which appeared conveniently in the late Pleistocene. Most of the most conspicuous animals of modern North America made the crossing at that time, including the bison, wapiti, and moose; so the megafauna of modern North American is essentially Eurasian. Human beings followed about 13,000 years ago and then spread steadily south to enter South America about 11,000 years ago.

So by the time we reach the end of the Pleistocene, about 10,000 years ago, human beings had colonized or at least set foot on most of the world's great landmasses. Indeed, only two large areas remained: Madagascar, which surprisingly escaped serious colonization until after the time of Christ, and perhaps as late as the fifth century A.D.; and New Zealand, which somehow escaped the first wave of Pacific exploration and did not receive its first human beings until the ninth or tenth century after Christ.

What these modern human beings did as they traveled, and the impact they had on their fellow creatures, I will discuss in chapter 8. For now I want merely to observe the obvious fact that the hominines as a whole have been a very successful group during some very trying times. They were direct descendants of tropical, arboreal apes who, extraordinarily, went on to endure the tundra of the Pleistocene and some of the harshest episodes of climate that the Earth has experienced, and have become the dominant life form in every land they have entered. In the next chapter I want to ask how they did it. In brief: "What's so good about us?"

CHAPTER 6

WHAT'S SO SPECIAL ABOUT US?

It is impossible to discuss what makes us successful without entering tricky philosophical territory, in which many modern biologists lie waiting to pounce. There are notions that have become extremely unfashionable, which the unwary enunciate at their peril. One such notion is that when one animal survives and multiplies while another goes to the wall, then the survivor is in some way "superior." By contrast, the fashionable interpretation is (a) that survival is almost entirely a matter of chance and (b) that all talk of superiority is tainted both by theological muddle-headedness and by vicious politics and is therefore out of order. In short, the moderns say, it is better to be minimalist: we are entitled to observe only that at time T animal A and animal B existed side by side; while at time T + 1 animal A was waxing fat and animal B had disappeared. The reasons for this reversal should invariably be ascribed to the environment. Thus, we are allowed to infer that animal A was better adapted to the prevailing conditions, but not that it was in any worthwhile sense "better." For example, if A was aquatic and B was not and the land flooded, then A would thrive while B drowned. But that does not make A superior. Merely "fitter"—more apt.

There is much to be said for this purist approach. We might, for example, assert that an oak tree is superior to a moss, and at first sight that seems self-evident. Oak trees are huge and complex while mosses are small and apparently simple. Oaks may dominate the landscapes of entire countries, while mosses cling to damp corners. On the other hand, mosses grow in more places than oak trees do, and if and when the next asteroid strikes, most biologists would put their money on mosses to survive it rather than oak trees.

Then again, the concept of superiority has been used for centuries, and probably for millennia, to justify injustice. Human beings in general have not only subjugated other animals but have been frankly cruel to them; and we have justified this by our own "superiority." Tribes have enslaved and murdered tribes, races have murdered and enslaved other races, with the same excuse. Imperialism in general employs this reasoning: the conquering group feels it has a right to conquer because it is better. Charles Darwin was at pains to distance himself from such nonsense, yet his ideas have been pressed into the service both of imperialism and of racism. Europeans of the late nineteenth century and white Americans through much of the twentieth century have used the idea that they are "biologically superior," or are "further evolved," to justify the maltreatments they had already been handing out to non-Caucasians for several centuries. Stephen Jay Gould excellently describes the many misuses of such notions in *The Mismeasure of Man*.

But still I want to invoke the notion of biological superiority. We must, of course, begin by defining exactly what we mean by it and what we do not. As I will discuss presently, the notion can legitimately be applied only to entities or phenomena that are measurable. It is inadmissible to apply value judgments and to suggest, for example, that if one animal runs faster than another—which is an objective statement—that it is therefore better in some general way. In particular, to suggest that the creature which is measurably superior in any particular respect is ipso facto morally superior is bad science (since it is not in the brief of science to make such judgments) and also bad moral philosophy (since this is merely a variation on the deeply suspect theme of "might is right"). In short, if people festoon the idea of biological superiority with value judgments and moral justifications, then they deserve to be attacked precisely in the way that Stephen Gould has so ably done.

But if we avoid such solecisms, as Darwin did and I hope to do, then it seems to me that the idea of biological superiority is not only useful but necessary. It is true, as the moderns assert, that chance plays a large part in determining which creature flourishes in any one circumstance. It is also true that when circumstances change, then the baseline statement must simply be that the survivors are the ones that are most apt—just as a moss may survive when an oak tree does not. Yet it remains the case that there are measurable differences between organisms; and these measurable qualities are at least pertinent to survival.

So despite the caveats, let us look again at the notion of biological superiority. In practice, I suggest, the expression has two acceptable and useful meanings. One refers to qualities that might be called "technical," and the other is "logistic."

TECHNICAL SUPERIORITY

The idea that animals are *simply* machines, as René Descartes suggested in the seventeenth century, is one I find deeply repellent and indeed plain wrong. Certainly, they behave in ways that are quite different from any machine. Yet animals are machines of a kind. They have mechanical parts, and they convert energy into action, just as a machine does. The components of a machine are measurable and, more to the point, the overall performance is measurable. Clearly, some machines perform better than others, and to that extent they are superior.

In practice, of course, we have to judge like against like. We might objectively observe that a BMW is faster than a combine harvester, but we cannot say that it is superior. Clearly, they are not designed to do the same job. But if we take two vehicles that are designed to do the same job, then we can make objective comparisons. We might, for example, prefer to own a vintage Bentley rather than a BMW, because the Bentley is so beautiful. But an engineer who was asked purely to judge objectively would declare that in virtually all measurable respects, the BMW is superior. For one thing, at any given speed, the BMW uses considerably less than half as much fuel as the Bentley; in short, it is measurably more efficient, and dissipates less of the available energy in the form of heat. Importantly, if your life depends on fast transport, then you should choose the BMW.

Animals and plants have thousands of characters whose function is obvious and whose performance is measurable just as the performance of a car is measurable. It is clear, for example, that a key function of the protein-lipid membranes that surround all living cells is to keep the inside of the cell in a state of chemical equilibrium. The membrane allows some things to pass through it, and not others. Some it positively repels. It is also clear that a double membrane—two protein and two lipid layers—can carry out this task with greater certainty than a single membrane could. We can therefore assert, it seems to me, that double layers are technically superior to single layers. On a larger scale, I suggested in chapter 4 that rumination is in general more efficient than hindgut digestion. The ruminant derives more useful energy per kilo of leaf digested than the hindgut digester does. That is objectively measurable.

The important question, of course, is whether the creatures with the measurable technical advantages survive better. If not, then we must simply say, "So what?" Well, to revert to our first example, it seems extremely likely that single-membraned cells evolved before double layers did; yet the only creatures that still retain single-layered cell membranes are bacterialike microbes living an extremely marginal existence. It does rather look as if the demonstrable superiority of the double-membraned cell has won the day. Similarly, it really does seem hard to deny that the technically superior ruminants have come into their own since the Miocene, when high-cellulose grass began to prevail; and it is the case that the ruminant bovids, rather than the equids, diversified after the Pliocene (and in some cases continue to diversify); and that the equids cannot apparently compete in the middle range of body size where difference in digestive efficiency is likely to become significant. *Nannippus* was an exception, and modern Shetland ponies are domestic artifacts.

The issue is not simple. For one thing, technical ability does not always correlate straightforwardly with technical complexity. The homoiothermy of mammals requires greater physiological complexity than the poikilothermy of reptiles, but poikilotherms often win precisely because they require less food and are therefore intrinsically more efficient. Mammals have to justify their complexity by working harder. Nevertheless, when you compare like with like, you do find that equivalent creatures do the same things with measurably greater or lesser effectiveness, and that in general the more effective ones win out. In short, I submit that "technical superiority" is a worthwhile concept and that it does indeed correlate positively, though obviously not invariably, with survival.

LOGISTICAL SUPERIORITY

In recent years ecologists have begun more and more to speak of "survival strategy." Effectively, it means, "What is the creature trying* to do, and how do its actions contribute to its survival?"

---

*"Trying" in this context need not imply any act of will. Many biologists happily speak of grass or trees "trying" to do particular things. The language conveys the meaning colorfully, but obviously is not meant to imply that the plants are consciously following a plan.

What truly makes such thinking worthwhile—and has placed the science of ecology onto a quite new footing—is that these strategies for survival can also be quantified. The key to quantification lies in the system of math that was developed in particular by the Hungarian-American mathematician John von Neumann in the middle decades of this century: that of game theory.

The basic idea is to assess the likelihood of success of any conceivable move in any conceivable game and hence to quantify the chances of winning in any particular line of attack. But this general approach can also be applied to the things that animals do in the wild. For example, ecologists like John Krebs of Oxford University have looked in particular at the attempts of animals to find food—their "foraging strategy." It is possible to demonstrate mathematically what the animals would need to do in order to obtain the best quality or the maximum amount of food with the least effort (or risk). Such thinking has revealed, for example, the point we have made earlier: that in general it is best for predators to attack prey of their own size—big enough to be worthwhile but not so big as to be risky (unless they cooperate like lions and hominids, when they can take on bigger prey). Indeed, it is possible to define objectively any particular animal's optimum foraging strategy. From this we can objectively decide how good a particular animal is at doing what it ought to be doing, just as we can objectively judge how efficiently a given herbivore digests a kilo of grass. We have to be careful, of course, and in particular to concede that the animal probably knows its own business better than we do, and that some of the perceived inefficiencies may be maneuvers to avoid problems that we do not appreciate—for example, a detour to avoid a particular predator.

By and large, though, the point applies. We can objectively assess how well animals carry out the things they ought to be doing. By the same token, we can assert that one creature may be carrying out the allotted tasks more efficiently than another—in short, that it is logistically superior.

The genus *Homo* had a sister genus, *Paranthropus*, "sister" implying that we both descended at roughly the same time from the ancestral *Australopithecus*. *Homo* now dominates the world. *Paranthropus*, in the words of Professor Bob Brain, former director of the National Museum of South Africa, had the ecological impact of a baboon, and then died out. Rarely have two sisters pursued such different paths. The notions outlined above, of technical superiority, but more importantly of logistic superiority, explain how the discrepancy came about, not simply be-

tween human beings and paranthropines, but between us and all the
rest of creation.

WHAT DO WE DO THAT IS DIFFERENT?

If we are truly to get to the essence of what we do that is different,
then we need to consider our abilities at three levels. First, there are
our physical attributes: in particular, the hand and brain, and the special development of the brain and the repositioned larynx and the extraordinary muscularity of the tongue and face that underpin language
and rapid speech. Then there are the practical abilities to which such
assets have given rise, such as farming, and speech, the use of missiles
and fire. But finally, to grasp what we really do, we should analyze our
own abilities in logistic terms, which in principle can be quantified by
game theory.

In the tour of our abilities that occupies the rest of this chapter, two
points are particularly striking. One is that the difference between
what we do and what other animals do often seems very small in technical terms. Consciousness is clearly not unique to us. Our kind of language, projected by speech, probably is unique; but even so, the
qualitative advance can be analyzed into components most of which
are not so different in kind from the attributes of other animals. In
other words, human beings do not seem to have invented the ability of
speech *ab ovo*, but to have combined attributes that are already present
in other animals. Darwin was keen to emphasize that there is continuity between human beings and the rest—that the differences are primarily matters of degree even when they appear to be absolute—and
Darwin usually turns out to be right.

But the second striking feature is that very small technical develop-
ments may translate into huge logistic advantage. When you boil it
down, the new life form that we have created rests on only a very few
extra techniques. Indeed, I often feel that we have beaten our fellow
creatures out of sight in the same kind of way that a fencing master will
invariably overwhelm a tyro. The master may not actually be particularly brilliant (and may well be old and arthritic to boot), but he knows
a few things that the novice does not; and a few crucial tricks is all it
takes to turn an even contest into a rout. The abilities that have made
us supreme could similarly be classed as tricks. If natural selection
could be said to work according to a plan, then it would simply be to
produce creatures that can survive for the next generation. But some-

times the ad hoc devices that favor such short-term survival have huge sequelae; new qualities emerge as epiphenomena, and whole new vistas open up. This is what has happened to us.

In the rest of this chapter I will look at our various abilities in turn. But in each case I will explore them at three levels: the physical underpinning, the practicalities, and the logistic advantage.

MENTAL TRICKS: THE POWER TO ANALYZE

It is absolutely not true to suggest, as René Descartes suggested in the seventeenth century, that animals do not think. It is undoubtedly the case, however, that human beings think more deeply and broadly than other animals. Brains emphatically do not work in the same ways as computers, but they do have things in common, and one of those things is that the ability to take in and process data—sheer power—is to some extent a function of size and complexity. Our brains are clearly much bigger than those of other animals (relative to body size) but we can presume that they are also more complex: they contain more neurones, which are connected to each other in more intricate ways.

Our superior thinking power has several connotations. First, it implies that when we look at the world about us we perceive many more factors than other animals do. We take more into account.

Second, there is the matter of hard-wiring versus software. Thus, many creatures have ad hoc mental abilities that go well beyond anything we can do. A nuthatch can hide 100,000 separate seeds for the winter and find every one of them when it needs to. Some of us find it hard to remember where we left our keys. But the ability of the nuthatch is like that of the pocket calculator: wonderful, but inflexible. Move one of the nuthatch's seeds six inches to the left and it is thrown into confusion. We should not blame the nuthatch for this. To waste time looking for seeds that are not where expected would be a very poor foraging strategy indeed. But the point remains: the nuthatch does what it does, and cannot deviate. Many other animals (and indeed nuthatches themselves, in other contexts) have to learn what to do in any one situation and then are able to respond in a variety of ways; in other words, they decide what to do. Thus, their responses are not hard-wired. One of the many surprises sprung by field biologists in recent years is that many wild animals are far less hard-wired than old-fashioned behaviorists liked to believe; for example, young cats need to learn how to hunt, and monkeys must learn to climb trees, and most animals have to learn social skills and the arts of bringing up children, and most creatures can respond in a far greater variety of ways to any one situation than tidy-minded biologists of the recent past liked to believe. Nonetheless, the fact remains that although much of our own behavior is hard-wired (more than we like to admit, probably), the proportion that is hard-wired is small compared to what is not; and in those areas where we do indeed make decisions, we are able in general to choose between many more possible options than most other animals. In short, to combine the first two points: we take more factors into account, and we are able to respond to any one of the factors in a greater variety of ways.

Third—and largely perhaps as a consequence of these first two abilities—we do not simply take the world at its face value. We have, it seems, an innate tendency to observe the patterns that underlie the behavior of the things around us, and behind the patterns to infer the rules that produce those patterns. Of course, in its extreme form this search for order has culminated in the notion that the whole universe operates according to universal laws, science being a way of searching formally for those laws. But long before our species hit on that way of doing things, our ancestors perceived, for example, that prey animals moved to particular places at particular times, that they congregated where the grass was green, and so on. Of course, other predators do this, too. But it seems reasonable to suggest that we—at least *Homo sapiens*—developed such an ability to an advanced degree. Here is one of the points at which a quantitative difference gave rise to a qualitative advance, for once a creature begins to understand the rules by which other creatures operate, its power over them increases by leaps and bounds. Once the rules are grasped, then anticipation can lead on to manipulation.

So that is the first of our mental tricks: a great power simply to analyze—what psychologists call cognitive ability. But it is the next refinement—really a set of linked refinements—that has really made the difference.

MORE TRICKS: THOUGHT AND CONSCIOUSNESS, LANGUAGE AND SPEECH

Descartes suggested that animals could not think because, he said, thought is dependent upon verbal language. Animals do not use words. Ergo, they do not think. Descartes was renowned for his logic.

The science of behaviorism, which dominated animal psychology from early in this century practically until the 1980s, was rooted to some extent in this Cartesian notion. Its other roots belong in positivism: the notion that nothing is worth thinking about except what can be directly observed and measured. Hence, the behaviorists regarded animals simply as machines that do not think, but merely behave, and regarded the notion that they do think as a piece of muddle-headedness generated by sentimentality.

In fact, biologists have studied animals more and more closely as the century has worn on—far more closely than ever in the past; it is one of the more pleasing aspects of our times—and the more they looked, the more it became obvious that the behavior of animals could not be explained simply by assuming they were machines. At least, if they were machines, they were incomparably more complex than any machine that any human being has ever devised. They are machines, in fact, only in the sense that we could say for purposes of discussion that human beings are machines. In practice, more and more biologists came to conclude, it is impossible to explain what animals do unless we suppose that they think. As Professor Herbert Terrace of Columbia University suggested at the Royal Society's two-day symposium on animal intelligence in 1984, the task before us now is to explain how animals think even though they lack verbal language.

So what's going on? Is it the case that human beings do think verbally (as Descartes supposed) and that animals think in some other way, or is there more to it than that? There is indeed more to it. The relationship between thought and verbal language has for the most part been misconstrued. But once we construe it correctly, we can see what it is that humans do that animals do not. We can see, too, that although the ability of animals goes far beyond anything that they were generally credited with in the past (except, of course, by animal lovers, who never lost faith in their abilities), nonetheless the extra verbal tricks of human beings give us a supreme advantage.

Point one is the observation that, when you think about it, it is obvious that human thought does not depend upon words. A key example has been offered by the American geneticist Barbara McClintock, who was awarded a Nobel Prize in 1983 for her discovery of jumping genes: the ability of some genes literally to shift from one part of the genome to another. Dr. McClintock describes how this extraordinary notion came to her, effectively, in a flash (see Evelyn Fox Keller, *A Feeling for the Organism: The Life and Work of Barbara McClintock*, San Francisco: W. H. Freeman, 1983). Later, she related the idea to her colleagues in the no-frills language of a specialist talking to specialists, and it took her two hours. How is two hours' worth of concentrated verbiage compressed into a "flash?" Yet we have all, in lesser contexts, had many such experiences. How else does repartee operate? Witticism follows witticism instantly. The joke is conceived long before its verbal manifestation has time to be enunciated.

Words, in short, are not the necessary raw materials of thought, as Descartes supposed. They trail behind the thoughts. The thoughts themselves flow in us like some dark river in the caverns of the brain (a very Coleridgean kind of notion), just as they must do in animals. Words merely describe those ideas. They are labels that we attach to ideas. Or as Ludwig Wittgenstein said, words point at ideas.

But to describe ideas in words, to point at them, is not a trivial thing to do. In the language of computers or of librarians, this ability enables us, as thinkers, to access our own thoughts. More: it enables us to monitor them, and to direct them. Thus, in the process of thinking, words are not the frontline troops, as they have appeared to be. But they are the rapporteurs, and they can be the commanding officers. In short, words give us a degree of control over our thoughts which is beyond that of all other creatures.

Why, though, does it feel to us (as it clearly did to Descartes) as if we think in words? Why do we need to be reminded, as Barbara McClintock's experience reminds us, of the central absurdity of such a notion? The answer surely lies in the phenomenon of consciousness, and in the circularity of arguments that inevitably surround consciousness.

Many have sought to describe consciousness, and to discuss its evolution. But in the end it is difficult to improve on the notion that seems intuitively obvious: that through consciousness we know that we think. Consciousness is the means by which we observe the dark rivers of thought, and direct them. Among other things, it offers, as psychologists tend to say, a way of directing the attention of the thoughts to matters that are most important.

But I seem to have said much the same thing just now about words. Is it the case, then, that words and consciousness are the same thing? Well, of course the answer must be no, for various reasons, one of which is that scholars of animal psychology are also increasingly of the opinion that animals are conscious, contrary to the wisdom of past decades; and perhaps we might adapt the comment of Herb Terrace and suggest that a further task is to explain how animals manage to be conscious even though they lack verbal language. It is true, too, that all of us think to some extent in pictures; the image of a blank and accusatory screen flashing across the mind may remind us that the television is on the blink. But once the image is established, the words tend to take over: "Damn! Must get the television fixed!"

Thus, although we may deny that thought in general is linked ineluctably (or even particularly closely) to verbal language, there does seem to be a strong relationship between consciousness and verbal language. Through consciousness we know that we are thinking. Through verbal language we tell ourselves that this is the case. So although it seems proper to admit that other animals are conscious (and impossible to deny that this is the case), we must nonetheless admit that our possession of verbal language makes us more conscious than other animals. We might indeed envisage another feedback loop: consciousness employs verbal language in the way that the brain in general employs the dexterous hand; and verbal language enhances the quality of consciousness just as the dexterous hand enhances the brain. Because words are labels attached to ideas, because indeed they point at ideas, we can survey the words as if they were the ideas themselves—lay them out for inspection like fish on a slab.

This is where the circularity comes in, and the illusion that led so many people (including Descartes and the behaviorists) to misconstrue the relationship between verbal language and thought. For the only thoughts of which we are conscious—obviously—are the conscious ones. These do not represent the bulk of our cogitations. They do, however, provide the means by which we access the bulk of our cogitations, and hence the means by which we recognize and "bring to mind" those cogitations. Furthermore, our conscious thoughts for the most part are verbal. So the thoughts of which we are conscious—or at least the ones over which we exercise the most control—are verbal thoughts. It seems to us, then, that the only thoughts we have of any kind are verbal thoughts. It is indeed beyond our imagining—how could it be otherwise?—that we can think seriously without words, even though it is obvious that we must; for how can we think about anything consciously—including thinking itself—without putting the thought into words? And if we do not put it into words, how do we know we are thinking about it? Among the many wise things that Marvin Minsky says in his excellent book *The Society of Mind* (New York: Simon & Schuster, 1985) is that our brains are evolved organs that have been selected to observe and cope with the outside world and emphatically have not been selected for the purpose of self-examination. In short, we are innately bad at introspection. Our failure to perceive the nature of our own thinking, or the nature of the consciousness that gives ac-

cess to that thinking, or of the words that give order to that consciousness, is surely a manifestation of this ineptitude.

So what is verbal language, and how did we come by it?

WORDS, WORDS, WORDS

Biologists have discussed at length whether other animals have language—and, of course, they do. What is language after all but the ability to convey ideas (some fact about the outside world or about one's
inner emotional state) by means of symbols that the receiver under-
stands? And of course animals produce such symbols constantly: sights,
sounds, odors—all signaling to the recipient, "Here is food!"; "Here I
am!"; "Watch out—danger!"; and so on. Such language is not merely
"functional." It often has a conversational mien, as when squirrel monkeys feeding in the trees chirp from time to time just to let the others
know where they are—and sheep do the same thing on the hillside.
Neither are the sounds random utterances. Monkeys such as vervets
demonstrably make a range of warning sounds—a true vocabulary—
one meaning "snake," one meaning "hawk," one meaning "leopard,"
and so on; each suggesting a different form of evasive action. Perhaps,
unlike us, animals do tend to refer only to what is immediate rather
than to abstract thoughts, but perhaps we have not looked at their
utterances closely enough to justify such generalizations. Probably,
they are less able to convey subtleties—subjunctives and conditionals.
But they clearly think conditionally: "If this, then this; and if that, then
that." In short, it can be difficult to pin down the absolute difference
between animal communication and human language. We seem to
have incomparably more vocabulary than other animals do. We clearly
have a greater capacity for abstraction than they do. But it seems cava-
lier in the extreme simply to suggest that we have language and they do
not; and if—as is the case—they make different sounds to suit different
circumstances, then it seems difficult to sustain the idea that we use
words and they do not.

Yet there are clear distinctions. For a start, we might simply observe
that our language, besides being verbal, is open-ended. Animals make
particular sounds—"words"—to pick out items on the central registry
of notions that is peculiar to their species. We do much more than this.
We can invent sentences that have never been coined before, and can
in principle express notions that have never been expressed before.
Furthermore, we can do this effortlessly. Even babies do it. They are

wonderful inventors of new thoughts and coiners of new expressions. Part of the point is our literally endless ability to coin new words to fit any circumstance—at least in principle; for we do have difficulty in finding words to express ideas that are quite beyond our experience, as theologians have found both to their cost and to their advantage.

But the larger point was pinned down in particular by Noam Chomsky. What our language has and that of other animals apparently does not, is syntax. Underpinning our languages are a few basic rules that enable us to manipulate the sounds—the words—in an infinity of ways; and by applying the rules of syntax we can convey an infinity of meanings and infer the right meaning when another speaks. The fine rules of syntax differ from language to language—from German to Chinese and from Chinese to Finnish—yet there seems also to be a deep syntax that is common to us all. By applying the syntactical rules, we not only convey meaning by word order (or by word endings, or both) but also are able to stack subsidiary clause upon subsidiary clause without getting lost. And again I stress we do this effortlessly; and so do babies.

Words are not necessarily spoken. Obviously, they can be written. But written language is obviously a secondary skill: a way of conveying visually, and permanently, symbols that are by their nature sound bites. In principle, presumably, human beings could have developed an olfactory language or a visual language. Many animals, including many insects, are able to convey extremely precise and detailed information by chemical signals. For example, the parasitoid ichneumon wasps, which lay their eggs in the bodies of aphids, are able to tell from chemical signals precisely how many eggs have already been deposited by other ichneumons, and the sex of those eggs. Aphids themselves employ a wide range of chemical signals, including alarm pheromones that warn others of danger. And so on. But primates have in general moved from smell to vision since their prosimian days. Vision offered an obvious route, and of course primates communicate a great deal by vision, from the blue base of the blue-based baboon to the hundreds of flickering expressions on the human face. But sound has its own advantages. In particular, it travels by night and around corners. Through sound, animals can communicate with other individuals who are not looking directly at them. That offers a huge logistic advantage, which almost all land vertebrates (and many aquatic ones) exploit.

Syntax is a way of manipulating words; and since words in practice begin as spoken sounds, it is in essence a way of manipulating sounds. The point is important if only because sounds for the purposes of speech (as opposed to those of orchestration) are issued in a linear series, one after the other. But ideas are rarely linear. This is the problem that beleaguers all writers of books: how to present ideas in a linear sequence even though the ideas themselves branch out or go around in circles (for the beginning generally cannot be understood until you get to the end). Ideas can be presented visually in flow diagrams and the rest, but if we commit ourselves to a verbal presentation, we must begin at the beginning—the beginning being almost invariably arbitrary—and work along a single narrow path. This is narrative. Syntax enables us to create an order in this inveterately linear sequence that preserves meaning, allowing us to insert subsidiary clauses and parentheses and emphases and all other tricks that we take for granted but which, if analyzed, are truly miraculous. Perhaps most miraculous of all is that we do take these abilities for granted. We must study the syntactical details of other languages, but syntax itself is innate. People robbed of language—slaves, for example, plucked from their own countries and effectively prevented from conversing with their new masters in their new countries—spontaneously invent languages known as creoles that are odd in vocabulary but are perfectly and complexly syntactical. Pidgin languages represent creoles in embryonic form, the first attempt of people to synthesize a new language by imposing syntactical rules on half-grasped fragments of other languages. Pidgins sometimes sound funny, but they are truly heroic inventions, a tribute to the human proclivity to create true language in the most unpromising circumstance. Babies have no problems with syntax. They effectively invent their own, and then modify their invention in accord with the special rules of their particular society.

Syntax is innately geared to the spoken word and yet it is possessed by people who have never heard words spoken. The sign language employed by people who are born deaf is perfectly syntactical, although—interestingly—the details of the signing syntax may differ somewhat from those of the prevailing spoken language. Perhaps the syntax of the prevailing culture is modified for visual purposes. Never mind: the fact remains that syntax is innate. In practice, our syntactical ability—at least in the form it takes now—surely could not have evolved except in association with a spoken language. We can imagine a feedback loop between articulacy on the one hand and syntactical subtlety on the other. Hence, deaf people in their signing partake of a syntactical ability that could not have evolved if their ancestors had themselves been deaf. But now that syntax has evolved, it has a life of its own.

From what, though, did this syntactical skill evolve? Chomsky simply asserts that there is nothing else like it in nature, that it is nonsense to speak of its evolution, that it simply arose. But that seems a terribly nonbiological idea. We might rather point out that the essence of syntax is the ability to place things in categories and hierarchies. The "things" in this case are words; the categories are the parts of speech— nouns, verbs, adjectives; and the hierarchies are the clauses and conditional clauses. Other animals cannot apparently order sound signals in this way, so they cannot convert their vocabularies of sounds into anything that resembles a human language. But they can and do order other things in comparable ways. Even pigeons can learn quite rapidly to tap symbols into a machine (in expectation of reward) in an order that in logical structure resembles that of words arranged according to the rules of syntax. Syntax undeniably represents another qualitative leap for humankind. But again we can at least find parallels in the mental abilities of other creatures. We need not assume that natural selection has conjured our ability out of the air for the express benefit of our lineage. Again, we can see the opportunism of natural selection: adapting an ability to impose order in general onto the special case of words. Sometimes opportunism is serendipitous in the extreme.

What of our articulacy—our ability to frame such a variety of words so rapidly? This is no mean thing. No other animal can deliver such a rat-tat-tat of varied sounds so rapidly and so precisely. It requires not only tremendous wiring at the deep level—the ability to formulate the thoughts into words—but also to deliver them: to move the tongue and lips with fine coordination. Again, we can find some precedents, although we have to let the imagination run a little. It has been suggested, for instance, that our marvelous ability to coordinate tongue and lips is comparable to that of some monkeys, who can peel fruit, swallow the flesh, and spit out the pips simultaneously and at breakneck speed, an ability certainly favored by natural selection when the food is limited and the competition is fierce and the pips are toxic.

Most intriguing of all, however, and peculiar to ourselves, is the position of the larynx. In most animals it sits high in the throat and serves as a valve. As the animal drinks, the larynx forms a continuous tube with the space at the back of the nose. Thus, the airway—trachea to nose—forms a continuous hose that runs up through the stream of fluid that is flowing down the throat into the esophagus. In other words, the animal can drink and breathe at the same time.

But the larynx of the human being has been repositioned halfway down the throat, as the Adam's apple in the adult male makes plain. So human beings cannot breathe and drink at the same time. Any attempt to pull such a stunt results in choking. That seems like an enormous disadvantage—except that the descent of the larynx leaves an enormous space above it, at the back of the nose and the top of the throat. The space acts as a sound chamber, like the ceiling of a concert hall. It gives our voices a resonance that other species do not have—or at least, if they do, then they must resort to other extraordinary means, like the throat pouches of the orangutan or the howler monkey. This resonance, combined with the extreme pliability of tongue and lips, gives us a verbal dexterity that indeed is comparable with and is as wonderful as our manual dexterity, and again is uniquely human. Incidentally, there has been much debate of late about the position of the Neanderthal larynx. Was it high in the throat as in all other mammals, or low-slung like ours? If high, then their voices must have had a nasal quality—a disappointing quack in such impressive creatures. If low, then their voices may indeed have matched their physiques, and resonated as one feels they should across the steppes and the glacier edge and along the Neander Valley.

When our lineage acquired verbal language remains controversial. Beyond doubt, that ability depends in large part upon the presence of unique regions of the brain known as Wernicke's area and Broca's area. Phillip Tobias, one of the discoverers of *Homo habilis,* claims to have found an indentation in the cranium of *H. habilis* that corresponds to Broca's area. So perhaps speech is as old as our genus.

Others suggest that according to the archaeological record the huge takeoff in human culture occurred only 30,000 years ago. So perhaps speech began then. My own guess veers more toward Professor Tobias than toward such lateness, though we surely have to envisage a long period of what we might call protolanguages. All existing languages are emphatically modern, with not a whisker to choose between any aboriginal language and the most highfalutin German in the subtlety of their syntax. A great deal must have gone on between the first stirrings in *H. habilis,* if that is what it was, and the first pontificating *sapiens.*

This concludes, then, our overview of the mechanics and the acquisition of language. What really counts, though, as always, is the logistic advantage language confers. Here there are two aspects, both of supreme significance.

HUMAN LANGUAGE AND COLLECTIVE INTELLIGENCE

With our kind of language, unique among earthly creatures, we can in principle convey any thought of whatever degree of detail, however abstract or concrete, from one individual to another. Obviously, with the medium of writing to help us we can pass ideas from generation to generation en masse through indefinite time. But even without the aids of writing (and the modern elaborations of sound recording and film), huge amounts of information can pass from parents to children over vast sweeps of time. We know that some of the folk memories of modern Australian aborigines are at least 8,000 years old; they tell of once-familiar landmarks that were submerged after the last ice age and have now been rediscovered by modern divers, just as the native Australians described them. Their memories of mythical beasts—bunyips and the rest—may well allude to diprotodonts and their ilk.

We have folk memories, too, that are just as ancient. All of us can draw after a fashion. This ability is not in our genes. We all learned to do it as children from our parents. They learned it from their parents; and so on back in time to its first invention, probably tens of thousands of years ago at least, and still to be seen in striking form on caves and rocks in all five inhabited continents. What of the needle and the button, the adze and the plough, weaving and the firing of pottery? They were all invented long before the invention of writing. They are all part of our collective knowledge. They were all passed on through the generations, partly by demonstration but also by word of mouth: "This is how it's done—and this indeed is how others do it!"

In short, through the medium of verbal language, which in effect means speech, all human beings may in principle partake of all the thoughts of all other human beings through all of time. Other animals transmit ideas. Old ewes are retained on hill farms to show the new generation of lambs where to graze and how to avoid trouble. Particular packs of wolves have their own ways of doing things—as biologists say, their own traditions. The technology of chimpanzees varies from place to place; they can truly be said to have their own cultures. Various groups of animals have even been observed to acquire new skills, which then spread through the whole population and pass to the next generation. Blue tits in Britain have learned to remove the foil tops from milk bottles delivered to the doorstep and then drink the cream; and the skill has spread. The macaques of Japan have learned to wash their food in the sea and to take hot baths in the geysers, and have acquired those skills over the past few decades. But no other animal approaches our degree of collective thought.

This collectivism represents yet another qualitative advance; and one again whose logistic significance can hardly be overstated. Thus, it strikes me that in intellectual competitions any one of us ought to have the advantage over, say, a dog. But in fact competitions between dogs and humans do not start fairly, because dogs can draw only on their own experience and that of their parents and immediate pack but otherwise must work things out for themselves, whereas individual human beings can draw upon all that they have ever been told and have read about, which could in principle represent all that all human beings have ever thought. Then again, there is a fictional character on British TV called Dr. Who, a Time Lord who can travel through time and may live, apparently, as long as the Universe. But it occurs to me that an individual person who lived as long as the Universe and yet spent his time alone would never in all that time get around to inventing the computer. Indeed, few of us alone in all that time would be likely even to get so far as inventing the wheel. Or indeed the needle. If any of us were alone, in short, then we would probably make no significant progress in a billion years. But for all kinds of easily perceived logistic reasons, two heads really are better than one; and 100 million heads communicating through space and time are in a different league from any individual, no matter how venerable or innately intelligent that individual may be.

Thus, the communicative skill of the human species represents a qualitative leap in personal ability. But it represents far more than that, for the collectivism of thought that such communication allows effectively turns us into a new kind of organism. Each of us has become a neuron in a global brain—a brain that thinks across time as well as space. The power of this global creature is, quite literally, overwhelming. Termites have the same collectivism, but with termites the individual creatures are automata, whereas each of us in theory is a free spirit. Our perennial problem, of course, is to reap the enormous benefits of such collective thought while retaining our individuality. Some societies have accomplished the trick much better than others, but few have managed to sustain the delicate balance for very long.

But there is a second logistical aspect that is even more fundamental, and truly gets to the essence of human success.

HUMAN BEINGS AS TURING MACHINES

John von Neumann's game theory may be seen as the mathematiciza-tion of Machiavelli, and its influence accordingly extends through all human affairs and, as we have seen, into ecology and evolution. The English mathematician Alan Turing has been equally influential.

For in 1937, Turing wrote a paper "On Computable Numbers" in which he showed that in principle a huge array of problems (though not quite all) could be solved by one universal method—or at least by a series of methods each of which was an algorithm. He also described a hypothetical machine that could, in essence very simply, apply all of those algorithms. In other words, he had described in theory an (al-most) universal problem-solving machine, the basis of the modern computer. Such a theoretical device is known as a Turing machine.

I suggest that the key feature of the human brain is that it functions as a Turing machine. The brains of all other animals are in practice analogous with the ingenious calculating devices of the seventeenth century on, which in effect could deal only with a set and often ex-tremely limited menu of problems.

Two caveats are worthwhile. First, I am not suggesting that the human brain is literally like a computer or that the abilities of animals should literally be compared to clockwork toys (as Descartes did). Liv-ing brains are qualitatively different from any machine that human be-ings have so far devised. This is merely an analogy: that the difference between human brains and animal brains is like the difference be-tween computers and ad hoc adding machines. Second, it is a terrible thing (though the mistake is almost always made) to underestimate the cognitive powers or the emotional resources of other creatures. Many other animals can get their minds around difficult and novel prob-lems; and it is both cruel and poor philosophy to deny the sufferings and moods of animals.

Yet the point remains. There is a logistic difference between *our* abil-ity to assess problems and that of other creatures, and this has opened a vast ecological gulf and established a new evolutionary principle. Human beings, not quite uniquely but to an unparalleled degree, do not have to wait until the chancy processes of natural selection equip them to cope with a new environment. They simply work out what the problems are and solve them. Indeed, they do not need to adapt—or at least only to a limited degree. They can just as soon alter the environ-ment to suit themselves.

For in practice this Turingesque all-purpose brain is abetted both by a powerful physique and, in the end more significantly, by a physique that has retained all the wonderful versatility of the primate, including in particular the whirling arms and dexterous hands. Nothing more is needed—at least in the present world. We do not need to develop extraordinary teeth to bring down mammoths, as the sabertooths did; or molars for crushing clams, like sea otters; or snouts for digging tubers, like pigs. With brains and hands we can make weapons and tools that can do all of those things, and a lot more besides, and in some cases more efficiently than the specialists.

In short, our universal brain and our generalized physique have made us the all-purpose animal that in principle can solve any problem that any environment can present us with, just by thinking about it. Ecologically, therefore, we are as significant as all other animals put together. None of this gives us rights over other creatures. None of this justifies our treatment of them. It is, however, the case.

The key question, though, at least as far as our fellow creatures are concerned, is what we have actually done with all this power. Well, one highly significant thing we have been doing ever since we were australopithecines is catching and eating other animals. By the time we had turned ourselves into culturally modern people about 30,000 years ago, we had become the most proficient predators the world has ever seen. The point is extremely significant to our own history, and to the fate of all other creatures, and is worth discussion.

THE HUMAN PREDATOR

Rob Foley suggests that hominids made the big swing to meat-eating with *Homo erectus,* and justifies that by a highly ingenious ecological argument. Thus, he says, the number of species in each genus at any one time depends on two features. First, obviously enough, genera that occupy large areas (relative to body size) are liable to contain more species than those that occupy less; and second, herbivorous genera are liable to contain more species than carnivores, because herbivores tend to be more specialist in their feeding habits and so to speciate more. I would like to add a third point, which is that no population can be viable unless it contains a certain minimum number of individuals, so any one population of predators is bound to occupy a large space because individual predators need far more space than individual herbivores of comparable body size.

Now, says Foley, australopithecines (including paranthropines) are known to have occupied a large swathe of southern and eastern Africa, but nowhere else. Though there are clearly some gaps in the australopithecine record, it seems as if the number of known species of australopithecines roughly corresponds with the number you would expect over such an area. The number of species relative to the area also accords with the notion, arrived at on other grounds, that they were carnivorous up to a point: perhaps obtaining about 10 percent of their calories from meat.

But the genus *Homo* quickly came to occupy most of Africa—perhaps all of it—plus Eurasia. If *Homo* had had the same kind of diet as *Australopithecus,* says Foley, then we would expect such a vast area to contain about thirty hominine species at any one time. But although there may well have been several different species at any one time (for example, a whole suite of different archaics), no one has ever suggested as many as thirty. Such a paucity of species relative to area occupied is best explained by suggesting that hominines from *Homo erectus* on were much more carnivorous than *Australopithecus,* that in fact they obtained perhaps 20 percent of their calories from meat. For the same kind of reasons, lions, gray wolves, and brown bears spread themselves throughout the world without significant speciation. As I suggested in chapter 5, the 20 percent meat intake of the hominines may still seem modest, but it would vastly increase their ecological options and chances of survival, and transform their ecological impact.

Foley's argument seems to me to be eminently logical. The archaeological record also shows directly that modern—Upper Paleolithic—*Homo sapiens* were formidable hunters, able to kill elephants, the largest of all their contemporaries on land and, indeed, as the modern Inuit demonstrate, able with very basic technology and huge bravery to kill whales, the largest animals of all. So I am sure we can take it that human beings have always hunted, and that hunting has been a crucial reason for our own success and for our impact on other creatures.

So now two questions are raised. First, how did we manage to be proficient hunters, given that we seem fairly feeble physically, compared at least with specialist killers like bears and sabertooths or even chimpanzees? And second, how have we apparently come to defy the ecological law which says that big-bodied predators must be rare? Why are humans such effective hunters?

To begin with, human beings are not outstandingly feeble. We obviously are deficient in teeth and claws, the standard equipment of specialist carnivores; and we are not as strong as lions, as agile as leopards, or as swift as cheetahs or deer. But we do have our assets. For one thing, we have enormous stamina, and we are better able than many to tolerate daytime heat. Louis Leakey is said to have wagered once that he could run down an antelope—and succeeded, simply by keeping going. Whenever the poor creature stopped to graze, there was Leakey, padding over the brow of the hill, starting it up again in the kind of exhausting and increasingly panicky dash that would work perfectly well against a leopard or a lion, neither of whom would follow up a chase once lost. This cannot have been a normal method of hunting; but Laurens van der Post has filmed bushmen in the Kalahari of the 1950s wounding an eland and then pursuing it over several days before it finally collapsed. The stamina and persistence of the human animal, and the mobility that enables us to cover twenty miles and more a day, can clearly form the basis of a hunting strategy that differs from that of lions or cheetahs but need not be less effective.

Second, of course, human beings cooperate to hunt; as do lions, wolves, chimpanzees, ground hornbills and a host of other creatures. Such cooperativeness clearly increases efficiency, but it also has another logistic advantage: it broadens the prey base. In particular, it enables the cooperative predators to hunt prey larger than themselves, which few single-handed operators are advised to do. Only a few specialists—sabertooths attacking mammoths, weasels killing rabbits—habitually do so. In general, though, cooperative wolves may take on moose, and lions may hunt mature African buffalo; but tigers, hunting alone, are less inclined to attack gaur. Humans in concert were clearly happy to take on mammoths.

Human beings, however, are not merely two-legged lions. We know that from at least 30,000 years ago, they were modern people, at least as brainy as we are. We can reasonably presume that their knowledge of where prey was to be found extended beyond that of lions. Modern humans had the capacity to know in some detail where any one species or herd was liable to be. Furthermore, they could cooperate to a far greater degree than lions are able to. The people of the Upper Paleolithic were already part of the collective human intelligence, able to communicate across space and time. Each group could build on the knowledge of other groups. Indeed, they were able to anticipate where favored prey was liable to turn up at some future time. Thus, the efficiency of their own foraging strategy was improved enormously, for while ordinary predators follow, modern human predators could simply lie in wait. Clive Gamble suggests in *In Search of the Neanderthals* that such a logistic advance in hunting efficiency, brought about by memory, analysis, and cooperation, largely accounts for the eventual triumph of modern humans over Neanderthals in Europe. We need not envisage direct confrontation. It is just that the moderns were more efficient; they got more out of the environment with less effort and less risk. Again, it looks much like the difference between red and gray squirrels—a conflict not of strength but of logistics.

The third huge predatory asset of human beings seems to me to have been underemphasized. Crucially, human beings developed the use of the missile. If the missile is a stone, then it provides significant advantage. Once the missile becomes a spear, it puts the assailant into a qualitatively different league.

Again, the true significance of the spear is logistic. With a spear a hunter can deliver a lethal blow without ever running the gauntlet of the prey's defenses. Other predators must close with the opponent, as boxers say. They must slip the guard. They must get past the flashing hooves or the slashing horns. But a man with a spear can hit and run. He can even afford to miss without reprisal, though a mistimed attack by a normal carnivore could result in severe injury. Armed with spears, in short, even single hunters can ignore the rule that proscribes the hunting of prey that is much bigger than the predator. In the forest of tropical Africa, even today a pygmy may kill a forest elephant (though the technique is not to throw the spear but to stab, as with a halberd). With spears, in short, human beings acquired the ability to hunt large and eminently rewarding prey effectively without risk. No predator in all the rest of nature has ever enjoyed such extraordinary advantage.

Again, we can trace the point through. Human beings can throw because they have such wonderfully mobile arms, and great muscles on the chest and back to deploy them with. They have extraordinarily mobile arms because they inherited the abilities of ancestors who evolved in the trees; and instead of abandoning the skill of the arms when they came to the ground as other bipedal creatures have often done, they turned those arms to new purposes. One of those purposes—almost an incidental one, it seems—is throwing. But that incidental skill has transformed the human being from a hunting ape into a predator of a quite new kind; one that has never been seen on Earth before and one that tips the balance absolutely in favor of the hunter. Here is a prime example of the principle outlined at the start of this chapter: that possession of a single trick can change the rules of the game totally, that it can turn combat into rout. Add the fact that the missile-throwing hunter also has the power to analyze, is also cooperative, and also has stamina and persistence, and we have a devastating killer indeed.

There is one further logistic matter that we have touched upon already but deserves further emphasis. Human beings are not out-and-out carnivores like lions and leopards but they are extremely accomplished predators, and only bears are comparably versatile. We can subsist on clams or witchetty grubs or bring down mammoths or giant sloths. Paradoxically, we can make greater inroads into individual prey species than a specialist can because we are so flexible and do not rely absolutely upon any one prey species.

We have seen how numbers of northern lynx depend entirely on the fortunes of snowshoe hares: as the hares diminish, so do the lynx. But hunters as flexible as we are can simply go elsewhere when any one kind of prey becomes rare. Of course, the rarer a species of prey becomes, the harder it is to find, and so the predation pressure upon it is somewhat relieved. But the smaller number of prey still has to contend with the same number of predators as before. Thus versatile predators can and often do drive at least some of their prey species to total extinction, while they happily subsist on others. This is how feral cats are able to wipe out Australia's small marsupials.

The principle clearly applies to paleolithic *Homo sapiens,* only more so. For human beings may subsist for a time on a plant diet even if the game disappears altogether—but they are there waiting when the prey returns. They can also kill very large animals, which are particularly vulnerable because their numbers are bound to be relatively low. Furthermore, rarity is less of a natural protection against human hunters than it would be against, say, lions. Lions would be bound to hunt a rare animal less than a common one simply because it encounters the rarer one less often; and on the whole lions are not too fussy about what they kill provided the prey is of appropriate size. But human beings understand the natural history of their prey. If a favored animal becomes rare, well, they still know where to find the remaining herds if they choose to. And if the rewards and the prestige of the kill are great, as we may assume was the case when mammoths were the target, then they will be tracked down. Human beings are opportunist hunters. But unfortunately for their prey, humans are too intelligent to be merely opportunist.

Thus, by their ability to switch from food source to food source, human beings could maintain relatively high populations even though they were key predators. But also, if they chose to hunt a rare beast, they had the knowledge and the persistence to hunt it, as other predators would not. They were able to break the rules, in short. And by these means, in the end, they broke the ecological rule which says that big predators must be rare.

So we can put the whole thing together: a cooperative hunter who not only cooperates but also analyzes and anticipates; a predator able to kill animals much larger than himself with little personal risk; a predator that can attack the entire spectrum of potential prey animals from clams to mammoths, and can readily switch from one prey to another—or even eschew game altogether for a time; and a predator with the wit, if it seems appropriate, to hound a favored creature even after it seems to have taken refuge in rarity. In the next chapter I want to discuss the thesis known as the Pleistocene Overkill: the notion that human beings destroyed huge swathes of creatures long before the modern age of obvious destructiveness. Many have attacked this idea because they say that direct evidence is lacking, but since all paleoanthropological evidence is scarce, and we cannot be sure what should pass as evidence and what should not, such criticism is not entirely convincing.

More cogently, perhaps, many have questioned the plausibility of the Overkill hypothesis. Paleolithic hunters just could not have wiped out entire suites of huge and powerful creatures, the critics say. It seems to me, however, that once we begin to apply a little ecological thinking— and in particular once we consider the logistics of human hunting, and compare human hunting logistically to that of other predators—then the Overkill hypothesis becomes eminently plausible. It is all too easy to see how creatures like us, with modern human assets and a professional interest in hunting, could transform the fauna of entire continents. That those fauna were indeed transformed as paleolithic hunters got into their stride seems to me perfectly to match expectation.

But of course human beings did not remain as other animals have— as creatures that simply live off the land, as mere hunters and gatherers. Degree by degree our control increased until hunting and gathering became husbandry. A key step in this progression was the harnessing of fire.

FIRE

Human beings evidently harnessed fire a surprisingly long time ago. Evidence of hearths in China and France dating from about 480,000 years ago indicate that *Homo erectus* made use of fire; but recent evidence from Ethiopia suggests controversially that the very first deliberate fire-makers may have lived as long as 1.1 million years ago. *H. erectus* is still the most likely pioneer fire-maker, though it has been suggested that *Paranthropus boisei* may have made fires even longer ago in Ethiopia.

It is easy to envisage the many practical contributions of fire. Late in our history the regular use of fire in a variety of venues would sooner or later have revealed to our ancestors that clay becomes hard by firing, which is the key to functional pottery, and that certain rocks when superheated dribble globules and rivulets of metal. Each of these discoveries opened new cultural vistas.

But long before our predecessors employed fire to create new materials, they would have appreciated its ability to transform food. That it may detoxify and disinfect may not have been obvious, though they would have reaped the benefits of sterilization nonetheless. They would certainly have appreciated the ability of fire to tenderize and hence to turn what was unchewable or unpalatable into pleasing and satisfying provender; and they would have realized, too, that cooked meat may be stored longer than raw. But again, the logistic point is what counts. Fire increased the range of the human diet even more widely; it enabled early people to live in even more places, and squeeze through in even harsher times, than they were able to already.

The overwhelming significance of fire, however, is the control it has given the human genus over landscape as a whole, and over all the plants and animals that live in it. Ecologists now appreciate that over much of the world—wherever it is not permanently wet, it seems—fire largely determines which plants will grow where, and whether forest or grassland prevails, and if forest, which trees that forest contains. The point has been beautifully if sadly illustrated since the 1950s by the endeavors of Rutgers University biologists to conserve the much-loved Mettler's Woods, which is the last remaining uncut upland forest in central New Jersey. They protected the oak trees well enough by restricting public access to a single trail with an expert guide. But the trees are now senescent and the forest is not regenerating. It is clear now, as it was not when Rutgers began, that oak forest of this type does not regenerate unless there are regular fires—roughly once every ten years, to judge by the scars in ancient timber. Presumably, if we want this forest to continue for a few more thousand years, we should set it afire (see Peggy L. Fiedler and Subodh K. Jain, eds., *Conservation Biology: The Theory and Practice of Nature Conversation, Preservation, and Management,* London: Chapman & Hall, 1992). The other side of the coin is revealed in America's national parks, such as the wonderful Yellowstone, which was designated by Congress in 1872 and is the oldest in the world, where conservation biologists, farmers, foresters, and local people have very different views on whether or not to try to quell the many fires that arise spontaneously (let alone the many more that are started by human beings). In general, the biologists take a laissez-faire attitude to the fires, which they know are ecologically normal, while the people trying to make a local living decidedly do not (see Robert B. Keiter and Mark S. Boyce, eds., *The Greater Yellowstone Ecosystem*, New Haven: Yale University Press, 1991).

In dry land, natural fires influenced the distribution and indeed the evolution of plants long before human beings learned incendiary arts. Outside the rain forest of Queensland in Australia, many of the continent's native plants are fireproof—as witness the rapid recovery of vegetation in New South Wales that has followed the enormous bush fire of 1993, a fire that at times generated temperatures of 1,000°C. Many of Australia's plants escape fire by hiding their storage organs and accompanying viable tissue below the soil surface. Australia's native eucalyptus, with five hundred or so species, has become one of the world's most successful genera of flowering plants precisely because it is effectively fireproof. Eucalyptus sheds its outer bark to form an innocuous tinder on the forest floor while maintaining an iron-smooth trunk on which the flames can gain no hold.

Even more obviously, grassland worldwide is largely controlled by natural fires. Natural fire refreshes the grassland by burning off the rank vegetation, letting in the light, returning minerals to the soil, and allowing fresh green pastures to come through.

Modern human beings employ techniques that imitate nature's own fire regimes. In swidden farming, swathes of forest are first burned off, and the exposed soil, temporarily nourished by the ash of the trees, is cultivated until its fertility wears off, at which point the farmers move on. Such swidden, or slash-and-burn, agriculture is sustainable and effectively harmless on a sensible scale, although it becomes grotesquely wasteful in the hands of some cattle barons who burn away entire forests to create fragile pasture.

Gamekeepers on Scottish moors burn the heather at intervals to restore its leafiness for grouse and to quell invading trees—which certainly works, though it creates an extremely limited flora and fauna where there once was enormous variety. Arable farmers burn stubble, a controversial practice that looks to me like sound grassland management.

More pertinent, however, are the techniques of Australia's aborigines, who deploy fire in several very astute ways, though primarily to freshen the vegetation in discrete patches and hence encourage small marsupials to recolonize and multiply and so provide the people with food. Australian aborigines do not farm in the conventional sense—which is odd, since Australia has at times been joined to New Guinea and the people of New Guinea are excellent horticulturalists who presumably were cultivating plants at the time of the last conjunction. But through their use of fire the native Australians have influenced the flora and fauna of the dry regions for the past 40,000 years.

The true logistic significance of such controlled burning has been summarized by Reese Jones of the Australian National University at Canberra. For the practices of aborigines, he says, amount to "firestick farming." Astute deployment of fire raises hunter-gatherers from the ranks of superpredators to those of the game managers. Fire-raising is not simply a precursor of farming, or a tool of farming; it is a form of farming. Hence, the deployment of fire has brought about what must be seen as the most profound logistic transformation of all.

Farming, then, must be the next issue to discuss. But as a final parenthesis, I am greatly impressed that Native Americans traditionally employed smoke signals, and that people in historical times made enormous use of signal fires, especially, it seems, in times of war, when their beacons signaled the approach of enemies. Clive Gamble stresses that much of the success of modern human beings in Eurasia may have depended on establishing good relationships with neighbors (see Christopher Stringer and Clive Gamble, *In Search of the Neanderthals,* London: Thames & Hudson, 1993). This indeed was one of the "tricks" that gave our own European forebears an edge over the Neanderthals. Neighborliness was achieved by diplomacy and trade, and the key to both is communication. I wonder idly (or perhaps not so idly) if people on that vast territory stayed in touch day to day by signal fires, conceivably just to say hello (which is a very useful thing to say) and inter alia pass information on the migratory herds of saigas, bison, and horses, which can quarter continents so effortlessly. This is pure speculation. But on the other hand, the Native Americans were Eurasians first.

Taken overall, then, the various manifestations of fire, possibly more than any other single innovation, bridged the gap between human beings as superpredators and human beings as controllers of all they surveyed. Australian aborigines still represent the transitional state; although perhaps in their harsh and unpredictable continent the native Australian approach is not simply transitional but is ecologically the most appropriate. After all, many a modern farming enterprise has come to grief in Australia, and it seems doubtful if large-scale agriculture could have gained a hold on that continent without constant re-plenishment from the established wealth and technique of Europe.

But over most of the world it has been farming—two logistic steps in advance of mere predation—that finally enabled human beings to break the rule which says that big predators cannot be numerous. Farming deserves a chapter to itself.

CHAPTER 7

THE END OF EDEN: FARMING

Everywhere you look: agriculture. That is not quite true of course, for there are still vast ranges of mountain and desert, and a great deal of tropical forest; in fact, the statistics show that agriculture occupies only a third of the world's total land. But the land that is occupied by agriculture includes the temperate valleys and former wetlands and places that once were deciduous forest. In short, agriculture already covers most of the most fertile areas of the world. It also, of course, fragments what is left of wilderness. So landscape is in reality dominated by agriculture to a far greater extent than the raw statistics suggest. Creatures that are not domesticated or cultivated now occupy the interstices. They survive on sufferance. A thinking animal with the human ability to take an overview would indeed exclaim, "Everywhere you look—agriculture!" And out of agriculture has come industry, which in turn has enabled agriculture to become even more productive and prominent—for example, to plough places with heavy machinery that before were left to rough pasture. Here is another positive feedback loop.

As with thinking, as with consciousness, as with speech, agriculture seems uniquely human. But when you look more closely, you see again that farming is a compound skill, built from components which, taken individually, can in many cases be observed in other animals. Thus, do farmers protect their crops: but then so do some coral fish, which stand guard over particular patches of algae. Farmers alter the genetic makeup of crops; but then, so do those same coral fish, because algae that are protected from predators at large are bound to evolve differently from algae that are not. Farmers sow seeds of the plants they favor. Well, take away the fruit bats and orangs from a tropical forest and the fruit trees dwindle, too, because those fruit-eaters are also the seed-scatterers. Farmers nourish their crops; but so do termites and leaf-cutter ants, which raise nutritious fungi on custom-built compost heaps. Farmers nurture animals of other species for food. So do ants, which guard herds of aphids and milk them for their honeydew. Farmers plough; so do elephants. Farmers create vast landscapes expressly for their own purposes. So, spectacularly, do beavers, which create entire beaver counties by the sides of obliging rivers, so successfully indeed that the Hudson's Bay Company that became so huge was founded on the sale of beaver pelts. Our special skill, then—our uniqueness—lies in our versatility; not necessarily to do any particular thing that other animals cannot, but to emulate everything that they do, and then to mix and match. Again, as the great generalists, we are ecologically equivalent to all the rest put together.

Since farming is so dominating; since it obviously confers such power, since it has given at least some of us a life of unprecedented ease (for all we have to do to eat is drive to the supermarket), it seems obvious that once human beings had realized the advantages of farming, they would thank their gods for the new enlightenment and get down to it with a will.

That is how traditional archaeologists envisaged matters. Modern scholarship puts the origin of farming in a very different light.

HOW FARMING BEGAN: THE TRADITIONAL VIEW

Archaeologists prefer to deal, as proper scholars should, with the hardest possible evidence; and the earliest hard evidence of true farming dates from the Middle East of around 10,000 years ago. There are caches of grains that seem different from their wild relatives. From somewhat later dates come the bones of sheep and cattle that appear domesticated because they are smaller than their wild counterparts, suggesting poor nutrition and, perhaps, the deliberate selection of more manageable runts.

The traditional conclusion has been, then, that formal agriculture did indeed begin in the Middle East about 10,000 years ago, and from there the signs of its subsequent spread through Europe are clearly discernible. Whether the techniques spread culturally like a sound wave through a cymbal among people who stayed in the same place or were carried from place to place by migrating farmers is still discussed, although most scholars seem to incline to the notion of cultural spread.

The picture is neat and seems convincing as far as it goes, although the archaeological record now reveals what appear to be separate foci of farming beginning in China, the Indus Valley, Africa, New Guinea and other areas of the Pacific, and North, Central, and South America. All foci, though, are taken to have arisen after 10,000 years ago. That date—10,000 years ago—is taken to be the end of the Pleistocene and the start of the Holocene, or Recent, Epoch.

Onto these archaeological observations, however, a mythology has been superimposed in the manner that Misia Landau described. Thus, it was traditionally taken to be self-evident that farming was better than hunting and gathering. It produced a more reliable food supply, did it not? It even produced surpluses, which gave people spare time and freed more and more of them from the chore of food production, and so created leisured and specialist classes who seem to be the sine qua non of civilization. The location and the timing seemed right as well. For did not civilization—at least in its western mode—begin in the Middle East? And is it not the case that an origin around 10,000 years ago would leave plenty of leeway, but not too much, to establish the emphatically agricultural societies of the Old Testament, which referred repeatedly to sheep, cattle, wheat, figs, grapes, and olives? Yet there are various problems with the traditional scenario. Presumably, to be sure, agriculture must have succeeded through a kind of Darwinian selection—and that seems fair enough at first since the farming societies we know about so obviously seem better off than the hunting societies of which we have direct knowledge. But life is not so simple. For as I stressed in chapter 3, Darwinian selection deals only with the immediate. It is not prescient. It does not have goals. Darwinian selection would not have said to the world's first farmers, "If you start planting crops now, then in a few hundred years people will be able to build great cities, with armies and temples and paintings and music." If farming did not bring immediate reward, then it would not be favored. Its first stirrings would be snuffed out. And the problem is that farming in its early days seemed to offer very little advantage indeed. In fact, more and more evidence suggests that it was ghastly.

Such evidence is of various kinds. First, the diminution in body size seen in the first domesticated animals is also seen in the world's first farmers. Along with that general enfeeblement there is often a bending and thinning of bone that suggests disorders of privation from rickets to tuberculosis. More specifically, at London's Natural History Museum, Theyer Molleson has now described skeletons from early Egyptians that show a peculiar and most horrible pathology. The bases of the toes are arthritic and deformed, as if the toes had been constantly bent back upon themselves, and the bones of the lower back are also deformed and arthritic. For a long time Dr. Molleson was hard-pressed to find an explanation, but finally she did; on Egyptian murals that showed young people (few lived beyond thirty in that debilitating age) working at saddle querns. In these diabolical devices an upper stone is rubbed back and forth across a lower one to mill the grain. The miller must kneel on the ground. Hence, the bent toes as the feet are folded—and hence the horrendous strain on the back. Of course, the growing of cereal is only the beginning of food production. Once grown, the grain has to be processed. It is clear, too, from the rickets that after all that effort the grain-based diet left much to be desired. Besides, living as they did on the edge of the desert, the Egyptians consumed egregious quantities of sand. This broke and wore down their teeth by their mid-twenties and produced abscesses that now can be seen as neat rows of holes in the jawbones even of aristocrats, who presumably had a better diet than the common people.

The Old Testament confirms again and again that the life of early farmers was harsh indeed. One thinks, for example, of Ruth at her gleaning, or David fighting the Asian lions, which perhaps did not abound at that time but were certainly present; or the mortal conflict of Cain the tiller of the ground and Abel the keeper of sheep—for arable and pastoral farming have always been at odds. (It is interesting to note in this age of gentle vegetarians that the arable farmer was presented as the murderer.)

Clear, too, in the Old Testament is a yearning for the old days of hunting and gathering. Eden is seen as the lost paradise; and although Eden is presented as a garden, there is no suggestion that Adam and Eve were called upon to cultivate. The benison was just there. Thus, when God expelled Adam from Eden, it was with a curse—the curse that Adam must become a farmer: "In the sweat of thy face shalt thou eat bread, till thou return unto the ground. . . ." (Gen. 3:19).

"In the sweat of thy face," incidentally, is so much stronger than "By the sweat of thy brow," which is how this withering dismissal is usually remembered. It has the venom of the modern American: "Get out of my face!"

The general impression is growing, then, that farming in its early days was unremittingly harsh; whereas the hunting-gathering days that it replaced were remembered with affection and nostalgia, albeit by a folk memory that by the time of Genesis was about 6,000 years in the past. This impression has been reinforced in recent decades by studies of modern hunting-gathering peoples. Thus, Professor Richard Lee of Toronto University showed in the 1960s that the !Kung bushmen of the

Kalahari made a perfectly good living by hunting just for about fifteen hours per week. The rest of the time they sat around and told stories. John Yellen of the National Science Foundation, Washington, showed in similar vein that the Hottentots, who essentially are the same people as the bushmen, occasionally kept goats, but then would decide from time to time to stop keeping them and revert to hunting. In other words, the Hottentots were perfectly good husbandmen, but to them husbandry was no big deal. Sometimes they decided it suited them, and sometimes not. There was no feeling, as traditionalists would suppose, that husbandry once discovered was immediately acknowledged to be superior. On modern Cape York, the pointed promontory that pokes out of Queensland, the aborigines are said to have an enviable time among the permanently resident shellfish and a constant throughput of migrating birds who stop by to feed.

Finally, it is clear that advanced hunter-gatherers (or people who seemed to live mainly by hunting and gathering) clearly could and did develop many of the trappings of civilization, including graphic art, trade (and presumably diplomacy), and indeed the beginnings of cities such as Çatal Hüyük in Turkey which may well date from preagricultural times. In fact, it could well be that the period immediately before the era of agriculture—the time remembered as Eden by the authors of Genesis—was a golden age.

So to the fundamental problem of natural selective advantage we must add another question: Why? Why should thinking and already civilized people, artistic and doubtless philosophic, have abandoned an apparently easy way of life in favor of one which, until the late twentieth century A.D., has remained almost unremittingly awful? What on earth has been going on?

WHAT REALLY HAPPENED

It is in fact impossible to know how agriculture really began, or why precisely it has come to be the dominating force despite innate unpleasantness that is alleviated only intermittently and only in a few fortunate societies. But again it is reasonable to speculate on the basis of common sense and of observations in other, relevant fields—notably in that of ecology, reinforced by the rigor of game theory. In practice, hunting and gathering are the complementary components of a double act, and their gradation into pastoralism on the one hand and horticulture (leading to arable farming) on the other require different if complementary explanations.

To begin with the latter: John Yellen suggests that the arts of cultivation probably began not 10,000 years ago but perhaps as long as 30,000 years ago, and that they probably began not in the Middle East with the sowing of cereals but in tropical forest, perhaps Africa, with the simple planting of favored trees. Why not? Again, we might stress that, 30,000 years ago, we are talking not about australopithecines or even *Homo erectus* but about people who, if they were alive today, would be surgeons or bus drivers or accountants or what you will, but in those days just happened to be full-time professional hunter gatherers. They knew their environment. They knew where things grew and, in a detailed if ad hoc fashion, how they worked. Of course, they knew that seeds grew into plants: they could see them doing so in the forest litter. Of course, they knew that a new plant would grow—almost without fail in a tropical forest—from a stick thrust in the ground. It is inconceivable that they would not have known how to spread the plants they favored at the expense of those they did not. Add to that their mastery of fire and we have cultivation.

Note just a few points, however. First, cultivation in such a fashion and on such a scale would not show up in the archaeological record. What would there be to see? A change in the frequency of a few trees in a few places in a vast forest. The archaeological record is not that precise. Note also, though, the point made in chapter 4 on the dates of the first origins of animals: that by the time they appear in the paleontological record, they are probably already common, and probably have already been in existence for thousands, if not many millions, of years. Thus, if clearly cultivated grains are found in the Middle East from 10,000 years ago, as indeed they have been, then this by analogy suggests that the true origins of cultivation probably stretch back for a huge swathe of time before that.

It is the case, however (as emphasized in chapter 6), that such an innovation—a slight increase in the reliability of a favored food source—can make a huge logistic difference. Creatures that did just a little cultivation would be marginally but crucially more likely to survive in the bad times than creatures that had to rely simply on what nature cares to provide. Again, the gray-squirrel/red-squirrel model discussed in chapter 3 suggests that such a tipping of the balance, hanging on or not hanging on in any one place at any one time, is the difference between long-term survival that may lead on to a flourishing population, and oblivion. In short, the ecological consequences of cultivation bear very little relation to its conspicuousness. Just a little touch can make the crucial difference.

The idea that cultivation is indeed this ancient seems to mesh with three other significant observations. First, the spread of human beings around the world after about 30,000 years ago—including way out into the Pacific—has an inexorability about it. We can imagine that the ability just to plant a few crops to eat after the emergency supplies had gone would have made all the difference. In the same way, the first act of the Pilgrim Fathers when they arrived in North America, albeit in a quite different age, was to sow. Second, a propensity for modest cultivation reinforces the vision of Pleistocene overkill; for cultivation implies the expropriation of land, usually the best land, and thus would increase the impact of invading human beings wherever they went. Specifically, the best archaeological evidence suggests that the first colonists of Madagascar were pastoralists; and although this was well after the Pleistocene (in fact, at about the time of St. Augustine), it shows what can happen, and what might therefore have happened elsewhere. Third, if cultivation of a kind did indeed arise 30,000 years ago, then a significant mystery is removed: why it was that agriculture seemed to arise independently in many different locations at different times after 10,000 years ago. The point is that different people were not reinventing the skills from scratch. They all knew perfectly well how to cultivate, and had known for tens of thousands of years. Cultivation and perhaps husbandry were skills which, like the modern Hottentots, they invoked when called upon. The apparent first appearance of farming in the archaeological record of any particular place indicates merely that cultivation had become economically desirable at that particular time and so was being practiced on a scale that was at last large enough to leave some trace.

It is easy to see, too, in a general way, how advanced paleolithic hunters drifted seamlessly into husbandry. The flushing of game by war whoops and fire becomes indistinguisable from herding. Where initially they might have driven the animals into corners for easy slaughter—or indeed over cliffs, for which there is evidence—they might soon have guided them into custom-built corrals.

Eventually, as their control increased, it extended even to the animal's reproduction. Then passive herding truly became husbandry.

On the animals' side we see two processes at work. The first and less subtle is the traditional view: that the first pastoralists chose the species that were of the right size—cattle rather than mammoths; were least agile—sheep rather than gazelles; and were least aggressive—although the popularity of cattle is not easily explained in this way, or indeed that of pigs. But whatever the species, we can imagine the first husbandmen selecting the individuals that were most passive and, probably, the smallest: the ones they could handle most easily. Temperament and size are to a large extent heritable, and so the first husbandmen quickly produced strains of manageable livestock that were distinct from their wild ancestors, just as the first serious horticulturalists quickly produced native varieties—landraces—from their most favored food plants.

However, this traditional speculation has been greatly enriched of late by Stephen Budiansky in *The Covenant of the Wild* (London: Weidenfeld & Nicolson, 1994). He suggests that the animals played a far more active role in their own domestication. First, he says, the phenomenon of commensalism is common in nature; that is, many species tend to tag along with other species because, in various ways, it suits them to do so. Thus, do egrets follow elephants to pick the insects from the kicked-up dust, and cleaner wrasse pick the teeth of predators who know when to strike and when to lie back and enjoy it. Human beings, active and invasive creatures that they are, attract many such commensals; from the gulls that follow the plough to the raccoons, possums, rats, and dogs that hang around human settlements worldwide. Also, many animals are natural followers—like sheep, who live on difficult mountain territory, and whose best chance of survival lies in following the most experienced animal; exhibiting no leadership skills themselves, they build mental maps for future reference. All in all, then, it is not at all fanciful to imagine that many of the first domestic animals, including sheep and pigs, essentially gave themselves up. A life of commensalism suited them. Then, bit by bit, their freedom to do otherwise was curtailed.

But here is the final twist. Modern evolutionary theory suggests that natural selection operates at the level of the gene. That is, we should envisage that individual alleles that produce advantageous characters—including characteristics of behavior—gradually (or quickly) spread through the population. This is what Richard Dawkins meant by "the selfish gene." In practice, says Budiansky, individual animals that lent themselves to domestication would, in fact, have enjoyed greater reproductive success than those that stayed on the outside. So the genes that made them compliant would have spread, while genes that produced standoffishness would have left their owners out in the cold. Hence, natural selection rather than human will favored domesticity in those animals who were by nature commensal. The proof lies in the modern statistics: domestic cattle and hens are among the commonest large species on Earth—comparable even with ourselves—

while many if not most of their wild relatives, who for one reason or another slipped through the domestic net, are now endangered. In short, it is not dignified for an animal to be domesticated. But at least for the time being, it works. And natural selection is not concerned with dignity—only with reproductive success.

We can easily envisage, then, that cultivation of plants and domestication of animals would have developed almost inevitably as human beings became more adept in their gathering and hunting, and as they began to cooperate more widely with each other and in general to operate more strategically. We can see, too, that people who cultivated as a hobby and cooperated with complaisant wild creatures to create the phenomenon of domesticity would have made life easier for themselves, tightened their grip on their own environment, and thereby been favored by Darwinian selection. Hottentots seem (or seemed until recent decades) to demonstrate the enviable halfway state, in which husbandry is conceived as a take-it-or-leave-it option. The people of modern New Guinea often demonstrate a slightly different model, combining life off the land with horticulture. So far, so idyllic. But why take the next step, apparently demonstrated in the Middle East around 10,000 years ago, into the drudgery and indeed the misery of virtually full-time agriculture? Where is the sense of this? In fact, there is no sense in it at all—not, at least, if we assume that to be sensible is to move toward greater contentment. Yet again the point is merely one of logistics. People who cultivate a little and who begin the transition from hunting to husbandry are more likely to survive. More that that: because they increase their own food supply ever so slightly and—perhaps more important—their chances of pulling through the worst of times, their population tends to increase. At the same time, the creatures and plants that have not been caught up in cultivation and domestication are being forced to retreat. Thus, the options for hunting and gathering at large are reduced. Thus, people who begin domestication and cultivation are likely to be forced to do more of it—partly because their population goes up, and partly because the options for doing otherwise begin to be taken away. Thus, they embark upon another positive feedback loop. The more they farm, the more their numbers rise, the more the wilderness retreats, and the more they are obliged to farm. They may not enjoy it. But that is not the point. Darwinian selection is not concerned with enjoyment, any more than with dignity. Survival is the game.

In practice, the transition from part-time farming to virtually full-time—from convenience to reliance—is not inevitable. The Australian aboriginals never took to serious large-scale cultivation. Some native North Americans were farming when Europeans arrived, but some were not, or at least relied mainly on hunting and gathering. Hottentots clearly managed to sustain the putatively ideal halfway state until the last few years. Presumably, people commit themselves to agriculture only if the conditions favor the switch. Clearly, though, the necessary conditions have arisen many times—for witness the apparently independent origins of large-scale agriculture at many different foci.

In general, I suggest, the conditions that favor the final commitment fall into two categories: general and specific. The general point is simply that people who do farm on a large scale enjoy an overall advantage no matter how hateful they may find their existence. For the whole point of farming is that it does indeed produce more food suitable for human beings in any given area. Therefore, population does increase. The point the traditionalists made is also true: the increased population is likely to include a leisured class. In general, then, a farming population is liable to be more powerful, in a crude military sense, than a nonfarming group. Thus, if your neighbors are farming, you had better start farming as well; otherwise you will be swamped. In short, farmers obliterate nonfarmers in exactly the same way that human beings in general tend to override existing faunas when they arrive on fresh continents. They exploit the land more efficiently. They turn more of the local resource into their own flesh. They outcompete the groups that are less efficient. A favorite theme of the movies was the farmers versus the bandits, a theme commonly set in Mexico or Japan. In the movies the bandits always do well for a time, but in the end they always meet their comeuppance. That is how it is in real life. The farmers are still farming and indeed are driving Range Rovers, while the bandits have mostly been hanged. It is sad, perhaps: the farmers in the movies are always miserable wimps while the bandits are given to curly mustaches and wild bouts of mayhem. But that's the way Darwinian selection works. This particular game is to the dour. Theory shows that the race is not to the swift but to the Calvinists and killjoys.

But there must always be special, local reasons why agriculture develops from a part-time convenience to a full-time obsession in any one time and place. In the Middle East, it seems to me, the evolutionary kick from part-time cultivator to committed farmer was ultimately delivered by the same kind of global force that prompted apes to become australopithecines and australopithecines to become humans: yet another shift in climate.

THE MIDDLE EAST: THE FARM AND THE ARK

There are marginally mundane reasons why agriculture took off in the way that it did in the Middle East. Among the local fauna, the sheep is particularly compliant—slower than the gazelle, possessed of a wonderful propensity to follow when led, and able to subsist on low-grade pasture.

Marvin Harris of the University of Florida suggests that Jews were emphatically forbidden to eat pigs not in fact because pigs are "unclean" (although this is the reason given in Lev. 11:7) but because they need a lot of water and are also omnivores, requiring a varied diet. For desert people pigs would be an economic disaster. Sheep are the thing; able to get by on wisps of grass and browse (see Marvin Harris, *Cows, Pigs, Wars, and Witches,* New York: Random House, 1974).

Then again, the hills of the Middle East still harbor a series of grasses whose seeds are particularly large and tasty, notably wheat and barley. Seeds are excellent food because they are nutritious and built to last— the plant produces them, after all, as a storehouse for its offspring.

Seeds of all kinds have played a key part in agriculture, including many kinds of nuts, a host of pulses, various chenopods such as fat hen and Good-King-Henry in pre-Roman Britain and quinoa in South America, and amaranth, also in South America. But the bigger seeds of grasses—the cereals—are unmatched. In any hierarchy, after all, something or somebody has to be at the top, and wheat, rice, maize, barley, and a few others just happen to be outstanding. In short, the cultivators and husbanders of the Middle East had promising raw material: sheep, wheat, and barley, plus the highly nutritious and esculent olives and figs. Embarkation upon agriculture was therefore liable to meet with greater reward than was sometimes the case elsewhere.

Even so, the final reason why Middle Eastern people became committed farmers was, as ever, because they were pushed into it. And what pushed them into it was the end of the last ice age, around 8,000 years ago. For with the end of an ice age comes flood. Water runs off the land, if not directly from melting ice then from increased precipitation, and rushes to augment the sea. People who live near estuaries are caught two ways: by the rivers flooding out, and the seas surging back in. And many people do live by estuaries— the Ganges, the Hudson, the Murray—because in normal times they can be extremely comfortable, with mollusks and crabs, a host of birds, easy navigation, and plenty to drink.

Thus it is that five hundred or so of the world's mythologies include accounts of flood. The area that gave rise to Sumeria, Assyria, and Babylon and is regarded as the birthplace of western civilization is Mesopotamia, which was and is particularly flood-prone. After all, Mesopotamia is placed between the Tigris and the Euphrates—its name means "midst of rivers"—and the marsh Arabs in what is now southern Iraq live there still, albeit under political siege and the constant threat of land drainage. The particular account of Noah, so graphically related in Genesis, is only one of many such stories from that region. Indeed, the same story has been traced to *The Epic of Gilgamesh,* Gilgamesh being a Sumerian king who became the world's first literary hero. Sumeria itself is the earliest known civilization of Mesopotamia, where Gilgamesh reigned over the city of Uruk sometime between 2,700 and 2,500 B.C.

The Epic of Gilgamesh contains the story of Utnapishtim who, like Noah, was warned in a dream of impending rainstorms and thence of flood; built an ark with family and livestock aboard; floated through the storms, buffeted here and there until the water subsided; and finally was able to hitch the ark to a reemerging mountaintop. The parallels with the account of Noah in Genesis are exact. The story probably came to the Jews via Abraham, who, so Genesis tells us, began his long journey into Israel from Ur, not far from Uruk: "And Terah took Abram his son, and Lot the son of Haran his son's son, and Sarai his daughter-in-law, his son Abram's wife; and they went forth with them from Ur of the Chaldees, to go into the land of Canaan: and they came unto Haran, and dwelt there" (Gen. 11:31).

Abraham and Gilgamesh were real people, while Noah and Utnapishtim may be symbolic figures; but of the fact that there were horrendous floods in Mesopotamia, a little before Sumeria came into being, there can be no doubt. The debris of those floods is still to be seen. At Ur itself there is still a 3-meter layer of sand and silt, which, to judge from the pottery immediately below and above, was deposited around 5,000 years ago. That is the spoor of a huge flood. At Fara (Shurippak) is a 70-centimeter layer dating from somewhat later—still huge, although less horrendous; and at Kish is the debris of yet another flood, somewhat later yet.

But what has this to do with the birth of Middle Eastern agriculture? Well, consider this hypothesis: that the memories of flood originated 3,000 years before Utnapishtim and Noah and the evidence at Ur and Fara; that the first memories date, indeed, from the end of the last ice age. After all, at the height of that ice age around 18,000 years ago—the last "glacial maximum"—the sea level dropped by 150 meters, nearly 500 feet. Because the shelves at the edges of continents have only a shallow slope, that fall translates into an enormous area, so that at the last glacial maximum there was 40 percent more dry land than there is now. When the ice age ended, that 40 percent had to disappear beneath the rising sea.

Now cast back to the first people in Genesis, Adam and Eve, and their banishment from Eden. Where was Eden? Genesis tells us (2:10–14):

> And a river went out of Eden to water the garden; and from thence it was parted, and became into four heads.
> The name of the first is Pison: that is it which compasseth the whole land of Havilah, where there is gold.
> And the gold of that land is good: there is bdellium and the onyx stone.
> And the name of the second river is Gihon: the same is it that compasseth the whole land of Ethiopia.
> And the name of the third river is Hiddekel; that is it which goeth toward the east of Assyria. And the fourth river is Euphrates.

A little geographical speculation is now required, but not too much. Havilah is in the southwest of Mesopotamia. Gold was mined there; and the aromatic resin bdellium can still be found. The Pison could be the present-day Wadi Batin, now a dry riverbed. "Ethiopia" in the beautiful King James Version, quoted here, is probably a bad translation. It probably relates to an area of southeast Mesopotamia, in which case the Gihon could be the present-day Karun, which before it was dammed carried most of the sediment out of the highlands of Iran to form the delta of the modern Persian Gulf. Hiddekel is the Tigris, and the Euphrates is the Euphrates. Trace these four rivers back and they converge at a spot that now lies several kilometers offshore in the Persian Gulf. But then, at the height of the last Ice Age, most of the Gulf was dry land. Eight thousand years ago, when the big melt really got under way, half of it was still dry. It would have been a fine spot for hunting—a gathering place of fish and clams, wading birds and antelopes. It would have been a fine place for casual cultivation, too: flat, fertile, and sheltered. This, at least, is the thesis of Juris Zarins at Southwestern Missouri State University.

But why the switch to serious farming? *The Epic of Gilgamesh* offers a clue. The narrative is down-to-earth and humane, with stories of forgiveness and tolerance. But also, intriguingly, it warns against the dangers of overpopulation—this at a time when the world population was probably only around 8 million, less than one–five-hundredth of its present level. But then, of course, it seemed to the people of Mesopotamia that the world began and ended in the Middle East. What mattered were the people in their own region, and the numbers relative to the local resource. We can imagine that in the original estuaries of the Tigris and Euphrates, way out in what is now the Persian Gulf, human numbers rose as people enjoyed the fruits of a fertile valley. This was Eden indeed. But we can also see how those people must have been squeezed as the rising sea forced them inland. Then the cultivational skills that they had indulged as a bonus had to be deployed as a matter of urgency. Thus, they were embarked on the rising vortex of agriculture and population from which, ever since, there has been no escape.

There has been no escape, either, for our fellow creatures. Agriculture has truly transformed the ecology of the whole world.

THE COUP DE GRÂCE

There are obvious and perhaps not so obvious physical reasons why agriculture has changed the world so completely—transformed its ecology and confirmed the status of *Homo sapiens* as a new life form. But as ever, the true significance of farming lies in the intangibles of logistics. Agriculture has changed the nature of the game.

Farming is obviously of physical significance because it occupies so much land, which previously belonged to other creatures. Indeed, as I will argue in the final chapter, some of the principal tasks for modern conservationists are to create farms that are more hospitable to other creatures, or else to make the farms more focused so that more of the wilderness is left alone. But perhaps less obvious is that the habitats of every wild creature are in effect mosaics, with many separate components that may play only an intermittent role and yet are essential. Agriculture is liable to remove components from the mosaic, as in the expropriation of a water hole, or introduce a discontinuity: many bats, for example, will not fly across open fields. (On the other hand, and tragically for all concerned, farms often provide essential components of mosaics that have otherwise been removed. Thus, the small farms of India now provide essential provender for thousands of elephants whose forests have been swept aside, and the confrontations between elephants and farmers are increasingly desperate.) More to the point, however, is that farmers have a relationship to their resources that differs from that of hunter-gatherers and of all other animals. Thus, other animals, plant-eaters or carnivores, are limited by the generosity, or the lack thereof, of nature. To an extent, some animals do tend to increase their own food supply as we have seen—scattering seeds or flooding fields—but in general an animal or a hunter-gatherer can eat only what nature chooses to provide. The populations of predators in particular are geared to those of the creatures they feed upon; and even the most versatile predators, with many separate herds to sustain them, remain relatively rare. But if a farmer wants more food, he merely has to work harder.

In fact, of course, populations of farmers are limited, too, and we can see what happens all around the world when the limitations are not recognized. The process of desertification, brought about by over-exploitation, dates from preliterate times. Yet in broad terms, the generalization applies. The whole point of farming is to convert the food source, and indeed to convert the landscape that produces that food source, until all the nutrients in the entire landscape are pressed to the service of the farmer. A rhinoceros in a forest needs tens of square kilometers, because only a small proportion of a few plants serve it as food. Modern cattle need barely half a hectare each of good grassland because all the grass is grown for their needs, and it grows far faster than any wild plant. If human beings eat the crops themselves, without feeding it first to animals, then with intensive cultivation each of us can in theory subsist indefinitely on the produce of 100 square meters. With farming, in short, the potential food supply per unit area increases not twice or ten times but perhaps a hundred or even a thousand times; and until the limits of the Earth itself are reached, the increase in provender is directly proportional to the work put in.

The crucial logistic point that follows is that hunting and gathering offer only modest rewards for extra endeavor, and will quickly stifle whoever is overambitious. The !Kung sit around most of the time largely because they have no sensible alternative. It really would not be a good idea to catch significantly more than they do already. Farmers, on the other hand, can reasonably work all the hours of daylight and then some, and they often have.

This is the sense in which the farmer plays a quite different game from the hunter. The hunter, whether a human being or a lion, has a strictly limited output, which in turn defines his sensible input. The farmer has not. Thus, to the point that farming increases the human population is added another key factor: that the ecological impact of each individual human being is increased perhaps ten, perhaps a hundredfold—simply because each individual works harder.

Game theory, too, when applied to the practical world, should include the concept of attitude: whether the player chooses to play with enormous effort, or to be relaxed; whether to seek to change the environment, or to lie back and enjoy what the environment offers; whether to experiment or to be conservative, and simply do what members of the particular society have always done. For hunters, human or animal, it does not make sense to do more than is necessary, and the attitude that Darwinian selection is liable to favor is one of relaxation, acceptance, and conservatism. But farming favors effort; it favors those who seek to exploit; and because it tends to provide food surpluses and so provides leeway, it also allows the farmer to take risks, and hence to experiment. In short, with farming the rules of Darwinian selection change. Vigor, exploitativeness, and experimentation take over from laid-back conservatism. To be sure, there have been plenty of laid-back conservative farmers; but game theory shows that sooner or later they will be swept aside.

When we put all the points together, we see how it is that human beings have apparently broken the ecological law which says that big predatory animals are bound to be rare. How is it that although we are big predators we contrive to be among the commonest of all land vertebrates? Quite simply, we have invented a new ecology, one in which, to an ever-increasing extent, all the resources of the world that previously nurtured many millions of species, are channeled toward just one.

Of course, it is clear that the expansion cannot continue forever. The world clearly could support the present human population of 5 billion–plus. The cause of the ever-present famines at present lies in politics and not in the physical limits of the world. But we can seriously doubt whether the world could sustainably support 10 billion, or 20 billion; yet those are the kinds of figures that would be reached within the lifetimes of our children if the present growth continued.

To call a halt we need to do more than change our day-to-day activities. We also need to alter the attitude of endeavor and exploitativeness that has been favored this past 10,000 years by natural selection and then by politicians. Such attitudes are appropriate to expansiveness. Those are the attitudes that enable a society to come to grips with a landscape, and to drive out the people next door. They are not the attitudes that can allow us to sustain ourselves into the indefinite future in a state of peace. But attitudes are hardest of all to change.

(For further discussion of some of the ideas on archaeology and geology in this chapter, see Charles Officer and Jack Page, *Tales of the Earth*, New York: Oxford University Press.)

CHAPTER 8

WHAT DIFFERENCE DO WE MAKE?

In the eighteenth century Jean-Jacques Rousseau established the myth of the "noble savage" who, whatever he lacked in European graces, was supposed to live in harmony with his fellow creatures. Only "civilized" people, Rousseau felt, have lost their respect for nature and despoil it so wantonly. The myth formed a powerful strut in the Romantic and indeed the revolutionary movements of the late eighteenth and nineteenth centuries. It persisted, albeit in self-delusory form, in Paul Gaugin's paintings of Tahiti at the end of the nineteenth century, and indeed was a fashionable component of anthropology and green politics in the 1960s and 1970s.

Of course, there is some justice in the Rousseau myth. The hunting and gathering peoples who survived into modern times clearly protected many of their prey animals by taboos: some could not be killed at all, while others could be killed only in special circumstances—exactly comparable with the laws that now protect game (and which also may have quasi-religious overtones, as Rudyard Kipling describes in his stories of Stalky & Co., which were published in collected form in 1899. Anyone who bagged somebody else's fox in the English shires without a yoicks tallyho became a social pariah). It seems obvious, too, that traditional hunter-gatherers could not be as destructive as modern people with their rifles, their heavy tractors, their civil engineering, and their all-pervading pollutants.

Rousseau knew nothing of anthropology, of course. He had not seen "savages" at first hand and certainly had not studied their way of life. All he had to go on were the stories of travelers who, in his adventurous century, were returning from the ends of the Earth with wild tales of strange people. Almost always those people, for all their gaucheness, seemed to live in paradise: surrounded by animals and plants of marvelous exoticism and undreamt of variety. Many of those animals—the ones that lived on islands—were ridiculously tame. Obviously, the people lived harmoniously with them, or the animals would not exist. Contrast the variousness and apparent tameness of those animals with the narrow suite of nervous creatures in agricultural and citified Europe.

Yet Rousseau was deceived. First, it just happens to be the case that the creatures of the tropics are innately more various than those of higher latitudes. The fact is not intrinsically surprising, but no one has explained it in a completely satisfactory way. It may have to do with the general opulence of the tropics—the never-failing, high-energy sun. Or it may be a historical phenomenon, reflecting the mosaicism created by the expanding and retreating forests as the tropics responded to the ice ages—mosaicism that encourages the emergence of new species, including our own. The point, though, is that the eighteenth-century explorers were the first Europeans with the leisure to study the natural variety of the tropics, and they encountered it in lands that also happened to be inhabited by "savages." So savages and natural ebullience seemed to belong together.

More cogently, however, Rousseau was quite unaware that the variety that was coming to light in tropical lands, and particularly on islands, was generally only a relict of what had existed 100,000, or 10,000, or, in some cases, only a few hundred years before. Of course, the tropics are productive. The vegetation was lush and the animals teemed accordingly. But what Rousseau and his eighteenth-century contemporaries did not begin to suspect, and what indeed is only now becoming apparent, is the sheer scale of destruction that had taken place before the Europeans arrived and recorded the remaining fragments. The travelers undoubtedly encountered pristine floras and faunas on some islands—those that human beings had not previously visited. But however wondrous they may be, the floras and faunas of islands never approach the sheer opulence of continents, and no one in historical times has ever seen a continental fauna in its proper state of glory. Only Africa, and then only in parts, remotely approaches the pristine state that was worldwide until the late Pleistocene. Ironically, some of the countries that are now regarded as the wildest and most natural are in fact the most degraded. The Highlands of Scotland have become a symbol of natural wilderness, though they effectively contain nothing but red deer and grouse for shooting; and yet those mountains and valleys would still abound with moose, wolf, lynx, beaver, and bear were it not for our nobly savage ancestors and the zeal of Tudor shipbuilders. New Zealand, which sells its wild landscape so energetically to modern tourists, has in a few brief centuries stripped out the moas, the most wonderful suite of ratite birds the world has ever seen—they and the great eagle that preyed upon them.

In short, the modern hunter-gatherers who were newly discovered in the centuries before Rousseau may have seemed to live in harmony with the creatures that still surrounded them, but those creatures were only the vestiges of former glories. Eighteenth-century "savages" lived harmoniously with them in the way that, say, the Saxons of England lived among the ruins of Roman Britain. It was picturesque, but it was devastation nonetheless.

But were our ancestors truly responsible for the deaths of the animals with which they once shared the planet? After all, the demise of the animals can be explained in other ways. Notably, there could have been changes of climate—changes that favored the spread of human beings but simultaneously compromised other creatures. So before we abandon the kindly, Rousseauesque view of our "savage" forebears, we should give them a fair trial.

OUR ANCESTORS ON TRIAL

The facts of the matter, only now coming fully to light, are first that a huge number of creatures all over the world disappeared during the late Pleistocene and continued to die out en masse well into modern times. Second, this die-off coincided with the spread of human beings around the world, first from Eurasia into the Americas and Australia and then into the islands of the Indian Ocean and the Pacific, including Madagascar and New Zealand, and those of the Mediterranean.

The charge is that the deaths of those animals was caused by the diaspora of modern humans—in other words, that our ancestors, willfully or inadvertently, killed the now extinct creatures. In fact, this putative scenario has been called the Pleistocene Overkill although it actually extended well beyond the Pleistocene.

The chief prosecutor in this trial is Paul Martin, professor emeritus of geosciences at the University of Arizona (see "Prehistoric Overkill: the Global Model," in Paul S. Martin and Richard G. Klein, eds., *Quaternary Extinctions,* Tucson: University of Arizona Press, 1984). Of course, the evidence he can bring to bear is bound to remain circumstantial but—just to anticipate—it has three powerful features. First, there is an awful lot of it: the sheer magnitude of the extinctions now being revealed by fossil and subfossil evidence is staggering (subfossils being the remains of dead creatures that have not yet had time to become completely fossilized). More and more creatures are coming to light whose existence was unsuspected. Second, the samples of extinct creatures are biased, for large species suffered far more in the putative Overkill period than small ones—large here being arbitrarily defined as 100 lb. (or 44 kg) and above. Yet as Martin argues, if climate caused the extinctions, we would expect the small creatures, like shrews and mice, to suffer more. Third, although there are often problems with dating and sampling (it is difficult or impossible to identify first events), it seems that the extinctions in each particular location generally occurred soon after the arrival of human beings.

So let us examine Professor Martin's thesis, continent by continent, and island by island.

CONTINENTS

North America

People arrived in North America via Beringia some time after 13,000 years ago. Their direct descendants are still there, now known as Native Americans. The large mammals that survived their coming, and are with us still, span twelve genera. There are moose, wapiti, white-tailed and mule deer, bison, musk-ox, mountain goat, bighorn sheep, and pronghorns among the hoofed animals; and pumas, wolves, and various bears (black, brown, and polar) among the carnivores. There are several feral types as well—domestic animals gone wild—including burros (donkeys) in the Rockies, left over from the gold rush days, and mustangs, scattered here and there over centuries. Thus, the native North Americans that are left to us form a wonderful lineup, particularly from the perspective of denuded Britain, where the red and perhaps the roe deer are the only large native terrestrial mammals that remain.

So where is the problem? Well, until at least 100,000 years ago North America had forty-five genera of large mammals, and several of those genera embraced several species. Those that have been lost include a glyptodont, *Glyptotherium;* at least four genera of giant sloths; the spectacled bear, *Tremarctos* (which still survives in South America) and the formidable short-faced bear, *Arctodus; Smilodon,* the sabertooth, and

Homotherium, the scimitartooth; *Acinonyx,* the cheetah (which of course survives in the Old World); *Castoroides,* the giant beaver; *Hydrochoerus,* the capybara, which still lives in South America, and *Neochoerus,* which is a capybara that is gone altogether; several species of *Equus,* both horses and asses; *Tapirus,* the tapir, which also has survived elsewhere; two genera of peccary, *Mylohyus* and *Platygonus;* the camel *Camelops,* the llama *Hemiauchenia,* and the shortlegged llama *Palaeolama; Sangamona,* the fugitive deer, and *Cervalces,* the stag moose; *Tetrameryx,* which was a pronghorn; *Saiga,* the saiga antelope which still abounds on the Russian steppe; *Euceratherium,* the shrub-ox; two genera of woodland musk ox—*Symbos* and *Bootherium;* the yak, *Bos,* still found in Tibet; and three proboscideans from three different families—*Mammut,* the American mastodont, *Cuvieronius,* which is a gomphothere, and *Mammuthus,* the American mammoth. Note, too, that in some of the above cases entire families disappeared from North America, although they survived elsewhere. These include the families of the glyptodonts and the giant sloths; the *Equidae,* the *Tapiridae,* and the *Camelidae*—all of which had originated in North America; and the three proboscidean families. The loss of the proboscideans represented the passing from North America of that entire order. Some of those losses may have taken place before human beings arrived, but many clearly occurred soon afterward.

Note, finally, that the large mammals that survived the advent of human beings were largely themselves of Eurasian origin. In other words, they were accustomed to human beings: they had already been hunted by them for tens of thousands of years. Many of the survivors, too, are either solitary, like moose, or given to unpredictable migrations, like bison and caribou. In either case, this makes them difficult to hunt in a concentrated fashion. These points seem to me to be highly significant—especially in the context of Africa, as we will see later.

South America

South America was hit even harder at the end of the Pleistocene than North America—harder, indeed, than any other continent. Now it has only twelve genera of large native wild mammals. Again, they are wonderful to behold and include llamas, guanacos, alpacas, and vicuñas plus various deer and the Brazilian tapir among the hoofed animals; with jaguar, spectacled bear, and maned wolf among the bigger carnivores; the huge capybara among the rodents; and the giant anteater among the edentates. Of course, there is also a host of other creatures, including the New World monkeys; but although they are equally wonderful, none of those others are large.

Until 100,000 years ago South America had no fewer than fifty-eight genera of large mammals. These of course included descendants of the true South American natives that had survived the Great Interchange of the Pliocene, plus the descendants of the northern Pliocene invaders. With these two mingled faunas, South America was even richer than the north, which is not surprising, as we would in general expect to find more species nearer to the Equator. But the southern continent lost far more than the northern. No fewer than forty-six genera out of the fifty-eight disappeared after 100,000 years ago.

Again, in the late Pleistocene the losses are biased toward large mammals. For in the Pleistocene as a whole South America is known to have lost thirty-two genera of small and medium-sized mammals, including marsupials, armadillos, rodents, and carnivores; but, as in North America, these smaller animals were mostly lost in the early Pleistocene. In the late Pleistocene, when the large mammals were hit so hard, South America lost not a single genus of mouse, though mice can be extremely sensitive. By contrast, too, the losses of large mammals in late Pleistocene South America greatly exceed those of the previous 2.7 million years. As Professor Martin says, "While South America became a refuge for some large mammals that had ranged widely in North America, such as tapir, camelids, and spectacled bear, it became a graveyard for far more." But again the fossil records show that the commoner types of large mammal had survived at least until 15,000 to 8,000 years ago.

Thus, the late Pleistocene extinctions of large South American mammals include about nine glyptodonts—and note once more that we are talking genera, not mere species; about ten giant sloths; the last of all the litopterns, *Macrauchenia*, which was the last of the order to have lived in Earth; and *Toxodon*, the last of all the notoungulates. *Neochoerus*, a capybara, disappeared from South America as it did from the north; carnivore losses again included *Smilodon*, the sabertooth, and *Arctodus*, the short-faced bear—the total end of two formidable creatures whose like we do not see today; at least five genera of camelid; four deer genera; two genera of peccaries; three genera of horses, including *Equus* and two hipparions, *Hippidion* and *Onohippidium;* plus a gomphothere and three mastodons. Again, we have been brought up to believe in a vague sort of way that these elephant relatives were ancient creatures whose day had come, like Triassic dinosaurs. Yet, we see that they were modern animals (albeit from ancient lineages) who belong in the modern world, and might be with us still.

So what can explain these losses? Once more, climate has been suggested. Yet the losses were of allegedly tropical animals who disappeared at a time when, so the pollen records show, the climate was getting warmer. The aftermath of the Great American Interchange has also been blamed. But this took place 3 million years before the late-Pleistocene collapse, and if invading animals have an effect at all, it is generally quick. Besides, the fossils show no signs of long-term conflict. Both the glyptodonts and the giant sloths generated several new genera in South America after the northerners arrived—and of course spread and diversified in North America as well. Litopterns and notoungulates in general were apparently in decline from the time of the Miocene, but some at least were doing well in the late Pleistocene, and remains of *Macrauchenia* are found alongside those of the North American camelid invader *Palaeolama,* which does not suggest that there was conflict between them. Then again, twenty of the forty-six genera that went extinct in late Pleistocene South America were themselves of North American origin.

So again, circumstantial evidence implicates human beings. Climate and competition between different species of animals seem implausible causes. Common observation suggests that only human beings could select large mammals to destroy. The bulk of the extinctions apparently occurred soon after human beings are thought to have arrived in South America, 11,000 years ago. There remains a lack of direct evidence—the archaeological record does not abound with the relics of butchery—and we have still to show that human beings could have wrought such destruction.

Australia

People came to Australia at least 40,000 years ago, and perhaps as long as 80,000 years ago. Either way, they were late-Pleistocene arrivals. As with the Americas, Australia also suffered huge losses in late Pleistocene and Recent times.

Australia began the Pleistocene with far fewer genera of large mammals than either of the Americas, and all of them were marsupial; so it had far fewer to lose. Yet it did lose thirteen genera of large mammals, and they were extremely speciose—so Australia may in fact have lost almost as many large species as either of the Americas, and the proportion of loss was even higher. In fact, only one genus of terrestrial mammal of more than 44 kg now survives in Australia—that of the large kangaroos, *Macropus*. This includes four species, of which only two, the red and the great gray, can truly be considered large.

The large mammals of Australia that went extinct in the Pleistocene included *Thylacoleo,* the marsupial lion, the most impressive of all the Australian marsupial carnivores; three genera of wombat; nine genera of kangaroo, some of which were fleet of foot like the surviving *Macropus* and some of which were more like giant sloths; and four genera of the ponderous diprotodontids, of which the type genus, *Diprotodon,* was like a two-ton but unarmed rhinoceros. The losses of mammals continue in Australia, of course. Most of those that are now disappearing are small and unimpressive to the nonzoologist. But as late as the 1930s, Australia lost the last of its thylacines, *Thylacinus,* which was the size of a small wolf and the only Australian marsupial predator that could truly have been a runner.

In Australia, too, more than in the Americas, nonmammals have played significant roles among the large terrestrial fauna (though we should not forget the *Phorusrhacoids* of South America—the giant relatives of the moorhen, discussed in chapter 4). Many of them also disappeared in the late Pleistocene. The losses include *Megalania,* the giant monitor lizard; *Meiolania,* a giant horned tortoise; *Wonambi,* a python-sized snake; and *Genyornis,* a giant ratite bird. Two big ratites do remain, of course: the emu and the cassowary.

Although the dates are far from clear, these Australian creatures apparently disappeared sometime before the late-Pleistocene collapses of the Americas. But then, people had arrived in Australia much earlier than in America. Present evidence suggests that in Australia as in the Americas, the die-offs peaked after the human invasion. Again, it is hard to ascribe those extinctions to changes in climate. In Australia there is no question of invasion from alien animals apart from humans. The evidence against human beings is circumstantial but again, as I will argue later, the cumulative case seems inexorable.

So Australia and the Americas show a similar picture: a rash of large mammal extinctions—though sparing the small ones—apparently following hard on the heels of the first human invasions in the late Pleistocene. The remaining continents, Africa and Eurasia, show a different pattern of Pleistocene extinctions: much smaller and much earlier in Africa; and much more spread out in Eurasia.

Africa

In the whole Pleistocene—from 1.8 million years ago to 10,000 years ago—Africa lost thirty-seven genera of large mammals, but no fewer than thirty of those losses occurred in the early and middle Pleistocene, which means they occurred before 130,000 years ago. In Africa as elsewhere there is a propensity toward the extinction of large mammals, for in the whole Pleistocene only ten genera of small mammals are known to have gone extinct.

The large-mammal losses in the early and middle Pleistocene of Africa included at least eight species of primate, including various baboons and the hominid genera *Australopithecus* and *Paranthropus,* though of course the former did leave descendants in the form of *Homo.* Three hyena genera disappeared in that time, including *Percrocuta,* and four cats, including the sabertooths and scimitartooths *Machairodus, Meganteron,* and *Homotherium.* Proboscideans lost the gomphothere *Anancus* and the true elephant *Mammuthus,* the mammoth, which of course survived elsewhere. But the deinotheres disappeared altogether, as *Deinotherium* went extinct. The only perissodactyl lost in the early Pleistocene was the last of the African chalicotheres, *Ancylotherium.* The mooselike giraffe *Sivatherium* disappeared, plus five types of pigs and six genera of bovids: a grand total of thirty.

But in late-Pleistocene Africa (after 130,000 years ago), in stark contrast to Australia and the Americas, only seven large mammal genera disappeared. These include the genus *Elephas,* which of course survives to this day in Asia; the three-toed horse, *Hipparion,* which left its footprints alongside those of Lucy's relatives; the camel, *Camelus,* which of course lives on in Arabia; the giant deer (Irish elk), *Megaloceros;* and three genera of bovids.

Some of these Pleistocene losses could have been due to competition between animals—for example, between the relatively diminutive hipparion and the emergent middle-sized ruminant antelopes. *Mammuthus* may have lost out early in the Pleistocene to *Elephas.* Overall, however, Africa gained in general during the Pleistocene, notably because the bovids continued (and continue) to radiate. The continent now remains as the richest in large mammals, with forty-two genera—of which nineteen are bovids, mostly antelopes.

Clearly, though, Africa has retained far more of its late-Pleistocene fauna than the Americas did, or Australia. How come? Several explanations have been proposed, all of which are attractive and all of which may in part be true. For example, the forest animals of Africa survived because they are solitary and elusive; while the surviving savannah species, like the survivors of North America, tend to be unpredictable migrators.

I like the notion that infection has played a large part in curbing the zeal of human beings in Africa, and hence of protecting the animals. A high proportion of major human infections have arisen in Africa, including the malaria parasite, yellow fever, and trypanosomiasis. The reason seems fairly obvious: the infectious agents themselves have to evolve, and they have had longest to evolve in the continent in which human beings arose. By the same token, infections in general arise initially by transfer from other species, and humans in Africa have been in contact with other animals that carry potential pathogens since before the time that they were human—notably monkeys. We know that human beings in Africa have evolved specific defenses against parasites even at great cost to themselves: notably a widespread tendency to anemia, which protects against blood parasites in general, and the specific condition of sickle-cell anemia, which clearly protects against malaria in particular. Such an apparent sacrifice of physical ability could not have come about except in response to enormous selective pressure. Even today African wildlife gains some protection from invasion from domestic cattle because of the trypanosomes that cause the bovine equivalent of sleeping sickness. I suspect there have always been places—many places—in Africa where human beings could not survive because of the sheer weight of infection, which means that other animals could. In general, ecologists are only now beginning to realize that infection plays a huge role in the organization of the natural world, that parasites can be among the leading players. The human beings who migrated out of Africa left many of their infections behind, just as domestic crops may do when taken to a new location, and sometimes the humans flourished more than in their land of origin, as many a crop has done.

But the third explanation for human and animal coexistence is surely the strongest: that human beings and the post-Pliocene African fauna evolved side by side. Prey animals can of course survive in the presence of evolving predators; if it were not so, then there would be no predators, because there would be nothing for them to prey upon. So the big African animals have been adapting to hominid hunting for 3 million years or more, and their various evasive tactics—including erratic migration—was probably shaped in large part by the hominid presence. But the human beings who first came to the Americas and Australia were full-fledged hunters with novel tactics, thrown suddenly among creatures who were entirely naive. As has been demonstrated time and time again in recent years, there is a world of ecological difference between predators and prey who have evolved side by side and have coadapted, and the clash of naive natives with introduced exotics. Thus it is that mice and rats have flourished in Eurasia in the presence of wildcats, while the descendants of those cats, tamed, transported, and returned to the wild, wreak havoc among the ecologically comparable marsupials of Australia. There is a similar contrast between the coevolution of hominids and other animals in Africa, and the sudden imposition of a full-fledged hunting human being among the innocent faunas of the Americas.

Eurasia

The recent fate of Asia's fauna is still largely unknown, but that of Europe is well understood. In Europe as in Africa there were steady losses (and gains) throughout the Pliocene and Pleistocene, but losses were not particularly concentrated in the late Pleistocene: just thirteen genera went at that time, of which eight still survive elsewhere (the tahr, the rhinoceros, the horse, the wild dog known in India as the dhole, the hippopotamus, the musk-ox, the spotted hyena, and the saiga). The five late-Pleistocene mammals from Europe that have disappeared totally are the scimitartooth *Homotherium; Megaloceros,* the giant deer; *Mammuthus,* the mammoth; *Palaeoloxodon,* an elephant (though this genus may be the same as *Elephas*); and *Coelodonta,* the woolly rhinoceros.

Overall, then, we have three continents in which widespread extinction of large mammals seems to have followed hard on the heels of human occupation: the two Americas and Australia. We have one, Africa, in which the extinctions were much more drawn out, and mainly focused on the start of the Pleistocene; but this is the continent in which humans were getting into their stride in the early Pleistocene, and in which animals and hominids had the chance to coevolve. And we have one continent, Eurasia, that seems intermediate, with a more erratic scattering of extinctions. But this was the one in which hominids arrived early—at least a million years ago—and which, through that long time, had other things to think about, including a periodic engulfment under ice. In short, the pattern of extinction worldwide is exactly what you would expect if human beings had been directly responsible for the animals' demise.

Of course, we need more evidence. It would be good to produce unequivocal traces of human assaults upon animals. It would be good to show that other factors—notably climate—could not have been responsible, or at least not entirely. And of course we should show that Stone Age hunters could have eliminated other animals on such a scale—that the overkill hypothesis is plausible.

I will come to all this. First, we should finish the litany of circumstantial evidence. For the pattern of extinction perceived on continents has been mirrored in microcosm, time and time again, on islands.

ISLANDS

Continents are huge melting pots. Space is not a problem. Whatever is biologically possible may occur. Lineages and ecomorphs let rip. Islands are much more confined. They are qualitatively different.

The key problems with islands are severalfold. First, islands are not big enough to sustain many viable populations of large animals—or certainly not of large land mammals, each individual of which needs a great deal of energy. Second, they are by definition cut off from continents.

The creatures that wind up on islands do so very much by chance. Some islands arise by splitting from continents, and their fauna evolves from whatever happens to be marooned or manages to float or fly in before the divide becomes too wide—as Madagascar budded from Africa with its cargo of protolemurs and, fortunately for the lemurs, without any monkeys. Some islands grow as volcanoes in the middle of oceans, and never have contact with continents. They acquire their faunas extremely chancily: for example, the partula tree snails of the high volcanic islands of the Pacific were probably carried in from ancient Gondwana on the feet of birds. In general, then, the fauna of islands is a rag, tag, and bobtail: a random scattering of creatures from which there is liable to be an absence of large terrestrial mammals, albeit with a commensurate flourishing of less-demanding reptiles. Furthermore, the assemblages of islands bear less and less resemblance to the fauna of the nearest continent the further they are from the shore.

On each island the random scattering of founding creatures evolves in sublime isolation. But islands impose a logic of their own; and the evolutionary paths on different islands tend to run in similar directions.

First, we find that if big mammals become stranded on islands, they either go extinct or else adapt to the reduced food supply by becoming smaller. The logic is simple: small individuals eat less than big ones, and it is easier to accommodate a viable population of small individuals than of large ones. Thus, as a homely example, we find tiny sheep and horses on the Shetland Islands. Far more intriguingly, we find dwarf elephants on islands all around the world.

On the other hand, because islands cannot support large mammalian predators, the prey animals no longer find it advantageous to be really small: after all, mice evolved their exiguousness principally to be elusive. So while the big mammals become small, the tiny ones tend to grow larger. Thus, Malta once had dwarf elephants and giant dormice; perhaps, with time, the two would have met in the middle. On the continents of Europe or Asia such ecomorphs would have made no sense at all. But islands impose their own rules.

Many reptiles, too, become giants on islands. Predatory reptiles need far less food than mammalian carnivores and in the absence of mammalian competition seem to come into their own. Thus, Australia—an island continent—once had an impressive suite of land crocodilians and a truly massive goanna. Today the largest lizard of all lives on the Indonesian island of Komodo, and we saw in chapter 4 that it may have evolved originally to feed upon dwarf elephants. Tortoises, too, in the absence of hyenas or bears that could bite their heads off, become giants on islands. Now they exist only on Galápagos in the Pacific and in Aldabra in the Indian Ocean, but once they abounded also on Madagascar and in the Mediterranean.

In the absence of mammalian predators, too, island creatures commonly abandon the apparatus and the techniques that ensure their safety on continents. We discussed a host of flightless birds in chapter 4. Many such birds and other island creatures have also lost the reflexes and the elaborate programs of behavior that once protected them. In short, they are often ridiculously tame. The dodo seems to have given itself up to seventeenth-century voyagers to Mauritius. The booby bird still exists. It is a fabulous flier—a gannet—but acquired its name from the sailors who used to invite it to jump into the cooking pot.

Island creatures, in short, are innately vulnerable. Their populations are invariably small and so they are always closer to extinction. They have nowhere to run when threatened. A fire, a hurricane, a flock of goats can wipe out their habitats altogether. They often lack the reflexes that might have enabled more streetwise creatures to adapt. It hardly seems surprising that scores of island creatures worldwide are

now on the endangered lists, including the kakapo and the Pacific par-
tulas and a huge variety of fruit bats and Madagascan lemurs. What is
surprising and indeed shocking is the mass of creatures that have already died out within the last few centuries, and are only now coming
to light. We should look at a few.

Islands of the Mediterranean

Mediterranean islands have the typical fauna of inshore islands that
may be joined to the continent when the sea level drops. Their faunas
look like a fairly random but much reduced sample of the mainland,
and the individual species show typical island modifications. Nowadays
the biggest mammals of the Mediterranean islands are the acrobatic
ibex; and connoisseur visitors on at least some of the islands delight in
the black vulture (which has now been reintroduced to Majorca). But
what remains in the Mediterranean is a sad relict of what was there
until just a few thousand years ago.

For the fossils and subfossils that are still coming to light on a total of
eleven Mediterranean islands, including the Balearics, Sicily, Malta,
Sardinia, Corsica, Crete, and Cyprus, reveal giant land tortoises, so typical of warm islands in general, and, wonderfully, a bizarre array of
dwarf elephants, dwarf hippos, and dwarf deer, including (ironically) a
miniature version of the formerly giant deer, the so-called Irish elk,
Megaloceros. At least one of the dwarf elephant species (or subspecies)
was less than a meter high when adult, an astonishingly inappropriate
ecomorph for such a small creature but proving, if proof were needed,
that natural selection is ad hoc and simply goes to work on whatever is
around. The youngest of the known dwarf elephants disappeared on
Tilos around 4,500 years ago, the time of the earliest pyramids.

In the last few thousand years, too—almost in historical times—
there have been dwarf proboscideans on islands off the coasts of
Siberia, Southeast Asia, and California. They came from several different lineages—mammoths, stegodonts, *Elephas,* and mastodonts. Elephants in fact, having gone to such great lengths to cope with
hugeness, seem very good at dwarfing. It would be interesting to come
back in 10,000 years and see what has happened to the erstwhile working elephants that are now living feral on the Andaman Islands, 1,500
km or so off the east coast of India; assuming, that is, that our descendants prove more enlightened than our ancestors, and allow them to
live.

Wrangel

Wrangel is a high spot of Beringia that now pokes above the Arctic Ocean about 200 km off the northeast coast of Siberia. It has long been known for its mammoth remains, which hardly seems surprising, for Siberia was the principal center of mammothdom. The first surprise has emerged since 1989 when Sergei Vartanyan, a geologist from St. Petersburg, showed that some of the tusks dated from the time of the pharaohs—indeed, that the latest of all are only 3,700 years old. Thus, mammoths survived on Wrangel for at least 5,000 years longer than on continental Eurasia. The second surprise came when Vartanyan and Andrei Sher from the Institute of Animal Evolution in Moscow showed that the later mammoths were dwarfs, again showing the bizarre elephantine propensity for miniaturization.

Again, we must ask what killed them. People? Or a change of climate and hence of habitat? Well, it does seem that vegetation change played a large part in the extinction of Eurasian mainland mammoths. But the mammoths survived in Wrangel precisely because the island retained the ice age grasses and sedges that suited them so well. People are known to have lived on Wrangel around 3,200 years ago, which seems too late to account for the mammoth's demise. But then Wrangel has not been intensively studied: it is too remote, too expensive to get to, and has been politically sequestered. The earliest-known people were probably not the first. It seems, rather, an extraordinary coincidence that the mammoths departed from Wrangel so suddenly after 5,000 years of adaptation at a time so close to the known arrival of human beings. It would be surprising, would it not, if humans were not involved. Besides, as cited in chapter 1, there is a picture on a wall of a pharaonic tomb that looks remarkably like a dwarf mammoth. There is no very good reason to doubt that it is.

Madagascar

Madagascar is the fourth-largest island in the world—large enough to be acknowledged, like Australia, as an island continent. It has a tremendously varied terrain, with upland, lowland, wet tropical forest, savannah, desert, and of course a rich and various coast. At the time of Christ it had a commensurately rich fauna—what Robert E. Dewar, writing in *Quaternary Extinctions*, calls "a magnificent bestiary." The fauna is still wonderful, and includes entire suites of endemic creatures, which means that they exist nowhere else in the world. There are endemic insectivores, rodents, carnivores and, outstandingly, the unique lemurs. But at the time of Christ there was also a dwarf hippopotamus and giant tortoises in the genus *Geochelone,* plus a range of giant and robust ratites, the Madagascan equivalent of New Zealand's moas, collectively known as elephant birds.

We do not know what happened to the Madagascan fauna more than a few thousand years ago or how it came into being except in the broadest terms, for there are no Pleistocene fossils on Madagascar—in fact, there are no Cenozoic fossils at all. All that remains are subfossils, taken from a narrow band of time. But as usual there is a bias in these subfossils toward the larger animals. They seem to preserve better, and on Madagascar, there is bias toward primates, the mammals that in most locations have generally preserved so badly. Thus, of the living Madagascan primates, only two small genera, *Microcebus* (mouse-lemurs) and *Phaner* (fork-marked lemur), are unrepresented among the subfossils. But of the seven genera of carnivores on Madagascar, only two have been recovered (one of which belongs to an extinct species). Only one of Madagascar's seven endemic rodent genera has turned up as a subfossil, and only one of the eight or nine endemic insectivores.

But let us focus on the primates, which in Madagascar means lemuroids: lemurs and a few similar prosimians. Now there are ten genera. Two thousand years ago there were seventeen. Since these creatures are endemic, a loss from Madagascar means total loss. Two of the genera that do survive, *Varecia* (ruffed lemur) and *Daubentonia* (aye-aye), are known to have lost species: *V. insignis* and *D. robusta.* Each was the largest in its genus.

In fact, all the extinct lemuroids were large by lemur standards, for the biggest of the living types, the indris *Indri indri* would just about have matched the smallest of the extinct species for size. The largest of all the extinct types was *Megaladapis edwardsii,* whose skull was about 30 cm long—getting on for a foot; the adult males must have weighed 50 to 100 kg. The existing lemurs are largely nocturnal, but to judge from the shape of their eye sockets all the extinct types were daytime animals. They moved in more varied ways than modern types. The huge *Megaladapis* was probably a ground dweller who also climbed trees, vertically and clumsily, like a koala or a bear. *Archaeolemur* was a quadruped, like a modern baboon, though it may have had a flexible snout. *Paleopiopithecus* probably brachiated, like a gibbon or a spider monkey. Between them, in short, the lemurs as a whole have mimicked the ecomorphs of the monkeys and apes on other continents—creatures that might have outcompeted them, if ever they had found their way onto Madagascar. In one sense they outstripped the monkeys and apes, however; for in places up to seven different genera evidently lived side by side. Nowhere else in the world do we find—or have we ever found—such a coexistence of so many different primates. Perhaps, Dewar suggests, the lemurs were even more varied than they otherwise would have been because there were no bovids or deer to fill the niches of the middle-sized browsers.

Three nonprimate mammals are also known to have gone extinct on Madagascar in the last few thousand years, though there were probably many more, given the obvious sparsity of the nonprimate subfossil record. They include the only two wild artiodactyls from Madagascar: a hippo in the same genus as the existing African type but only two-thirds the size, and an extinct bushpig. There was also an endemic aardvark, *Plesiorycteropus madagascariensis*. Finally, there was a civet in the same genus as the modern fossa, *Cryptoprocta*. The fossa is a handsome beast, remarkably catlike, right down to the pointed ears and the retractile claws: a striking piece of convergent evolution. The extinct *Cryptoprocta* was bigger, like a low-slung puma. A sad loss.

There were two genera of the giant elephant birds, *Aepyornis* and *Mullerornis*, with six to twelve species between them. The biggest, *A. maximus* was like a massive ostrich, almost 2 meters high, the bird that was probably the "roc" of Arab mythology. The smallest was *M. betsilei*, also robust, but the size of a small emu. Their robustness probably reflects their life in the forest, or the general lack of large predators, which reduced their need for speed. Perhaps most striking of all, however, were their eggs, the biggest of which held 11 liters—almost 3 gallons. The elephant birds were grazers and browsers, largely filling the gap occupied elsewhere by deer and antelope.

Finally, two giant tortoises are known to have disappeared from Madagascar: one with a shell about 80 centimeters long, and the other reaching 1.2 meters. Smaller members of their genus, *Geochelone*, still exist.

Carbon dating is appropriate to these subfossils, and shows that most of them had gone extinct by about 900 A.D., though there are odd historical accounts from as late as the seventeenth century of huge birds, hippos, and a calflike creature with a human face, which could have been *Megaladapis*. A few, then, possibly survived until three or four hundred years ago.

It is clear, though, that many and probably all of the extinct types lasted until human beings arrived, and died out soon afterward. For archaeological records show that people did not settle in Madagascar before the time of Christ, and possibly not before 500 A.D. Indeed, the oldest known sites whose age is beyond question date from the ninth or tenth centuries, but these already show a varied culture—one group were coastal fishers, one gathered shellfish, and one was pastoralist—so the initial settlement was almost certainly sometime earlier. The people probably came from East Africa, though the language and culture owes much to Southeast Asia.

Even with this evidence of a flourishing ecosystem reduced to fragments within a few centuries of human arrival, some scientists have suggested that a change of climate killed the animals. But there is no evidence for such a change, and besides, on such a varied island the animals would surely have had room to maneuver. Then again, the plants of Madagascar are as wonderfully endemic as the animals. They include many unique members of the *Euphorbiaceae,* and one entirely endemic family of peculiar cactuslike trees known as *Didieraceae.* Stands of these plants have evidently stood in the same places for thousands of years. There is no evidence of climatic disturbance here. Again, then, powerful circumstantial evidence implicates the human invasion.

Hawaii

Hawaii has provided the stage for one of the classic conservation endeavors of the twentieth century: to save from extinction the native goose, alias the ne-ne, or *Branta sandvicensis.* In 1952 only thirty-two individuals remained, and they were under siege from introduced mongooses. Then Sir Peter Scott and his colleagues at the Wildfowl Trust (now the Wildfowl and Wetlands Trust) at Slimbridge, Gloucestershire, undertook to begin breeding them in captivity. Now there are thousands. They can be found in parks and reserves worldwide. Although the final phase has yet to be fully realized—to reestablish a self-supporting, viable population of ne-nes on Hawaii—this has truly been a brilliant conservation success.

What Sir Peter could not have realized, however—what no one knew until recent studies of subfossils—was that the ne-ne was merely one of a whole suite of waterfowl that recently included several other geese and at least six ducks, some of them flightless and all now extinct. Besides these, too, there were flightless ibises, flightless rails, a new genus of long-legged owl, a species of sea eagle, a number of perching birds that included honey-creepers and at least one crow, and a new species of petrel. In short, the richness of Hawaii's birdlife was once truly astonishing; the ne-ne was just a lucky survivor, which happened to be able to live (though not very happily, it now seems) high on the Hawaiian volcano. All of these birds, so the evidence increasingly suggests, disappeared after the Polynesians first arrived on Hawaii around 500 A.D.

The same kind of story is repeated again and again throughout the Pacific and Indian Oceans: the archaeological evidence increasingly suggests that the native birds and other creatures apparently disappeared soon after people first arrived. There are a few islands, too, that were not colonized by prehistoric people, and so can serve as controls. Lord Howe Island, for example, remained uncolonized, and it still contained two flightless birds, a wood hen and a rail, when modern Europeans first arrived. But two comparable islands at the same latitude that had been colonized, Norfolk Island and the Kermadecs, had lost their flightless species.

Indeed, then, the evidence that implicates human beings in a Pleistocene and post-Pleistocene overkill is circumstantial, but every year the pile of evidence grows deeper and it all points in the same direction. The best-known example of all can close our case—from a country as large as Great Britain that escaped colonization until the European Middle Ages.

New Zealand

Until 1,000 years ago New Zealand contained one of the finest suites of creatures that has ever evolved in any country, one that shows what evolution can do, given a promising gene pool and a free hand. That fauna consisted of the moas, somewhere between thirteen and twenty-seven species (depending on who is doing the counting) of flightless ratite birds that ranged in size from a big chicken to a very large and heavy ostrich. These birds were the principal herbivores and hence the driving motor of the ecology on both islands.

Now of course they are all gone. Gone, too, from those halcyon and not-so-far off days are various rails, of which several were flightless and one was a giant; a flightless coot; a swan and at least three geese, of which two were flightless; some ducks; a crow and a large shag; two raptors generally classed as harriers, which in fact may have been goshawks; and the giant short-winged eagle that preyed upon the moas in what must have been among the most dramatic confrontations in nature.

Besides the ones that are gone forever, a whole range of creatures that formerly held sway on the mainlands are now marginalized, which in many cases means literally banished to islands. These include the re-markable kakapo and the even more remarkable tuataras, reptiles that look like large lizards (up to 2 feet long), which in fact belong to an ancient order that is as distinct from all other living reptiles as the dinosaurs were. It seems probable, too, that albatrosses once nested on New Zealand, but now they do not.

As ever, the subfossil evidence continues to mount, and as it does, it increasingly suggests that these extinctions followed hard behind the first human colonization by the Polynesians known as Maoris, beginning in the tenth century A.D. In this case, in fact, the extinctions were not quite as rapid as might be expected. Indeed, there are reports of moas as late as the nineteenth century, and although it now seems that these were simply mistaken, some moas at least may have survived for five centuries or so after the Maori arrival, say until the time of the Tudors or the Stuarts in England. But then, although there is clear evidence that the Maoris hunted moas—indeed for a time they had a "moa culture"—they were not full-time hunters. In the main, they were horticulturalists, fishers, and gatherers. As I have suggested, such part-time hunting to some extent favors the prey species since the hunters do not focus upon them exclusively; but in the long term it probably ensures that their elimination is the more inevitable. Maoris also compromised the moas and other creatures through the fires with which—intentionally or not—they cleared much of the forest.

In one striking respect the extinctions and diminutions of New Zealand contrast with the first wave of destruction in the Americas and Australia. On those continents only the big animals apparently suffered. In New Zealand it is clear that some marine fish and shellfish, plus crayfish and a range of giant flightless insects, also suffered badly. But then, the Maoris were gatherers of crayfish and shellfish. And, crucially, they brought with them the Polynesian rat, who winkled out the small creatures that would have escaped the attentions of humans.

It is for similar reasons that the invasions of modern Europeans have often been even more devastating than those of prehistoric peoples, though now of course there is less to devastate. The creatures we have carried with us, which in Australia have included rabbits, cats, foxes, camels, water buffalo and cane toads, may be even more destructive than the human beings themselves. Indeed, speakers at a symposium held at Britain's Royal Society in February 1986 concluded that "biological invasions"—of exotic creatures introduced by human beings—have now become the second most significant cause of species extinction, second only to habitat destruction. The Australians with

their dingo, around 5,000 years ago, and the Maoris with their Polynesian rat, were among the first to demonstrate that humans plus hang-
ers-on are even worse than humans on their own. Hawaii, incidentally,
was originally free of ants; but now several species have inadvertently
been introduced and have transformed the insect fauna, wiping out
scores and possibly hundreds of species, including some that were essential as pollinators of native plants.

Overall, then, the circumstantial evidence that indicts our ancestors
seems powerful indeed. The dates do not always correspond as neatly
as a prosecutor might like, for it is not always possible to say that such
and such a creature or suite of creatures definitely disappeared within
a few years or centuries of human arrival. The principal problem here
is not technical—events of around 1,000 to 15,000 years ago can often
be dated extremely accurately—but the usual one of logic. The archaeological evidence cannot tell us precisely when human beings first colonized a particular place, but only that they must have arrived by a
certain time. The paleontological evidence cannot tell us precisely
when the last animal lived; it can tell us only that a particular animal
was still extant at the date of the fossil or subfossil. In general, though,
the evident pattern of extinction coincides remarkably with the evident pattern of human colonization; and whenever human beings
came to new places, among naive creatures, the shock of novelty seems
invariably to have been devastating.

But no jury should convict on circumstantial evidence alone. All
other accessible lines of inquiry should be pursued. In this instance
these are of three kinds. First, we should seek alternative explanations
for the Pleistocene extinctions, among which a change of climate and
hence of habitat is overwhelmingly the most obvious. Then we should
seek direct evidence of human guilt: where are the signs of slaughter?
And third, and I think most important, we should address the issue of
plausibility. Could our ancestors, who were not particularly numerous
and were armed only with stone-tipped spears, really have wiped out so
many creatures in so many places?

ALTERNATIVE EXPLANATIONS: CLIMATE

To explore the notion that climate killed the creatures of the Pleistocene, we might reasonably focus on elephants. Thus, Andrei Sher
(who revealed the dwarfing of Wrangel's mammoths) believes that

change in climate and hence in habitat was at least the principal reason for the final demise of mammoths on mainland Siberia. Fifteen thousand years ago when the mammoths abounded, the habitat was tundra steppe: grassland over permafrost. When the ice age ended, the permafrost melted and the grass gave way to a misty, ghostly landscape of lakes, moss, and a monocultural forest of larch, albeit with an understory of birch and blueberry. Mammoths evolved as mass consumers of grass and sedge, and as Dr. Sher says, "We don't see any source of forage for woolly mammoths in this forest." We know, too, as related in chapter 2, that when the ice ages ended, they could, in any one place, end fast. Twenty years could see a 7°C (44.6°F) rise in temperature; the difference between a frozen landscape and a temperate one.

The Siberian mammoths did try to make a getaway. They moved north, to the high Arctic. But, says Sher, "By 9,500 years ago larch forest and wet tundra spread almost everywhere, and I believe this was the main reason for mammoth extinction." Wrangel Island, far to the north even of northern Siberia, happened to be the place where the old tundra steppe stayed longest, and the mammoths chewed on for another 5,000 years.

The general picture to emerge from mainland Siberia is that the mammoths were creatures of the cold, as indeed is suggested by the anatomy of the occasional frozen corpses that still come to light, with their thick layers of fat and big pumping hearts. They are adapted also to the grass and sedge that such cold promotes. Warmth is not their thing at all.

But the point does not apply universally. Near Oxford, England, Adrian Lister of Cambridge University has discovered late-Pleistocene mammoth bones from a former riverbed in association with the shells of mollusks—mollusks of species that now live only in the Mediterranean or the Nile. Here, in short, the mammoths flourished in a climate that was clearly more balmy than today's. On the Colorado Plateau, too, Larry Agenbroad and Jim Mead have discovered mammoth dung from around 13,500 to 11,600 years ago; and found that it is rich in the remains of grass, sedge, spruce, sagebrush, and saltbush. In the place where this dung was found these plants no longer exist, for the climate now is hotter and drier, and so we may conclude, as Andrei Sher has done in Siberia, that the elephants disappeared because their favored plants had gone. But the Colorado Plateau is not flat, and as you go uphill it becomes cooler: a change of altitude is like a change of latitude. And the plants that the last of the Colorado mammoths were eating, 11,000 or so years ago, are still to be found just 4,000 feet up the slope. Whatever those elephants died of, then, it was not lack of food—or that at least could not have been the sole cause.

In short, although Siberia was the principal home of mammoths, it may ecologically have been a special case. The mammoths flourished there because there were miles and miles of flat grassland of precisely the kind they preferred. But when the grass retreated, the vegetation that replaced it happened to be particularly unhelpful, for larch is not exactly succulent and it is deciduous, so there is nothing to eat in winter. Because the land was flat, there were no hilltops to repair to; we can imagine the grassland retreating uniformly over a vast area, like the tide going out over a flat beach. In short, the flat uniformity that had allowed the mammoths to flourish also brought about their demise, for there were no microclimates to duck into or—as it happened—any lush new plants to enjoy. In the intricate landscapes of England or America, enterprising mammoths could clearly scratch a living even when the comforting cold was taken from them, if only something else had not added to their woes.

Neither can the demise of dwarf elephants on the islands of the Mediterranean be readily explained by climate. As Paul Martin points out, there were no known changes at the time of their going that would have had such an effect. Besides, at the trough of the last ice age, 18,000 years ago, the surface of the sea in the eastern Mediterranean is known to have cooled by more than 6°C (43°F), which in ecological terms is huge, and yet the elephants and their miniaturized neighbors came through those harsh times. "Ironically," says Martin, "the last European elephants appear to have (occupied) an environment viewed by biogeographers as especially prone to the hazards of natural extinction"; and, "Holocene extinctions in islands in the Mediterranean would be quite mysterious without the likelihood of overkill."

There is more to say about climate, for it clearly was a factor in the late-Pleistocene extinctions. But taken alone, except here and there, we seem to need some additional cause. But if that cause was indeed overkill, where is the evidence?

DIRECT EVIDENCE

In science, direct evidence is always a good thing. The overkill hypothesis would gain from middens of butchered bones, crushed skulls, and arrowheads trapped between subfossil ribs.

But in practice, evidence of this kind is sparse and what there is is equivocal. In Siberia, for example, Olga Soffer from the University of Illinois and her colleagues are excavating houses that were built around 18,000 to 12,000 years ago, not out of wood—for of wood there was precious little—but of mammoth bones. One of them, at Mezherich, is ingeniously cobbled from more than four hundred large bones, including ninety-three lower jawbones: an awful lot of mammoths. Here was a mammoth culture comparable with the moa culture of New Zealand, and as in New Zealand, the apparent profligacy implies carnage.

In practice, though, Dr. Soffer suggests that people and mammoths lived side by side for several thousand years on mainland Siberia. This does not seem to me to offend the general overkill notion. For one thing, as we have noted, the animals of Eurasia were exposed to hominids in a low-level way over at least a million years and had time to adjust: *H. erectus* and *H. neanderthalensis* before *H. sapiens*. For another, 18,000 years ago represents the height of the last ice age, and human numbers could not have been great, although mammoth numbers could have been very high. As we will see later, a high initial number does not protect an animal from extinction once the tide starts to flow against it, but it does take longer to wipe out a lot of animals than a few.

We might argue, too, as I will argue in more detail later, that the people of the late Paleolithic were not at all stupid. If they relied upon mammoth bones for housing, then they might assiduously have avoided killing them. Then again, they could find all the bones they needed just lying about. In fact, there were places where the bones accumulated, and as Dr. Soffer says, they used these as lumberyards: "They just settled next to a lumber yard and used that lumber for construction, for fuel, and for tool-making."

In North America, where continental overkill seems least equivocal, direct signs of human attacks on animals have been found at five sites. However, while some of these signs—including, for example, stone arrowheads lodged between the ribs of elephants—do give pretty convincing evidence of human involvement, other alleged evidence now seems somewhat shaky. For example, crushed mammoth bones, from young animals, have been found, with peculiar scarring and spiral fractures, for the all the world suggesting butchery, with a bias toward the most amenable prey. But archaeologist Gary Haynes has studied piles of elephant bones in Zimbabwe that resulted from deaths caused by drought in the early 1980s. At that time the animals gathered around the dwindling water holes, and as their desperation grew, their social structure broke down and the smallest animals, normally cosseted, were pushed to one side and were the first to die. Then the frantic animals trod on their fellows' corpses, producing precisely the kinds of marks and breaks that have been seen in North American mammoth heaps. There is every reason to suppose that at the end of the North American Ice Age mammoths would have gathered around the dying streams and fought and panicked just as they sometimes do today in arid Africa. We should not drag in notions of human butchery gratuitously.

Yet I feel that both sides, the prosecution as well as the defense, are equally entitled to treat such evidence cautiously. We all know how sparse the archaeological record is, and the further back you go, the sparser it becomes. Do we really expect to find a cave marked BUTCHER'S SHOP? Any direct evidence at all seems to me a bonus. Only a few flint spearheads have been found in skeletons, but a few, in this context, seems an awful lot.

So we must say that climate alone certainly does not seem always to explain the deaths of Pleistocene animals—certainly not of mammoths, who were among the most cold-adapted animals of all. The direct archaeological evidence, as ever, is equivocal: either side can bend it to their advantage. That leaves the last great issue—plausibility.

Could Stone Age hunters really have killed so many animals? The theoretical case for the prosecution—the one which simply says that Paleolithic and recent hunter-gathers could have killed vast suites of large animals—rests on three main points: that large animals are much more vulnerable than they seem; that human beings are even more destructive than they seem; and the modern science of theoretical ecology, including computer modeling, which increasingly suggests that surprisingly modest pressure of the kind that humans can bring to bear can all too easily drive sensitive creatures to oblivion. We should look at these points in turn.

BIG ANIMALS ARE FAR MORE VULNERABLE THAN THEY SEEM

Three outstanding biological features militate against big animals. First, there is the one we have met throughout this book, the importance of which can hardly be overstated: that of population. Animal populations need to be big if they are to be viable in the long term. Populations smaller than a few hundred seem almost bound to go extinct in centuries if not in decades through physical accidents such as epidemic, or statistical accident such as a skewed sex ratio; and in the longer term, lack of genetic variation will prevent their further adaptation when conditions change. In reality, "a few hundred" may mean several thousand. Thus, Tom Foose, when he was with the World Conservation Union, calculated that unless wild populations of black rhinoceroses contain at least 2,500 individuals, they are liable to go extinct. A population that size requires vast undisturbed space: on normal territory, perhaps up to 25,000 square kilometers. Just a little disturbance here and there can reduce the effective size of the territory below the size needed to support viable numbers.

Furthermore, and second, large animals are generally reproductive K-strategists: that is, they produce only one or two offspring at a time, and only at longish intervals, in the expectation that most will survive. They stand in contrast with r-strategists, like mice and flies, which produce big or even vast litters at short intervals, in the expectation that only a few will live. K-strategy reproduction is commonly associated with slow maturity and a long life.

It can pay to be a K-strategist. In Australia, wedge-tailed eagles can override the droughts that commonly last five years or more simply by stopping breeding until the good times roll, but the small rodents and marsupials have no such option because they do not live that long. The slow maturation of the K-strategist leaves plenty of time for learning, so that clever animals like elephants and apes have time to hone their survival and social skills. Human beings demonstrate the ultimate merit of K-strategy.

But K-strategists pay a huge price for their carefulness. When their populations are reduced, they are slow to recover. If the stress that reduced their numbers continues in the years when the population is low, then the animals may not recover at all. Continuous pressure, even apparently low continuous pressure, can erode them little by little until, inexorably, they crash. Modern orangutans illustrate the point beautifully—or tragically. Their reproductive rate in the wild seems astonishingly low: only one offspring every eight years or so. Even human beings in a state of nature improve on that, with one every five years or so (and a total of five in a twenty-year reproductive life). Simple calculations show that if only one in twenty female orangs of reproductive age is killed each year, then the population will crash.

Finally, and most obviously, big animals are conspicuous. They are easy targets for spears or (later) arrows. They do not need to die immediately: the hunters are so well rewarded that they do not resent a two-or-three-day follow-up.

In short, we may begin with a fairly small population, innately vulnerable; we pick off just a few easy targets; and because the animals have so little spare capacity and are quite unable to rebound as mice or herring may do, then little by little, or indeed quite rapidly, they dwindle to nothing. Once you apply a little theoretical biology or actually observe what is happening now among elephants, rhinos, orangs, big whales, and slow-breeding birds like cranes and birds of prey, you see how easily the most conspicuous and impressive creatures can be brushed aside.

Why has this not been obvious? Why have ecologists only recently framed such ideas? In part, I suggest, it is for the logical reason that the big animals that now remain to us are, for one reason or another, the least vulnerable. The more vulnerable ones have already gone. We have already explored various specific reasons why creatures might escape immediate extinction: the solitary elusiveness of the moose, the erratic and rapid migrations of bison and caribou. In general, the ones that survive now are the ones that were least easy to wipe out. This is obvious. But it also gives a false impression. The creatures who are long gone were far more sensitive than the survivors—and far more typical.

There is a second, more subtle general point. Unlike some modern biologists, I do believe that competition is a real and powerful phenomenon. Of course, animals in an ecosystem eventually settle down and live far more amicably as neighbors than the metaphor of "nature red in tooth and claw" would have us believe. Yet each kind lives within bounds set by the others. The corollary is that when two creatures are competing for the same resource, the removal of one can allow the other to expand. This, I am quite sure, is what happened after the late-Pleistocene extinctions of North America. With the removal of horses, elephants, camels, and a host of antilocaprids, the animals that were left—the deer and bison—could spread themselves as never before. The first creatures that European naturalists encountered in North America were these luxuriously expanded survivors: tens or even hundreds of millions of bison. It seemed implausible indeed that Stone Age hunters could have wiped out such herds, and of course they did not. But the Pleistocene creatures that the Stone Age hunters encountered were not like that. They were assemblages of many different creatures each of which had a far smaller population and was far more easily pushed over the edge. The apparent swarm of survivors is an artifact—like a rash of ragwort on a bombsite. The ragwort would hardly get a look in if the other wild plants had not been wiped out first.

In short, big animals are vulnerable. In general, the survivors are the less vulnerable but also the less typical. Now add to that the fact that human beings, who are obviously destructive, are even more destructive than they seem.

HUMAN BEINGS ARE EVEN MORE DANGEROUS THAN THEY SEEM

The hunters of the Upper Paleolithic—the late Pleistocene—were obviously proficient. But it is difficult to judge just how proficient they were. Were they simply good predators, like wolves or lions, but with a few extra tricks thrown in? Or were they more than that? Might they have qualified, in modern parlance, as game-managers? There is a spectrum of possibility. The notion that our ancestors functioned just as another predator is the conservative view, while the idea that they were game-managers is clearly more bold, and I have not heard it stated so explicitly. Some scientists feel that it is innately virtuous to be as conservative as possible, and would not consider the game-manager hypothesis until the simple predator idea was exhausted. But actually I think that is bad philosophy. It is at least as acceptable, and in the end more economical, to consider the whole spectrum of possibility together. So let us ask: were our late-Pleistocene ancestors—those of around 30,000 years or so ago—merely predators, or were they perfectly competent game wardens? Well, even if they were "merely" predators, they clearly would have been very adept indeed. We explored their assets in chapter 6. We know they could think, cooperate, remember, and share experience. It is not foolish to imagine them lying in wait for prey whose habits they understood well, or dividing themselves into beaters and killers, as many hunting-gathering people do today (and which is the technique employed by shooters of pheasant on Norfolk heaths). It is sometimes said that we should not extrapolate too readily from present-day experience to the past. But the same point can be made at least as cogently the other way around. We know from their anatomy and from all other signs—including their weapons and their paintings—that the Upper Paleolithic people of the late Pleistocene were the same as us. They lacked the thousand years of experience and accumulated infrastructure that we have now inherited, but as people they lacked nothing. If they were alive now, they would hold down jobs as builders and tycoons, bank managers, scientists, and coaches of tennis. Well, in those days circumstance obliged them to be full-time hunters. I see no reason to doubt that they practiced the crafts of their time as efficiently as their descendants practice those of the present. In short, we know that they were hunters; and there is absolutely no reason to doubt that they were extremely good at it.

This general aptitude was augmented, I suggest, by one technical attribute that is undeniable, and one ecological possibility that is highly speculative (I have never heard it seriously discussed) but is at least worth thinking about. The technical attribute was of course the missile, as discussed in chapter 6. The ability to hit and run—to strike the mortal blow with little or no personal risk—is unique among large predators and rare in all of nature. Here was an innovation indeed: a "trick" that transformed the nature of the game almost as profoundly as that, say, of speech.

The second ecological possibility, discussed in chapter 7, is that from about 30,000 years on (and perhaps before) our late-Paleolithic ancestors were already practicing horticulture, not on a scale large enough to show up in the archaeological record, but certainly enough to make a crucial difference to their survival in any one place and at any one time. The art of survival is not simply to flourish in the good times, but to get through the bad times. We have discussed the paradox in early chapters—that if those Stone Age hunters were cultivating, then their efficacy as hunters would be greatly enhanced, simply because the alternative food source would allow them to maintain their population when the wild prey became rare.

This principle applies to late-Paleolithic hunters even if we do not admit that they might have indulged in cultivation. For the key to human ecology is our versatility. If the supply of mammoths fails, then we can get by on crayfish. If the meat fails altogether, we can at least subsist on plants. The same argument applies, or seems to apply, to true Carnivora versus Creodonta. The Carnivora on the whole are less committed to meat, and so can survive in more circumstances than the creodonts, and so in the end were bound to oust them. Human beings have a more varied diet than any carnivore except the bears; and if we cultivated, which at least was possible, then we would have outstripped them as well. Besides, the use of fire and of stones for bashing sinews and bones would have made us more efficient than bears even at meat-eating.

The fact that human hunters are supremely clever gives them a crucial advantage that is denied to all others. In general, other predators are merely opportunist. They may have some preferences, but in general they take whatever is suitable and most available. They cannot afford to waste time seeking out what is rare; and this creates a balance of a kind because the species that are overhunted are protected by their scarcity and have a chance to recover. But we can imagine that human hunters were not such slaves to opportunity. If mammoths were rare, so what? The hunters were astute enough to know where the few remaining herds could be found. And it is doubtful if hominids ever regarded hunting merely as a source of food. There were always kudos attached to it—just as is obviously the case among present-day chimps. There would be kudos indeed in killing a creature that was not only huge and powerful but was also hard to find. In short, it is eminently reasonable from all that we know about hunting people to assume that our late-Paleolithic ancestors were uniquely capable of hunting rare animals to extinction; or—which is much more to the point—of hunting them until the populations were too small to be viable.

There is one final ecological scenario. All animal populations fluctuate in the wild. Species experience good years and bad years; and on a grander scale we can probably envisage that elephants experienced good centuries and bad centuries. Thus, we see that at any one time animals go locally extinct; which means that they disappear from part of their range but hang on in another. When times improve, they begin to spread again from their refuges and reoccupy the places they have vacated.

But an extremely versatile animal like the human being is less liable to be driven out of any one place by any one set of circumstances. In the same way—demonstrably—gray squirrels are less likely to disappear from any one spot than the more persnickety reds. Competition, however, is preemptive: in general, it favors the creature that is already in residence. Thus, the red squirrels found it harder to get back into the places from which they had been made locally extinct, because the grays had never left and were already in residence.

I find this kind of picture, as demonstrated by gray squirrels and red, eminently capable of explaining the erosiveness of human invaders in North America. People occupied spaces but then, in the bad times, had little need to vacate those places because they were so resourceful. But the less versatile creatures were obliged to leave from time to time. When they tried to get back in, however, the human beings were already ensconced, ready and waiting. This would not happen every time, and the human beings would not always win. But they would win often enough. The process worked as a ratchet. Little by little, the human range and population grew, and the animals diminished. Furthermore, because human beings are so versatile, the same scenario could be enacted in a hundred different habitats: in the forest, by the river, in the mountains, on the plain. The importunate primate was always there.

Finally, modern observations show that it is not necessary to kill an animal directly in order to make its life intolerable, and hence cause it to retreat, and hence in the end drive it to extinction. There are two main points. First, animals construct their habitats as mosaics: a collation of essential components. If any one component is made unavailable, then the whole habitat collapses. The mosaic exists in four dimensions—time as well as space. If you make it difficult for an animal to drink in peace at the waterhole, then its whole habitat is compromised. If you occupy some favored spot in which it feeds at one particular time of the year or in which it hibernates—an estuary or a cave—then again, the whole territory becomes inhospitable. If parts of a territory become inhospitable, then the population can be fragmented; and if it is fragmented then each of the divided subpopulations may be too small to be viable and again the whole population collapses, fragment by fragment. This is what seems to be happening today with Asia's wild elephants.

Most animals, too, are extremely sensitive to disturbance. In the Pleistocene, orangutans occupied a vast swathe of Southeast Asia; indeed, they must to a large extent have coevolved with giant pandas, which in those days were similarly extensive. Now the pandas are confined to a few hillsides in China, while the remnant orangs are thousands of miles away, on Sumatra and Borneo. But the modern orangs on Sumatra and Borneo are now demonstrating the extreme vulnerability of animals that are large and apparently formidable. Their reproduction is egregiously slow. They are also outstandingly standoffish. When human beings move in, they tend to move out, not necessarily because they are molested, but simply because they are accustomed to have the forest to themselves. I wonder if this is because the males stake out their territories by a call that for all other orangs is terrifying; and find that human beings call their bluff by ignoring them, and thereby announce their own dominance. Whatever the psychology, we know that orangs vacate vast areas that they could perfectly well inhabit, simply because they resent having neighbors. They cannot afford to be so choosy, but choosiness is built into them.

In general, human beings are far more disturbing than any other predator. Lions may walk in full view within a hundred yards of antelope without disturbing them. Provided the antelope can keep them in view, they know they are safe. But prey animals are not stupid. They know that human beings have missiles. They know that, uniquely, human beings can kill from a distance. So human beings cannot walk openly within 100 yards of wild antelope. Indeed, on walks in agricultural land if you keep your eyes open, you can see crows and pigeons take off from trees half a mile away. You usually do not notice them. But they notice you, and they act accordingly. They would not respond that way to a fox or even an eagle. But then, the fox or the eagle has to get in close in order to kill. Indeed, old Africa hands such as David Houston of Glasgow University and Brian "the Lion" Bertram, tell me that you can stride up to lions at a kill and shoo them away. I cannot believe that lions find human beings innately terrifying. But they are experienced, and although wild animals lack the detailed communicative skills of human beings, they do have traditions. They know, therefore, that human beings have tricks up their sleeves—if not spears, then guns. The safest option is simply to give our species a very wide berth.

In short, it seems reasonable to suggest that our late-Paleolithic ancestors not only killed their fellow creatures but also made their lives difficult: upsetting their patterns of feeding and of reproduction and generally compromising their habitats to a far greater extent than an ordinary predator would do. For large, vulnerable, slow-breeding species, that could be all that is required.

Yet all we have assumed so far is that human beings were proficient predators: the conservative hypothesis. If they were more than that—if they were protomanagers of game—then their influence could have been even more acute.

PALEOLITHIC GAME MANAGERS

The notion that late-Pleistocene hunters veered toward game management must be seen to be speculative, yet in part seems largely a matter of definition or of degree. We could argue, after all, that people who are conventionally considered merely proficient hunters—and yet are anticipating the movements of their prey, both in the long term and the short, and are employing beaters to drive their prey into corners or into the paths of their partners—are managers of a kind; just as we can argue that people who stick twigs in the ground in the expectation that they will grow at least have their foot on the ladder of horticulture. But it seems reasonable to me to speculate that Upper Paleolithic hunters were more organized even than this.

First, we can again learn from present-day hunter-gatherers, and in particular from the aborigines of Australia. For they are, above all, astute deployers of fire. In their early days they may well have used fire to drive diprotodonts and giant kangaroos, and now they use it to freshen the vegetation and attract small animals for prey. As we have seen, Reese Jones of the Australian National University at Canberra suggests that this is a fine way to manage that vast, uncertain continent. I see no reason to doubt that people of the Upper Paleolithic all over the world could have used fire to create amenable landscape: less forest, more open space. It seems eminently likely, too, that they knew perfectly well how to adjust the landscape to direct and manipulate their fellow creatures.

Why should we doubt, either, that those advanced Stone Age hunters preferred some creatures to others, and knew how to advance the cause of those they liked and undermine the ones they did not? It is easy to imagine such people imposing closed seasons on their favored creatures through laws of the kind that we choose somewhat snootily to call taboos, which were designed to keep the favored ones in good heart. On the other hand, if they decided that some other creature was bad news, why doubt that they might have targeted them? They knew perfectly well—they were modern people after all—that it is the young females who give birth; and if you want to eat meat sustainably, then it is wise to kill only the aging males. Contrariwise, if the animal is of a kind you feel you would be better off without, then you should kill the ones with the swollen bellies. Perhaps—only perhaps, but the idea is at least worth considering—the first colonists of North America set out to eliminate the mammoths. Perhaps they decided that smaller animals were of more use (since they did not need the elephant bones for timber). Perhaps they decided, too, with even greater sophistication, that they would be better off without large rival predators, and set out to remove the principal prey base of those predators. Get rid of mammoths and ground sloths and you eliminate sabertooths and the giant running bears. In a similar vein many modern historians now argue that the government of nineteenth-century America set out to cleanse the prairie of the bison—the most triumphant survivor of the Pleistocene pogrom—so as to destroy the economy of the "Indians," who were themselves the descendants of the Pleistocene invaders. I see no reason to suppose that the original Pleistocene invaders were less astute than the American government of a few thousand years later.

As I have already said at least once too often, however, the idea that Upper Paleolithic hunters were game-managers—that they consciously manipulated their prey and encouraged some at the expense of others—is speculation. It merely represents one point, the extreme point, on the spectrum of possibility and therefore should at least be considered. But what seems to me to clinch the case for the prosecution, and show beyond reasonable doubt that our ancestors certainly could have perpetrated the late-Pleistocene overkill and therefore (in the absence of a more convincing culprit) probably did, is the kind of theory that emerges from computer modeling. This shows that big and therefore vulnerable animals can be wiped out by hunting even when the hunters are merely predators—even when they fall short of game management.

In particular, Steve Mithen of the University of Reading has devised a model to show what would happen to populations of mammoths when they are subjected to various stresses. The population dynamics of the animals—the rate of birth, the interval between births, the age at sexual maturity, etc.—is based on that of their living relatives, African and Asian elephants in the family *Elephantidae*. The stresses Dr. Mithen invents for them are hypothetical but nonetheless realistic: a changing climate that diminishes the food base, and hunting pressure.

The model shows, as we have already noted is the case with orangs, that remarkably little hunting pressure quickly drives mammoths to extinction. Dr. Mithen first posited a reasonable population of human beings in North America and then showed that if one mammoth were killed per year for every twenty people, then, sooner or later, the mammoths would go extinct: especially if, as seems likely, they were already under climatic stress. The original size of the population does not affect the issue. It takes longer to eliminate a big population, but it will crash sooner or later. In the interests of conservatism, however, he proposed that the rate of hunting would diminish as the mammoth population went down, so that the rate of hunting never exceeded 2 percent per year of the mammoth population. Still, it dwindled to extinction. A mammoth population under climatic stress cannot increase at 2 percent per year, and so it cannot withstand a loss of 2 percent. Perhaps, too, as I have already suggested, the model does not need to be so conservative. Determined or bloody-minded human hunters with something to prove could well have hunted mammoths into the ground even after they had become rare. Many a modern creature—great auks, dodoes, passenger pigeons, moas, gray wolves, and many a bird of prey over much of their range—have similarly been hounded past the point of rarity and into the grave. Besides, rarity by definition implies a small population; and small populations tend to dwindle on their own account.

So we know that in the late Pleistocene many large animals died out on three great continents; while in the late Pleistocene and Holocene entire suites of creatures disappeared from islands worldwide. So we also know that the creatures we see today, wondrous as they are, are a shadow of the faunas of comparatively recent but "prehistoric" times. We know that the extinctions tended to follow the incursions of human beings. Did the animals simply fade away? Or did we kill them?

Of course we did. Or at least, our immediate ancestors did. I do not know whether they did it inadvertently as highly competent hunters who were also insouciant; or whether they did what they did with regret; or whether they set out, as protomanagers of game, to weed out the species they found least helpful and leave the others to flourish, a scenario that does not seem too fanciful. But whichever way they approached the task, the jury must find them guilty.

The evidence that implicates our ancestors is flawed because it is circumstantial, and yet it seems overwhelming. The direct evidence—of arrowheads stuck in mammoth ribs—is sparse, but the wonder is that it exists at all. The issue of plausibility seems to me to be open and shut. It would be amazing indeed if predators as competent as our ancestors had not had a tremendous impact. All that has become apparent these past few years in ecology and population dynamics, including the mathematical modeling that shows how creatures like orangs and mammoths will decline under what seems only moderate pressure, proclaims that for big, slow-breeding animals, extinction is always imminent. Other factors did play a part in the Pleistocene and post-Pleistocene extinctions, of course. Climate alone might have done the trick here and there. But the *coup de grâce*, and often the sole operant, was ourselves.

And now the pace of extinction has increased. Now the world is so arranged, if "arrangement" is the word, that the existence of every other creature is to some extent in our hands. The animals that serve us directly are overwhelmingly successful, if the criterion of success is indeed the replication of their genes, for cattle and chickens are far more numerous than they would ever have been in untamed nature. By contrast, the species for which we have no use are pushed aside— unless they discover some human niche, as aphids and rats have done, and become "pests." So now we have the world at our feet—where do we go from here?

CHAPTER 9

THE NEXT MILLION YEARS

Throughout the Cenozoic most species of mammal lasted roughly a million years and then went extinct or evolved into something else. There are no innate mechanisms to limit the span of species, at least that are known about, but that is the way things have tended to turn out. After a time conditions change and species must also change or go to the wall; and the cycle of change averages around a million years.

So what are the chances of *Homo sapiens* lasting a million years? Of course, the question is arbitrary because there are no rules. There is no a priori reason why our species should survive for so long. On the other hand, there are some very good biological reasons why we might last in recognizable form until the dying of the planet, because in theory there is less cause for us to alter and also less opportunity. So at least in the spirit of intelligent provocation, we might ask: What are our chances of lasting a million years? But at the same time we must ask: And what will happen to our fellow creatures along the way?

Already, of course, we have had some of that arbitrarily allotted span. But not much. The species *Homo sapiens,* as distinct from the archaic *heidelbergensis* and the Neanderthals, has been anatomically modern for about 100,000 years. So if we define ourselves in anatomical terms, which after all is the criterion that paleontologists apply to other species, then we can reckon that we have had only one-tenth of our time at most. Although our impact has already been frightful and for many other lineages has been decisive, we are very much at the beginning of our run. We still have 900,000 years to go; as near to a million as makes no difference.

Mao Tse-tung commented that a journey of a thousand miles begins with a single step. Our journey of a million years must begin with the next century or so. This will be a particularly testing time.

THE NEXT FEW HUNDRED YEARS: THE HUMAN SPECIES

All times are unique; all times are special. But some are more special than others, and the age we live in now is crucial. The next century or so—the next 10 centimeters in the kilometer of time we still have to run—will show whether or not it is possible for the human species to regulate its own numbers by means that are voluntary and benign, and whether in doing so we can preserve at least a fair proportion of our fellow creatures. If not, the coming century will produce a collapse of humanity and of environment that in ecological terms is entirely comparable with the death of lemmings, or with algae in a pond when the nutrients are gone, but on a far, far greater scale than either. Nature's rules are universal, and nature has no taste. We may find our own collapse unthinkable simply because it is so horrible to contemplate, but repugnance provides no protection at all, any more than incredulity protects against flood or the encroachment of ice.

Populations of all creatures grow exponentially. "Exponential" does not mean "fast." It simply implies growth by compound interest—not by a fixed amount each year but by a steady proportion of whatever was there before. In practice, with most creatures most of the time, this exponential tendency is countered by the hazards of life, of which limited resource is the principal one, either because the resource simply is not there or because some extraneous factor, like a rival species, prevents access. Eventually, even the most favored and rival-free populations must stop growing as they exhaust the capacity of their environments to provide more. But when everything is going well—in halcyon days when there is no practical limit on resource—populations grow maximally, the only limit being imposed by the reproductive capacity of the animal, which in practice usually means the female animal. The maximum expansion of elephants, who mature in their teens and produce only one infant at a time after a two-year gestation, is many times less than that of mice, which can produce litters of six or more at age six months, after a gestation of about three weeks. The maximum reproductive capacity of human beings is somewhere between those two.

Until human beings were finally obliged to commit themselves more or less fully to agriculture, about 10,000 years ago, the restraints on their population growth were all too obvious. Resourceful as they undoubtedly were, they still had to compete with bears, and the supply of mammoths and gazelles was all too fragile. It seems, indeed, that by about 10,000 years ago the human population had reached a maximum of around 5 to 10 million.

But the final commitment to agriculture removed the restraints. After all, the whole point of farming is to divert the highest possible proportion of natural output into human food, and up to a point—a point that even now we have not yet reached—more food can be produced simply by expending more effort. Thus, by farming, human beings can increase their effective resource by ten, a hundred, or, with modern intensive methods, by at least a thousand times. So by the time of Christ, after a mere 8,000 years of large-scale, full-time agriculture, human numbers worldwide are estimated to have reached somewhere between 100 million and 300 million.

After this, with the baseline already substantial and farming techniques already advanced, the exponential growth of the human population finally entered its rapid phase. The billion mark was passed by about 1800 A.D. By 1900 A.D. the world population had reached 1.7 billion. By 2000 A.D. numbers will exceed 6 billion. Thus, agriculture has allowed a thousand-fold increase in numbers over 10,000 years. In short, in recent millennia the exponential potential of human beings has more or less been unrestrained. Indeed, at times, as in Kenya in recent decades, human population growth has approached the maximum that is conceivable for our species—around 4 percent per year, a rate that doubles population in less than twenty years. Worldwide, the numbers now added each year are equivalent to about twice the current population of the United Kingdom.

The question is not whether this growth must stop, but when, and how great the numbers will be by the time it does, and whether we can call a halt by means that are voluntary and benign, or whether the eventual restraint will be out of our hands. If things simply get out of hand, we might ask how much damage is liable to be done as we enter our maximal phase and then collapse out of it, and whether, or how much of, that damage could be repaired in the following centuries. Soil erosion might be countered to some extent, for example, by fresh weathering of rock, but the loss of elephants will be permanent. Those who feel, as some clearly do, that matters can be left to chance or to God might simply consider some statistics that emerged from Princeton University in the 1960s, when the proportionate increase in population was somewhat higher than now (although the absolute number added each year was smaller). The Princeton demographers calculated that the growth rate then current would produce a world population of 17 trillion within seven hundred years; a trillion being a million million. Such a figure is clearly fantastic—several thousand people for every one that now exists. But seven hundred years is a trivial period of time—two-thirds of a meter, in our time-into-distance scale. Something has got to happen between now and then, and either that something will be a tight and clever policy, benignly expedited, or it will be very nasty indeed.

In practice, however, there are some clear and perfectly acceptable routes to population control that could bring about the necessary curtailment within a half century or so and level the human population at around 10 to 12 billion. That number would still be uncomfortably high, putting more strain on world resources than truly seems desirable. But those same policies, if maintained, would then bring the world population steadily downward to whatever figure our descendants feel is desirable. However, there is in populations the property of momentum: that is, trends that fall short of massacre take several decades to show a serious effect. For example, at least two-thirds of the people now on Earth are under sixteen years old and have yet to reproduce, so the world population will increase markedly even if they all decide to have only one or two children per couple. Taking momentum into account, modern demographic projections suggest that it would take about five hundred years for the world's maximum population (10 to 12 billion) to fall again to current levels (5 to 6 billion), if reasonably benign policies were put in place.

Once population begins to drop, then people can decide the figure to which they want to fall, and at which, roughly, it should be maintained. I have asked various conservation biologists what kind of world population they feel is reasonable and their answers ranged from 2 billion down to about 300 million. But all the biologists brought the same principle to bear: that the population should be small enough to be sustainable indefinitely, and leave plenty of leeway for ourselves and other species, but should be large enough to sustain a variety of healthy civilizations. Three hundred million may seem draconianly low (it is barely more than the present population of the United States), but it is the same as the world population at the time of Christ, and at that time there was certainly no shortage of genius or of cultural variety (any more than there is in present-day America). Incidentally, too, this "low" figure is 10,000 times higher than the present population of Asian elephants and 10 million times higher than the present population of northern white rhinoceroses. But whatever figure our descendants consider desirable, it could be reached within a few thousand years from now—more than one thousand but less than three thousand—simply by applying policies that most people would consider eminently benign.

But of course, there are "pronatalists" who maintain that anyone who seeks to reduce human reproduction must in some sense be "anti-humanity." To prevent the birth of possible babies is, they say, to "deny life." Some religious people argue, too, that all babies are born "for the glory of God" and that more babies means more glory. Such arguments can be answered even by crude statistics. Suppose, for example, that we do survive the looming demographic crisis; and suppose that our descendants finally decide that a world population of around 1 billion is a reasonable, sustainable target. With such a population there is no reason to doubt that our species could last a million years. In such a case, the human species will enjoy 1 billion $\times$ 1 million = $10^9 \times 10^6 = 10^{15}$ person-years. But if we allow our population to rise to, say, 20 billion, then we must surely doubt whether we could survive in recognizable form for more than another 10,000 years or so. If we faded after ten millennia, then our total presence through the time still to come will have been a mere 20 billion $\times$ 10,000 = $20 \times 10^9 \times 10^4 = 20 \times 10^{13} = 2 \times 10^{14}$ person-years—at most. In other words, if we exercise restraint, then the total number of human beings who will have trodden this Earth could be at least five times greater than it would be if we allowed populations to run away with us. Who, then, are the misanthropes? What, though, are the benign policies that could reduce human population so dramatically, without pogrom or epidemic or ecological collapse? Well, common sense reveals that if each couple (which effectively means each woman) elects to have only two children, then the population cannot grow (except for the first few decades, because of momentum). In practice, if two children per family was accepted as the preferred maximum, then the population would eventually fall, since some people would die before they reproduced, and some would elect to have only one child or none at all. So long as the average number of offspring reaching reproductive age is less than two per mother, then population is bound to level out and then to fall. Recent experience—for example, in West Germany, before the reunion with the East—shows that when people are rich they commonly do elect to have only two children. Large families are associated either with particular fashions (the Kennedys initiated a minor baby boom in the United States) or, more commonly, with deprivation: that is, large families are com-

monly seen (a) as a status symbol for women in societies in which females can gain respect only through motherhood; (b) as a source of
support in old age; and (c) as a buttress against high infant mortality.
Thus, the only sure as well as the only benign route to small families is
via increased wealth and security, and reduced infant mortality. It is
good to know that the only policies that have the slightest chance of
working in the long term are the benign ones.

In practice, common sense plus the experience of the past few
decades shows that several preconditions must be met if the two-child
family is to become the norm worldwide, all of which are difficult in
practice, but are conceptually undramatic. First, all efforts must be
made to minimize infant mortality. People must know that two children out of two are liable to survive. Second, everyone worldwide
needs a pension, so that they do not need to rely upon their children
when they stop working. Third, the trend in rich countries toward earlier and earlier retirement must be reversed, for if people retire earlier
and the birth rate goes down, then within a couple of decades or less
we will find there are too few young recruits for the job market and indeed that only a small minority of the population is actually working.
In many countries this is already a problem. We therefore need to employ people over sixty in ways that retain their dignity but do not (as in
the past) hold up promotion and clog the heights with a gerontocracy.
Finally, we need to ensure that the women of the world who have already declared an interest in the technologies of family planning
should have that technology available, and preferably should control
that technology.* As modern family planners say, the point is not to co-
erce but to empower. Coercion is obviously undesirable, but modern
experience shows that it is also unnecessary. People do not actually
want to have more children than they can reasonably support, or than
they need for their own security.

In short, the policies that could enable the human species to keep its
numbers within sustainable bounds are clear in principle, and they are
all benign. Indeed, the pleasant truth is that no policy that is not benign can possibly succeed. High infant mortality, for example, tends to

---

*Of course, contraception for men is desirable, too. But for good physiological rea-
sons, and not simply for reasons of male chauvinism, male contraception is a far more
difficult trick than female contraception. Fertility in men is far more closely inter-
linked with secondary sex characteristics than is fertility in women, and most male pills
produced so far have tended to emasculate if not frankly to poison. So far, vasectomy
remains the only safe male contraceptive option, but it is not fully reversible.

be countered by high birth rate, and leads not simply to misery and sadness but also to biological disarray and extreme wastefulness. War, famine, and pestilence have often produced dramatic blips in populations, but again, people invariably respond to these setbacks by breeding their way out of trouble. Wars are commonly followed by epidemics and also by baby booms.

Finally, we must of course acknowledge that the growth in human numbers is only half of the equation. The other half is to place a ceiling on human acquisitiveness. Thus, people on the receiving end of family-planning policies are wont to point out that a well-regulated, well-planned nuclear family in Los Angeles, with Mom, Pop, and two kids, consumes about twice as much as the average Bangladeshi village. We cannot all aspire to live like middle-class Californians; and indeed such a standard must be seen to be anomalous, just as the pending world population of 12 billion must be seen to be anomalous. A world standard closer to that of, say, a Mediterranean village is more realistic; a way of life that can be very agreeable indeed and which has produced more than its share of geniuses, from El Greco to Jesus Christ. We need have no fears for the future of civilization if we managed to achieve the material standards of, traditional Greek farmers. But we can improve dramatically on traditional standards and certainly on cultural horizons simply by adding electronics. Energetically speaking, electronics is cheap, and so is sustainable indefinitely. The Internet could be run many times over on solar power, if such was considered desirable. To live with the physical simplicity of a Greek villager and yet to be in touch with the culture of the whole of the rest of the world seems, at least to me, to be a very reasonable prospect. I think a lot of Californians would agree. Many, indeed, already aspire to live that way.

In short, if we were to address our own biology seriously, then we could secure a future for ourselves that could last indefinitely—though whatever we do, the next few centuries will be difficult. What, then, of our fellow species?

THE NEXT FEW CENTURIES: OUR FELLOW SPECIES

The survival of other species depends, to an extent that is hard to overestimate, upon ourselves. Of course, we do not exercise absolute control. Of course, there will be some that are liable to survive even though we may do our best to eliminate them—like mosquitoes and cockroaches. Others seem likely to slip through the net simply because we are too remote from them—such as the creatures that swarm around the hydrothermal vents in the ocean floor. But the creatures of which we are most directly aware, the grand survivors of the Pleistocene overkill—elephants, rhinos, tigers, lions, cheetahs, bears, camels, antelope, deer, koalas, kangaroos—none of these have a prayer unless we take their cause seriously. Some of them, like tigers, have no conceivable long-term future in the wild even as things are.

WHY SHOULD WE CARE?

In fact, there is no space in this book to address this issue at proper length, although I have discussed it more fully in *Last Animals at the Zoo* (Washington, D.C.: Island Press, 1992). But in general there seems to me a hierarchy of reasons why we should care, extending from the materialistic to the spiritual. The materialistic argument I feel is the least interesting, although I now think it is weightier than I once did. This simply says that other creatures—or, more generally, the genes that they contain—are a resource which, increasingly, may be pressed into human service; for drugs, food, or what you will. The materialist argument also suggests that biological diversity provides some protection against ecological collapse; but although this point seems intuitively obvious, it is actually quite difficult to defend. The most obvious weakness among several in the materialist argument is that it might tempt us simply to seek to preserve those species that we perceive to be most useful; and although the list of perceptibly useful creatures might extend to tens of thousands, it would fall several orders of magnitude short of the number that now live on Earth. Thus, we could conserve the "useful" creatures yet perpetrate a mass extinction nonetheless. In short, if we want to conserve more than a tiny fraction of present-day creatures, then we have to conserve them whether or not they are of obvious use to us.

So what other reasons are there? Next in the hierarchy of reasons is aesthetics: other creatures are beautiful, and they bring us pleasure. This point is far from feeble. What a shock it would be to contemplate a world in which there were no more tigers! The fact that *Homo sapiens* has already dispatched so many fine beasts of comparable beauty would not lessen the sadness. But aesthetics does not seem a strong enough reason. If nothing else, it is too anthropocentric. Again, it suggests that we suffer other animals to live only in so far as they bring pleasure to us. There should be deeper reasons.

So then we have to make this a question of ethics. It is simply right for us, as the world's most powerful species, to look after the others. If you ask, "Why is it right?" then a great list of arguments may follow, but in the end we would simply be forced to say that it feels right. This may seem feeble, but ultimately this is the true reason behind every ethical stance. All ethical positions rely in the end upon conviction. Moral philosophers offer reasoned arguments, but these arguments in the end merely justify the underlying conviction. They do not provide the foundation of that conviction. David Hume made this point in the eighteenth century, and although many have tried, including Immanuel Kant and J. S. Mill, no one has yet significantly improved upon it.

But mere convictions—feelings—tend by their nature to be inchoate. They need to be encapsulated in verbal form so that they can be discussed and conveyed. A useful heuristic device is to embed the underlying feeling within a narrative. The narrative that works for me is the traditional notion that runs through most of the great religions, that we and all other creatures on this Earth are part of a Creation. I do not want to argue that there is a literal Creator, or to argue that there is not. I merely want to suggest that the idea of the Creation, and the idea that we are part of it, provides the framework we need on which to hang the notion that we ought to seek to conserve our fellow creatures. With this established, we need to add just one other principle: that of noblesse oblige. We are the most powerful creatures within that Creation, and as such we must accept our obligation to look after the rest. It is our proper role to act as guardians and not as plunderers.

There is one final kind of argument of a more pragmatic nature, which some may find easier to accommodate. This is simply to evoke the principle espoused by the World Conservation Union: that of "minimum regret." The meaning is self-evident. We do not know what effect the elimination of elephants will have upon the future world—upon its ecology, or the well-being of our descendants. But there are many conceivable reasons for preferring a world that contains elephants to one that does not. At the very least, it seems perverse to substitute the impoverished world for the richer one, if we have a choice. We do have such a choice, and we surely would have less cause for regret if we endeavored to save elephants than if we did not.

It has been suggested, however, that the kinds of arguments presented in this book make nonsense of all conservation. For even if we do precipitate ecological disaster, something is liable to survive; and whatever survives will evolve again to provide new life forms, some of which will refill at least some of the niches that will have been vacated. If any mammals are left, even if only rats and mice, then their distant descendants will surely include some huge terrestrial forms that will emulate the elephant, even including the proboscis. But throughout this book I have also emphasized the role of chance. There is a fair chance that elephantine creatures would appear again, but no guarantees. It is certain, too, that history never repeats precisely. If present-day elephants disappear, then we can be sure that the Universe will never again see a creature that is precisely comparable.

Then again, we have seen that life on Earth has been interrupted at least five times by mass extinctions, possibly occasioned by collision with asteroids; and we can be reasonably certain that such a disaster will strike again. So what will it matter then if the elephants have already gone? They would be wiped out anyway.

But this argument fails on at least two counts. First, it is clear that some creatures do survive mass extinctions, for if it were not so, then there would be no animals now. But nothing comes through a period of mass extinction that did not enter it. In short, if there are elephants around at the time of the next mass extinction, then their descendants might come through to help repopulate and enrich the ecosystems that come after. But if they have already disappeared, then of course they will not. In short, the elimination of elephants could alter the ecosystems of the world, until the Sun that warms our planet finally dies, and all life stops. There need be no time limit on our depredations.

Finally, in the first chapter I mentioned John Maynard Keynes with disapproval, specifically his dismissive statement, "In the long run we are all dead." Well, according to Jack Sepkoski and Dave Raup, who first suggested that asteroid collisions occur at 26-million-year intervals, we are not due for the next one for another 13 million years. Even I would admit that that is "the long run." I just think it is worthwhile to think about the next million. During that time—brief or vast, depending on how you look at it—elephants should be significant players, provided that we let them live now.

That at least is a lightning résumé of the reasons for caring about other animals. If we do want to conserve them, however, then we must do a very great deal more than we are doing now.

WHAT MUST WE DO TO SAVE OUR FELLOW CREATURES?

We have to think first of ourselves. Any policy of conservation that does not take full account of the reasonable needs of the human species is doomed. The human activity that affects the planet most is that of food production. So we need to devise systems of food production that meet our own needs, and leave room for our fellow species.

I first started thinking seriously about world food production in the 1960s, but my thoughts achieved some coherence in 1974 when I attended the World Food Conference in Rome, which was convened after a series of particularly vicious famines in the late 1960s and early 1970s. First, it became clear that the famines of the time (as now) were caused not by an innate inability to produce food, but by misdirected policy. Then I began to realize that most of the nation states at the conference were not actually addressing the issue of food production, but were simply defending their own position: "It's not our fault"; "We cannot afford to help"; "We do enough already." That was fairly shocking: I was more naive then. But then the final realization dawned, which was even more astonishing: that the agricultural systems of the world are not actually designed to feed people.

Well, of course that is a slight exaggeration. Feeding people overall is what agriculture is for. But in general, agricultural systems throughout the twentieth century have been designed primarily to fit in with prevailing economic norms, or to justify some political conceit or other. Thus, western agriculture is designed in the end to maximize profit, and the agriculture of the old USSR was intended to show that collectivization is best (which in the context of agriculture was shown, spectacularly, not to be the case). If the prime concern of the human species was to feed people, then we would do things very differently. In this there are two main considerations, both of which would increase our own security and at the same time leave more room for our fellow species. The first, which in fact has been much bruited over the past two decades, is simply to produce less livestock. We do not need to contemplate worldwide vegetarianism; but we certainly should not be producing megaquantities of grain and pulses specifically to feed to farm animals, as at present. By doing this, we effectively double the population we need to feed: ourselves, plus the livestock that is supposed to be feeding us. There is certainly a case for feeding some cereal and pulse to livestock, but the true ecological niche of farm animals is to eat things that human beings cannot eat, such as grass, and to live in areas where we cannot grow cereals and horticultural crops.

The second requirement is to apportion the land surface of the whole world more efficiently, using some for highly intensive food production (which takes up less room), some for extensive agriculture (combining food production with wildlife conservation) and designating some specifically as wilderness, devoted principally to wildlife. This, I believe, is the essence of future conservation. But there is much more to it than meets the eye. The difficulties are such that if we really want to conserve more than an arbitrary smattering of our fellow creatures, then we need to dig very deeply into all of present-day politics, economics, and attitudes to the world at large.

THE NEXT THOUSAND YEARS OF CONSERVATION

It seems all too obvious that animals and plants should live and hence be conserved in their natural habitats in the wild, and this implies wilderness. But "wilderness" these days is a much-compromised concept. At best, it tends to imply a protected area, or reserve. Reserves range from esoteric patches, which in Britain are called Sites of Special Scientific Interest, or SSSIs, and may contain just a few recondite bushes or a pond or two, up to national parks that in some cases, like South Africa's Kruger or America's mighty Yellowstone, are the size of a small country.

Unfortunately, no reserve in the world can ever enjoy cast-iron and permanent legal status. Reserves are created by governments, and governments fall. Besides, as Tom Paine pointed out to Edmund Burke at the end of the eighteenth century (they were discussing the French Revolution), no society has the right to lay down rules that bind future societies—which means that there are sound reasons of law why no reserve can ever have permanent status. Even within its own time no government has absolute power, which in general we take to be a good thing, for the obverse is tyranny; but again, the limitations of government imply that they sometimes bow to pressures that are not benign. So it is that tropical forest may be sold for logging, or islands in the Great Barrier Reef for quarrying. Sometimes reserves are compromised for reasons that may seem benign or at least temporarily necessary—for example, when traditional pastoralists are allowed to graze their cattle in woods and on grassland which, on the map, are marked as wilderness. Many a "reserve" in Asia contains more cattle than wildlife. Even at its best, then, the legal status of reserves is precarious, temporary, and in various ways compromised.

Even the reserves that are given the best possible legal protection that their governments can provide are unsafe. In the early 1990s the conservation of black rhinoceroses of Zimbabwe seemed to provide a model for the world. Wildlife provided wealth for the country, so it was in the people's interests to protect them; and protect them they did, with teams of dedicated wardens whose optimum distribution was planned by computer. But between 1992 and 1994 about half of the rhinos apparently disappeared, which means they fell to poachers. The more space the rhinos are given, the better; but the more widely they can roam, the harder they are to protect. The animals of North America are safer. Many may be poached for meat and trophies, but none provides the incentive of a black rhinoceros, whose horn may be worth a lifetime's wages to the poacher—the difference between lifelong security and lifelong privation. The protective system of law in the United States is more intricate, more multilayered, than in most countries of the tropics. There is more wealth around to enforce that law. But space is at a premium in the United States. Even Yellowstone, which, with more than 3,000 square miles, is about one-fourth the size of Holland and is the biggest U.S. national park outside Alaska, is apparently too small to sustain a viable population of grizzly bears. In fact, very few national parks anywhere in the world can truly sustain viable populations of large carnivores. If tigers lived only in the wild—if there were no backup populations in zoos—then we could seriously doubt if the species would be with us in a hundred years. Yet tigers are prestige animals that many people, including the rich and powerful, would like to preserve.

The principal mistake that nonbiologists make is to assume that a national park created in the midst of a continent can be an ecological microcosm of that continent. In practice, the boundaries of the park might embrace some populations in their entirety, but they will cut through the populations of many others. The number of any one species that is left inside the reserve may simply be too small to sustain itself. The effects will not be felt at first, for if the species is long-lived then it may take a decade or so before the animals that are inside the reserve start to die off. Close observation will reveal, however, that as time passes too few youngsters are being born, and eventually, and often quite suddenly, the senescent population will just fade away. Many of the deceptively healthy flocks of cockatoos in Australia consist mainly of old birds, waiting to die. At least half of Asia's remaining elephants are liable to die without issue. We can expect a crash early in the twenty-first century.

For the fact is that a change in size of habitat alters the ecological rules. If an entire island is declared a reserve, fair enough: the island animals continue as before. But the designation of a reserve that is surrounded by farmland and cities effectively creates an area with island conditions to which continental animals, especially large continental animals, simply are not adapted. Then we must add a simple fact of geometry: a small patch of land (where "small" is a relative term) has a very long margin relative to its total area and to its diameter. For these reasons, small reserves are subject to "edge effects." Weeds, pollutants, and alien animals such as cattle, people, and feral cats encroach from all sides and do not have far to go before they reach the very center of the habitat, places that before might have been many miles from the madding crowds. Put everything together and we see why it is that reserves—even the biggest and best-run of them—are liable to succumb to "species relaxation," a quaint and almost cozy term that describes the steady loss of species over time.

What can be done about this? Well, in practice our impact is bound to become even greater than it is already as our numbers rise, and species relaxation continues to take its toll. So it is all too easy to give way to despair or to indulge in the pretense that the collapse of other species is somehow inevitable or "natural"—that indeed it is simply a manifestation of natural selection, as "inferior" life forms give way to ourselves. Despair, however, cannot be allowed onto the agenda. Hope, as St. Paul suggested, is one of the great virtues of humanity; and in the context of conservation it is a sine qua non. In a positive vein, then, there are two prime routes to conservation: first, to focus upon landscape and habitats as a whole; and then, complementarily, to concentrate on particular species, which should probably be those of outstanding ecological significance, and those that represent the larger branches of evolution.

First, then, we have to conceive all the world's landscape as a whole and as a mosaic—that is, not simply as an arbitrary patchwork, but as a patchwork in which the different elements complement each other. Some land, of course, must be city and road. Some should be preserved or re-created, and then managed, in a state as close as possible to that of natural wilderness. Some must be devoted to extensive agriculture, designed both to produce food and to maintain wildlife. Some should be apportioned to intensive farming, its task being to produce as much food as possible in the smallest space. It is wrong to assume, as so many "environmentalists" have apparently assumed in recent years, that intensive agriculture is innately wildlife-unfriendly. To be sure, wild creatures are not welcome among the crops themselves, and intensive farms are certainly unfriendly if they are allowed to pollute their surroundings. But intensive agriculture is not innately polluting; and if agriculture is practiced truly intensively, and cleanly, then it is benign precisely because it releases land for other purposes, including that of wilderness. The current policy of set-aside in Europe shows this principle in action. In truth, it is extensive agriculture that needs to justify its existence; and it can do so only if it does indeed benefit particular wildlife. This can be the case—as in the south of England, where traditional grazing of sheep preserved the short grass of chalk downland, and allowed a suite of ice age species to survive that otherwise would have been swamped by advancing forest.

Perhaps the most vital requirement, however, is to create "corridors" between areas of wilderness so that small and inviable populations are, as far as possible, linked into larger units that might collectively be viable. In Britain and Australia farmers are now encouraged to leave or create hedgerows, to guide woodland bats across open fields. In India biologists dream of creating broad "jungle corridors" to link the sequestered pockets of elephants.

Some reserves are designed with particular species in mind; and so it is now intended to establish large prairie-dog "cities" in the Badlands reserve of South Dakota, with a view to reintroducing black-footed ferrets. Increasingly, in all parts of the world, animals must be and are translocated from areas where they are temporarily too crowded to places where they are too sparse. Some otherwise inviable populations will perhaps need constant reinforcement. Increasingly, the wild populations must be backed up by captive populations, which, in the case of tigers, already outnumber their wild counterparts. The objection is often raised that captive populations inevitably drift from the wild state, both genetically and in behavior, that indeed they become adapted to a captive state and cannot then be reintroduced. Certainly, this often happened in the past, but as I discuss in *Last Animals at the Zoo,* this is avoidable. If the reproduction of animals is properly controlled (arranged marriages are necessary), then it is possible to maintain almost all of the genetic variation contained within the founders more or less indefinitely. Furthermore, if the animals are kept appropriately, they can maintain most of their original behavioral repertoire, and more and more species are being returned successfully to the wild. The objection which says that such a return is impossible is simply out of date. To be sure, reintroduction has often proved a great deal more difficult than was at first appreciated, but the difficulties are there to be overcome. Species returned successfully to the wild now include several that were once considered impossible, including "higher" primates such as woolly monkeys and orangutans. Increasingly, indeed, wild and captive groups must be managed together as single "metapopulations," with a regular flow of individuals and genes between the two. It is effete to be purist and object to such practices on aesthetic grounds. The next few centuries at least will be desperate times for all of us; and desperate measures are called for, unless we choose simply to give up and allow the tide of extinction to engulf the few big animals that are left to us, and a great many others besides.

Yet the measures discussed so far, difficult though they will be to carry out, address only the problems of the present. We know—it is perhaps the principal lesson of history—that the world will continue to change. So the task is not simply to correct present ills but to create resilience: a system that can cope with future pressures forever more, whatever they may be. This is a very tall order indeed, and requires new depths of thought.

A WORLD BUILT FOR RESILIENCE

The lessons of history are inevitably broad, for no set of circumstances can ever be repeated precisely. But the lesson that resonates through all of geology and the historical science of paleoclimatology is that the world changes, and is bound to change in the future. In particular, we know that the world is liable to warm in the immediate future as rising carbon dioxide gives rise to a greenhouse effect, just as the fall in carbon dioxide that was brought about by the rise of the Himalayas has left us with an icebox world. We cannot tell in detail how any one place will change—in a new greenhouse world Antarctic ice may expand and Britain could theoretically freeze. Even if things turn out as simply as possible, however, and relatively benignly—a Mediterranean climate for the north of England, for example—living things will still need to adjust. In high latitudes they will have to face peculiar combinations of high temperatures and dramatically changing day lengths; a pattern without precedent during the previous 100,000 years. At some time in the future we may reasonably expect another ice age.

History tells us, too, that animals can survive such radical changes in three ways: they may move, and in particular shift latitude; they may evolve into something else; or they may alter their behavior. The trouble is that even the best of modern conservation measures, at least as now enacted, remove at least two of these options. So present policies, even at their most enlightened, provide no more than a stay of execution, and that, perhaps, for no more than a few decades.

For how and where will the animals move as the world warms, or freezes? I do not see wildebeests migrating across Nairobi. It is hard to envisage reindeer crossing Paris. Lions failed to reinvade northern Europe after the Ice Age not because they are innately tropical animals but because human beings and their farms were already too populous. Lions would now be living in France if it were not for us. The present reserves are essentially traps, and however beautifully we manage them, we will condemn the creatures they contain to certain death if we do not allow them to move out as the climate alters. If we truly care about our fellow creatures, we must create north-south corridors of far grander proportions than are now envisaged. Either that, or we will have to translocate the animals on a truly fabulous scale, and indeed to re-create entire ecosystems in different latitudes.

As for the animals' second option—to evolve, and hence to readapt to the changing conditions—that will generally prove impossible in the foreseeable future for at least two reasons. First, the changes that we are liable to cause would generally be far too rapid. We might cause significant greenhouse warming in a few decades and we are reducing the size of habitats by the day, so there is no time for the evolution of dwarf races. The second reason is probably even more decisive: present populations, particularly of big species, are generally too small and/or too uniform genetically to provide adequate raw material for natural selection. The kind of pressures that might produce greater adaptation in big resilient populations are far more likely to drive the precarious herds of the present into oblivion.

This leaves only the third option: the animals might change their behavior. Here there are crumbs of encouragement, as some species at least have shown that they can live far more closely to human beings than seemed the case only a few decades ago. If they are left alone long enough, they become more trusting and so may reduce what they consider to be a reasonable "fleeing distance." The more that other animals can learn to tolerate human beings, the greater the space that is theoretically available to them. Thus, vixens in modern-day Britain play with their cubs on suburban lawns. A few decades ago they would rarely be seen by day, and never in cities. Carrion crows and wood pigeons, famously aloof in the countryside, are at home in modern suburbs just as foxes are. In a more romantic vein, parts of the United States are seeing the return of the bald eagle; the national bird, which Americans used to like to shoot, is now to be seen around many a picnic area. We need a few more decades of enlightenment before such tolerance becomes the norm, and some creatures will always remain standoffish. Marsh harriers, for example, simply cannot hunt when there are people around. But in time, if we learn to value the company of wild animals, they will increasingly surprise us with their complaisance. After all, wild animals do not generally run from each other the way they flee from us. We are the only creatures that kill at a distance.

However, having a few more animals around human settlements is small compensation for the threatened losses elsewhere. Clearly, if we want our fellow creatures to survive in worthwhile numbers—if we want more than the odd zoo, and we want zoos to be more than living museums—then we have to take conservation very seriously indeed. It would be a huge leap forward if we were able to bring our present-day wildlife reserves up to scratch, to make sure that there were enough, of the right size and in the right places, to serve present-day creatures. But in a world that we know is changing, we need to do far more than that. If we truly took our fellow creatures seriously, we would have to design the landscape of entire continents with them in mind—not only to designate areas of wilderness, but to provide corridors between them. That, of course, would mean a huge economic and political commitment; far greater than anything that is so far on the agenda.

WHAT POLITICS? WHAT ECONOMICS? WHAT ATTITUDES?

Is it possible, though, to redesign the world in ways that will enable wild animals not simply to survive in viable populations, but also to migrate from place to place as the climate changes? Is that possible in a world that begrudges even the meager reserves that we now accord to wildlife, and in which every postponed marina or diverted motorway is seen as a "major triumph" for the conservationists? Possibly not; but then, despair cannot be allowed onto the agenda. We must begin with the Sisyphean conceit that all is possible, and then ask what would need to be done.

The central trouble, it seems to me, is that people who are seriously interested in the conservation of our fellow creatures are all too easily laughed out of court. I have tried arguing the conservational case in influential circles and found that the arguments are generally dismissed as unrealistic—even, I have been chastened to find, within a society that includes the word conservation in its "mission statement."

The problem is that, as things are, the Jeremiahs are probably right, at least in global matters. There are a hundred competing claims on any one piece of land, and the idea that a patch of India, say—a patch that might be the size of an English county—should be left as a "jungle corridor" for elephants can seem not only ludicrous but inhumane. As it is, the farmers try to make a living from half a hectare; and they are the well-off ones. What happens to them when their last scrap of land is given back to the elephants? In western countries the argument takes a different form, but the end result is much the same. Why keep a mangrove swamp for crabs and manatees when it could in the short term be used to farm jumbo shrimp at $20 a kilo and in the longer term could be drained for hotels and casinos? One may attempt to seize the moral high ground by saying, "Well, I don't think we should simply be indulging the whims of people who are too indulged already!" But this kind of high-mindedness is easily countered. Hotels and casinos provide jobs. In theory, they create "thriving communities." They offer an alternative to the village life that can no longer be sustained. It seems just as inhumane to oppose the casino as it would be to create the elephant corridor. In both cases, the argument goes, poor people would suffer, and the trouble is that, as things are, that argument is probably true. Ergo, radical attempts at conservation do indeed seem unrealistic. You may put a notional fence around a desert or a mountain that nobody can make immediate use of and call it a reserve, but as soon as somebody finds oil or sees a chance to build a ski resort, then the conservation arguments go out the window. The idea that we might actually increase the area devoted to wildlife, and indeed increase it radically and in some cases by an order of magnitude, seems nonsensical indeed, morally as well as economically. What hope is there then? I have written about this before, and editors and publishers have sometimes tended to say, "Well, what's the answer? If you cannot provide an answer, why raise the matter at all?"

This seems to me most unfair. For the reasons behind the problem I am seeking to identify are of two broad kinds. First, there are the biological realities—the ecological and evolutionary facts which ensure that our fellow creatures are even more precariously placed than they so obviously seem, and need far more space and protection than we have thus far afforded them. So far, only professional ecologists have appreciated that this is the case.

But the other set of reasons embraces the point I made above in the context of agriculture. It is that the present political ideologies and economic systems of the world simply do not acknowledge the existence of any species apart from our own. This is equally true of all of them, in different ways. Thus, although there are pleasant socialists with sandals and beards who espouse animal rights, the powerful mainstream of socialism, for example as represented by Karl Marx or indeed by the British trade unions, has been fiercely anthropocentric. Conservation has rarely been on the agenda. China is a conservational disaster, while animals survive in eastern Europe mainly by default— basically because the agriculture has been too backward to wipe them out.

Apparently in complete contrast, the essentially feudal squirearchy of England boasts of its "stewardship." The gentry justify their ownership of land by the principle of noblesse oblige. These days they tend to acknowledge that to own entire counties or thereabouts may seem a little anachronistic, but claim nonetheless to have created and maintained a landscape in which everyone can delight. This is true. The traditional British landscapes are man-made; the landowners can claim much of the credit, and at its best in the Scottish highlands and the English lowlands that landscape is indeed glorious. I chauvinistically suggest that there is none more beautiful in the world.

Yet this is the glory of devastation. Britain is the northern temperate equivalent of denuded Crete, whose beautiful pastel rocks shine in the sun, but which, just a few thousand years ago, was covered in forest that might have harbored dwarf elephants and hippos. The soft purple hills of Scotland, with their melancholic accompaniment of pipes, were formerly smothered in ash, pine, and birch in the warm, damp west, and oak, pine, birch, and rowan in the Grampian Hills. The lush green fields of England were once covered by open oak forest—the temperate equivalent of australopithecine country—practically from border to border. In fact, Britain retains less of its pristine forest than any country in Europe. We also retain fewer of our large animals. There were wildcats around London in Roman times; the bear survived in Wales until the Middle Ages; there were beavers in Loch Ness in the sixteenth century; there were wolves in England until the sixteenth century and in Scotland until the eighteenth; and the white-tailed sea eagle, the biggest and once the commonest of Britain's larger birds of prey, was wiped out from its last Scottish stronghold in the nineteenth century to make way for sheep. Other natives, like the red kite, the black kite, and the otter, have been all but wiped out in recent decades.

We even managed to drive out the osprey and almost eliminated the peregrine, two of the most widespread birds of prey in the world. About half of our bats hang on by a whisker. On the other hand, most of the most conspicuous animals in Britain were imported, like the rabbit and the pheasant, while vast areas of "wilderness" are maintained for the traditional pleasures of the squirearchy: hunting, shooting, and fishing. The rich wildlife is reduced to a virtual monoculture of respectable targets: pheasant, grouse, deer, and imported species of trout. In short, Britain's conservational record is possibly the worst in the world, although we are rapidly being overtaken by Australia. So much for noblesse oblige. So much for stewardship.

Finally, of course, traditional landowning feudalism is now being overtaken by modern capitalism which, I would say, is no worse or better than squirearchy or collectivism. Like them, it is merely insouciant. Oddly, the arch exemplar of capitalism, monetarism, could be adapted to serve the needs of conservation very well. After all, one key difficulty at present is that the things the conservationists treasure are outside the mainstream economies; as the adage has it, "The best things in life are free." But so long as they are "free," they are abused, because they simply are not taken into account. Monetarist conservation does have its grisly side; an accountant behind every tree. However, so long as we live in a world in which the only things that are truly valued are those that have a discernible price, it is worthwhile at least to attach some prices. I often wonder, whimsically, why people pay thousands for Fabergé eggs and tread on beetles. Looked at closely, and objectively, the beetle is more beautiful.

But if the conventional economic systems cannot provide long-term security, what can?

ECONOMICS WITH WILDLIFE

For the next five hundred to a thousand years, while the human population rises, stabilizes, and then with luck begins to slide down to a sensible level, everything we do to protect wildlife must in effect be seen as an emergency measure. The prime concern must be to retain as many different species as possible, and within those species to maintain as much genetic diversity as possible, in the hope that the remaining lineages can pick up the threads again after the biologically anomalous human bloom has subsided. In practice, we have to modify the desire to keep as many species as possible, and decide priorities. But that is merely a conditional clause. The overwhelming requirement is to hang on to what we can in the face of appalling odds. Measures that we might undertake now, for short-term purposes, do not necessarily represent ideal long-term solutions.

In the short term—the next five hundred to a thousand years—there seem to me to be two overwhelming economic desiderata. Both have been discussed at length, for which we may be grateful, and are still being discussed. But neither has yet been taken seriously enough by the politicians who are actually in power. Just to be parochial, I have never heard Mr. Major or Mr. Clinton speak seriously on either issue.

The first is the overwhelming and urgent need to integrate wildlife conservation into the concept of development.

Most species of animals and plants live in the tropics, which means that most endangered species live in the tropics, and most of the tropics is occupied by countries that collectively form what Pandit Nehru referred to as "the Third World," and almost all of them have horrendous problems, among the chief of which is human poverty. The survival of tigers and rhinoceroses, let alone of mice and parrots and snakes, is rarely high among their perceived priorities. Yet wild creatures have little hope of survival until the people of developing countries are in a position to see them not as rivals, or as irrelevancies, but as assets.

In practice, of course, this nettle has been grasped, not least by the World Conservation Union, and all around the world we now find groups of native people earning their living from crafts that are based upon their natural resources. These activities tend to form only a marginal component of the national economy (which of course makes them extremely vulnerable) but ecotourism as a whole, which is another aspect of sustainable exploitation, can make a huge contribution. Thus, for Kenya, tourism is the greatest single source of foreign exchange, and at least half of the income from it can be ascribed not to the beaches but to the wildlife. Zimbabwe's farmers—unique in the world—now find it more profitable to return their farms to wildlife than to raise cattle.

But the problems are huge. In Kenya the swarm of tourist buses threatens to destroy the creatures they are supposed to support. There is deep ideological and hence practical conflict between Kenya and Zimbabwe; the former banning all hunting and trading in elephant ivory, the latter feeling that the sale of tusks from superannuated bulls is a potentially huge resource that could fuel the economy and hence give the people the reason they need for keeping elephants. Few countries can show off their wildlife as easily as Kenya can, for no other country has their wealth of savannah, where the animals line up to be photographed. You could spend a lifetime in Indonesia and never see either of its two native rhinoceroses (and indeed people have done Ph.D.'s on Sumatran rhinos in the field without actually seeing one). Ecotourism in tropical forest, which is disappointingly uniform when you get close to it and is nothing like the floral floor of Harrod's, is extremely difficult to sustain. Perhaps such countries must simply create superzoos on the margins of the forest proper, to show what the animals inside are like. Once the novelty wears off, western tourists also object to the long forest treks and the lukewarm evening shower in a hut. First-class hotels are needed to go with the superzoos. But this is all expensive.

In short, it will not be at all easy to integrate wildlife into growing economies. Yet it is vital. People must have a reason for looking after their native creatures. A live hyacinthine macaw must be seen to be worth more than a hyacinthine macaw that is shot and barbecued. Overall, this is not a problem that developing countries can solve by themselves, if only because the start-up costs can be so great (not least the cost of establishing ecotourism of a kind that would actually attract westerners). First, world cash must be injected. The Brazilians are quite right to accuse the British of hypocrisy as we lecture them on the joys of tropical forest and wild animals. "What have you done with your own wild animals?" they tend to ask, and with absolute justice. In the parts of the world that are now so beleaguered we (and the Americans and Australians) could make some amends for the devastation we have wrought at home.

The second prime desideratum, of course, must be to devise economies that are truly sustainable. In the short term—the next few hundred years—we do have coal and oil, but they will not last forever. After Chernobyl, we must at the least treat nuclear power with caution. Perhaps it is premature to smother the landscape in windmills, or to fit suburban houses with solar panels. But if we are serious about the future, we should regard the next few centuries' worth of fossil fuel largely as start-up money: providing the means to create technologies that will enable future generations to live safely on renewable power. There is no panic on this. Indeed, it would be a great error to panic: it might lead us, for example, to build too many precipitate dams. But the development of renewable energy must be firmly on the agenda.

More broadly, however, the notion of sustainability requires new kinds of economics: economics that can broadly be called "green."

There are university departments and institutes of green economics worldwide. Strength to their arms and brains. We will know the world is making progress when a Nobel Prize in Economics is won by someone who can show convincingly how the world can live well, and can indulge its ambitions, without the constant goad of material growth.

Put that together with sustainable energy and—provided we also control our own population—there is some hope of surviving the next 1,000 years in tolerable and recognizable form, and hence of thinking seriously about the following 999,000 years that our descendants should live to enjoy.

Still, though, the Jeremiahs will cry, "Unrealistic." And still, as things are, they will probably be right. Everyone knows—surely—that we need a sustainable world and a controlled population. Everyone knows that the world as a whole cannot live in the style of California, so that we must either ask the Californians to modify their lifestyle or accept the extreme inequity. Everyone knows these things, yet nothing serious is done to bring about the necessary changes. Why not? In the end, the Jeremiahs say, it is because present political and economic systems respond to the needs and aspirations of human nature. The Californian lifestyle is acknowledged to be the great good, and present-day economic systems are geared to the Californian goal because that, in the end, is what people want.

But this, surely, is not true. Nobody I know who has any knowledge of the world seeks to be poor, because poverty really is not ennobling, unless you can live in the glorious and fundamentally wealthy simplicity of some Italian Franciscan monastery (for who needs to be rich when the sun is shining and everything is taken care of?). But equally, nobody I know wants to be filthy rich. "Comfortable" is the kind of word most people use. "I just don't want to have to think about money" is what the thinking classes tend to say.

So in truth (I submit), the California lifestyle is not a universal aspiration or even, probably, particularly widespread. If it was, then nobody would become a schoolteacher, a nurse, or a musician; yet thousands and thousands of extremely intelligent people choose these careers. People go along with economies that seem to hold wealth as the prime ambition largely because, well, such economies seem to work, at least up to a point. In reality, people who do not care too much about money survive by hanging on to the coattails of those who do—or that at least is what the people who sport those coattails hold to be the case.

Nonetheless, the notion that the world's dominant economies are based on the aspiration to live like Californians does contain a significant germ of truth. Only a minority in any one society may hold that ambition overtly; but that minority drives the economy. Even if the intelligensia do not espouse the notion overtly, they still acquiesce in systems that are propelled by it. It is true enough, too, that the dominant economies of the present day reflect at least a component of "human nature." They reflect a widely held attitude, and attitude is all-important.

In the end, then, we are obliged to admit what the Jeremiahs maintain: that the present-day economies prevail precisely because they reflect the underlying attitude of a significant proportion of society. If we feel that the economy is unsuitable to present needs, or at least to long-term needs, then we have first to change the underlying attitude. Any attempt to alter the economy without changing its attitudinal underpinning can indeed be considered unrealistic. To change our underlying attitude it is necessary first to understand it. The history of the human species—the proper history: the five-million-year overview—can again provide worthwhile insights.

HUNTERS AND FARMERS: ATTITUDE

Psychologists like to devise matrices by which to define personality. Such matrices invariably include an axis that extends from extrovert to introvert, and sometimes another from assertive to compliant. Some qualities (like assertiveness and extroversion) seem naturally to go together, while others (like assertiveness and introversion) seem at odds. I have no views on whether such matrices truly help us to understand ourselves and our fellow human beings, but I would like to steal the technique. The approach can usefully be applied to the life strategies of animals, including the human animal.

One highly pertinent axis on the life-strategy matrix would extend from "acceptance" to "manipulation." Snails (and mosses, among plants) are accepters: effectively, they put up with what the world has to throw at them. If it rains, they come out and feed and reproduce. If it is dry, they hide, and if it stays dry for too long, they plug the aperture of their shell with mucus and hope for the best. Human beings are at the manipulation end of this axis. We do not put up with the weather, or simply hide when we do not like it. We dress ourselves appropriately and outface it. Indeed, we go further: we create controlled environments in which the weather does not matter. More even than that: some people (like the Israelis) are already seeking to control the weather itself, and most governments would probably elect to do so if they could. With control of course goes exploitation: we try to make the world do what we want it to (just as the English squirearchy have created what is essentially a garden). On the acceptance-manipulation axis, most animals fall somewhere between the extremes of snails and human beings.

Another highly pertinent axis would run from "conservative" to "experimental." Duck-billed platypuses would be close to the conservative end of such an axis (along with snails and earthworms and a great many other creatures), while human beings would be close to the experimental end. Duck-billed platypuses behave exactly the way their ancestors did in the Pleistocene, and indeed in the Pliocene and probably in the Oligocene if we but knew, while humans are very different from their Pliocene forebears.

We could also set up such a matrix purely for our own species. Even today we can see that some societies tend to accept what the world throws at them, while others seek to change whatever is inconvenient. Some behave as their forefathers did, and indeed make a virtue of it, while others constantly try new approaches and equate novelty with progress.

In practice, these two particular axes tend to run more or less in parallel. Acceptance seems to go with conservatism, while manipulativeness and exploitativeness obviously demand some experimentation.

All life strategies have their advantages and disadvantages. The advantage of the exploitative-experimental approach is obvious. We see it in the triumph of our own species. We do not simply dominate the world in the way that the dinosaurs are said to have dominated. We are not merely numerous and ubiquitous. We directly manipulate the environment in ways that ensure our populousness and ubiquity: notably by farming. The disadvantage of the exploitative-experimental approach is also obvious. Exploitation is potentially damaging, in a way that acceptance is not; and experimentation, by definition, is likely to produce unforeseeable consequences. Contrariwise, accepting-conservative creatures leave the world just as they found it, which is the safe thing to do, and operate on the assumption that whatever behavior allowed their parents to survive has a fair chance of ensuring their own survival as well. This policy is not foolish. Snails, one might point out, have been around a lot longer than human beings and are likely to be around in various forms long after we have gone. In short, acceptance-conservatism is a safe policy, while the exploitative-experimental approach is innately risky. On the other hand: nothing ventured, nothing gained.

So now we can ask, in the manner of a game theorist, when is it likely to pay to be accepting and conservative, and when is it better to push the boat out? The general answer is that the exploiters and experimenters need leeway. They need to be able to make mistakes and then recoup. They also—crucially—need a resilient environment that is resistant to overexploitation; although "resistant" of course can never mean "invulnerable." Overexploitation is equivalent to an experiment that fails.

When we apply this particular exercise in game theory to the history of human beings, a whole number of otherwise confusing issues begin to fall into place. We can begin by asserting, commonsensically, that all human beings have the capacity to behave conservatively and acceptingly, or to be primarily exploitative and experimental. The question then is, which circumstances most favor which approach? The broad general answer, it seems to me, is that hunters and gatherers really have to be conservative acceptors. This is true whatever their individual personality or mien may be; however fiercely they shake their spears. If they try too hard, if they invent ways of killing that are too efficient, then they simply wipe out their prey. In the end, so long as hunters remain hunters, natural selection will favor the lazy ones who tell stories and hunt the way their forefathers have always hunted.

But it is in the nature of farming to be exploitative and experimental. The whole point of farming is not to accept what nature has to provide. The whole aspiration is to manipulate other species and eventually to till the ground so that the environment provides more than it otherwise would. Furthermore, the harder the farmer works, the more he or she is rewarded. Farming, in short, turns an unimprovable resource into one that in principle can be increased indefinitely, just by stepping up the effort. In addition, by increasing output, farming societies create the surplus food and hence the leeway that allows them to take even greater risks. So here we have yet another positive feedback loop: experimentation favors farming that provides greater scope for experimentation.

Thus it is that over the past 30,000 years, which we may take as the rough date of the first deliberate cultivators, natural selection has increasingly favored the exploitative-experimental approach. Thus it is, too, that the advantages of this approach have become more and more apparent as the feedback loop has begun to bite, until nowadays we take it for granted that farming must be backed by agricultural research, which in turn partakes of the highest technologies. By contrast, however fancy hunting may become, with the infrared sights and helicopters of the millionaires, it will always be a marginal pursuit. The resource—the population of deer or grizzly bears—is simply too fragile. Hunting becomes economically significant only when—like grouse and pheasant in Britain—the prey animals are essentially farmed, and fed to the guns.

But the party really is over. This really is a special time in history—a critical time. The first full-time farmers of 10,000 years ago were perfectly capable of overfarming (we can see evidence of ancient soil erosion), but even so, they could effectively regard the world as a whole as a limitless resource. There were entire continents still to be discovered, and it would take another 10,000 years to find all the fertile spots and dig them up. But now we have had that 10,000 years. The fertile spots have been dug up, and we know there are no more continents to be had, that indeed, as the world warms, the land will shrink. Agriculturalists can argue about details—can we feed 10 billion people sustainably? or could we manage 20 or 30 billion if we really put our minds to it?— but no sensible person can seriously doubt that the finishing post is now in sight. The world is not indefinitely large. We cannot simply hurl ourselves at it with the abandon of the past 10,000 years.

In short, the attitude that has been so appropriate this past 10,000 years, and has allowed the most exploitative-experimental people to rise inexorably if fitfully to the top, has simply ceased to be appropriate. Yet our economies are geared to the exploitative-experimental approach, and so are our political systems. So all of a sudden, or so it seems, our political and economic institutions and philosophies are out of synch with the biological and physical realities of the planet. It might be unrealistic to devise new systems that are radically different, with a radically different motivation; but if we do not do this, then we cannot seriously contemplate long-term survival. Surely it cannot be the case that the only "realistic" course is to head pell-mell for disaster? Is that what the level-headed, sober-suited people are arguing? Our position seems not merely precarious, but ludicrous. It invites parody. It brings to mind an image that appears in all the best cartoons: the one where Tom or Jerry runs over the edge of a cliff, seems for a moment to defy the laws of physics, realizes something is amiss, turns to the camera, gulps, holds his nose, and is seen from above plummeting to the depths of what looks like Death Valley, to carve a perfect outline in the all too solid rock below.

So how do we get out of this mess? I am inclined to say, "Don't ask me." If you think that is a cheat, I can only say that the first stage in solving any problem is to state the problem clearly, and that is what I have tried to do. In fact, Jesus Christ expressed the essence of the issue almost 2,000 years ago when he said, in his Sermon on the Mount, "The meek shall inherit the Earth." It would be nice to believe the "shall." But it is surely true to say that if anyone at all is going to inherit the Earth, then it can only be the meek. On the other hand, St. Peter identified the snag, just as the modern Jeremiahs do: that we live in the here and now, and in the here and now the meek get bashed. Well, it is time to readdress the problem. The task is to create economies that are "conservative" with a small c and yet are robust, and can withstand the pressures of short-term aggressiveness. More broadly, the task is to develop what might glibly be called the politics of reverence, as opposed to the politics of exploitation. It is a tall order, just as it was when Christ preached to the five thousand. It is also necessary.

However, the practicalities of the next thousand years are only the immediate problems. There are broader issues.

WILL THE EARTH ALWAYS BE BENIGN?

The Earth has existed for the past 4.5 billion years, which is probably almost a third of the total life of the Universe. Latest evidence suggests that life has existed on Earth for almost 4 billion years. The earliest life forms must have been radically different from today's, although there are many creatures with us still, broadly classed as bacteria, which demonstrate at least in part how the earliest types must have lived. Thus, like many modern-day denizens of swamps, the earliest organisms obviously thrived in an atmosphere free of oxygen. They must have fed not by photosynthesis or by eating other things but by various forms of chemosynthesis, as is still practiced by some modern-day microbes. Ultraviolet must have poured through the ozone-free upper atmosphere; so the early types either stayed out of its way or else evolved protective mechanisms, like the pigments of modern-day corals and slime molds. Such pigments were presumably a prerequisite of photosynthesis, for unpigmented organisms could not have exposed themselves to the Sun in the first place. We may indeed suppose that chlorophyll itself, which now serves to convert electromagnetic solar energy into chemical energy, evolved originally as a protective pigment. If this were so, it would be another pleasant example of nature's opportunism. Some of the early organisms, too, may have been able to withstand extremes of temperature, like the thermophilic bacteria that nowadays live in hot springs and deep-sea vents.

But however tolerant those ancient creatures were, we can be sure that the conditions on Earth, although extreme and in several ways lethal by the standards of modern animals and plants, were modest by the standards of the Universe as a whole. Living forms in general operate at maximum efficiency at around 40°C—roughly the core blood heat of most mammals and birds. Survival is possible at much lower temperatures, and some organisms easily survive deep freezing. It is not a particularly difficult trick to pull; indeed, we can argue that living creatures of some kind or another could survive the coldest conditions of which the Universe is capable: the "absolute zero" of −273°C. But as the temperature falls below 40°C, their metabolism slows.

Heat is more of a problem. As the temperature rises above 40°C or so, proteins start to break down and indeed to denature, which effectively means to cook. The metabolism of thermophilic bacteria runs very fast to stay in the same place: the thermophiles contrive to make fresh proteins as fast as the old ones are broken down. Even so, these organisms do not survive for long at much above 100°C. But 100°C is gentle by the standards of the Universe. The core of the Earth itself is far hotter than that. The surface of our nearest planetary neighbor, Venus, can reach several hundred degrees centigrade.

In short, it is a minor miracle, or at least a matter of extreme statistical unlikelihood, that the Earth has maintained conditions suitable for life through 4 billion years. To be sure, the conditions have been very different at different times, requiring an almost absolute change in living personnel when the atmosphere first began to acquire oxygen. But always the temperature and moisture content of the surface—these in the end being more important than the details of chemistry—have been within the very narrow compass required even by the most tolerant of organisms. There are trillions of stars in the Universe, and many of them have planets, and many of those planets have moons. Only a tiny minority are likely to provide conditions suitable for living things, but that minority must still number many millions. The raw material of life is ubiquitous. Carbon, the key element, is one of the commonest materials in the Universe, and space contains many billions of tons of complex organic molecules of the kind from which living things are compounded. It is inconceivable, then, that Earth is the only planet with living organisms. Indeed, as Christian de Duve comments in *Vital Dust* (New York: Basic Books, 1994, p. 284), "Life is an integral part of the Universe." There are probably millions of planets with living creatures on board. De Duve speaks purely as a scientist: a Nobel Prize–winning biochemist who knows better than most what life entails.

Even so, earthly conditions are unusual, and, relative to the total number of heavenly bodies, must be extremely rare. Life seems inconceivable on a planet without an atmosphere, but only planets of a particular size have gravity enough to cling on to one. Liquid water seems vital, but water is liquid only within a band of temperature, 0° to 100°C, that is ridiculously narrow by the standards of the Universe. Temperature is extremely labile. The greatest influence on temperature, at least within the narrow range of circumstances that are relevant here, is the chemistry of the atmosphere. A little maneuvering either way can send it way beyond the compass of earthly life, or indeed of any conceivable organic life. The awfulness of Venus results not from its proximity to the Sun (though it is closer than we are) but from the greenhouse gases in its atmosphere.

We have seen how the rise of the Himalayas since the Eocene has steadily leached the atmosphere of its carbon dioxide and so produced an icebox world. We have seen the sudden catastrophic falls along the way which have demanded and prompted the appearance of entire new suites of creatures—including ourselves. History shows us, in short, the power of atmospheric chemistry, and also its subtlety, for the changes that have caused such radical shifts in the ecology and evolution of living creatures do not seem spectacular: a halving perhaps of an atmospheric gas whose concentration is measured only in parts per million.

We have also seen the phenomenon of catalysis, through which minute intrusions into the chemistry of the atmosphere can have huge consequences. The effect of CFCs on ozone and hence on the radiations that reach the Earth provides a spectacular example. Yet until a few years ago it would not have been possible even to detect CFCs in the atmosphere. We have seen, too, how chemical and physical changes, once set in motion, can take decades or even centuries to unfold. It will take decades for the oceans to absorb all the heat now being supplied by our mildly greenhouse atmosphere. It will take decades to leach the last of the CFCs from the stratosphere. Finally, history has revealed the phenomenon of catastrophe: steady changes can produce sudden effects. As we watch the changes taking place, we might bear in mind that in theory, as some threshold is reached, the world might lurch into some quite new state far more quickly than we can cope with it; and certainly more quickly than our fellow creatures can come to terms with, trapped as they are in their reserves.

It has come as a shock to realize that we, puny creatures though we sometimes like to think ourselves, can influence the atmosphere of the whole world. Perhaps it should not have been too much of a surprise; after all, it has been obvious for several thousand years that we could transform entire landscapes, not least by cutting down the trees. Yet I am not sure that we have been shocked enough. For although our influence is huge, our control is minimal. Having set the warming of the oceans in motion, there is not a thing we can do about it. Having initiated the breakdown of the ozone layer, we can only sit and wait, and hope that the chemists have calculated accurately, and that their understanding is now complete, and that the ozone will recover before we are all radiated literally to death.

Overall, history tells us that the beneficence of this planet cannot be taken for granted. It has always been hospitable to life of a kind, but not always to our kind. Conditions on Earth do not have to be the way they are now. The laws of physics will always apply, but when those laws operate in particular places in particular circumstances, they can produce many different kinds of outcomes. We absolutely cannot assume that in the future the Earth will always provide the kind of conditions that suit creatures like us. Given that our requirements are so narrow and that our influence is far, far greater than our control, we have at least to consider the possibility that we ourselves could generate global conditions that we, and creatures like us, would find impossible. In short, by our own manipulations of the atmosphere we could create a climate that was suitable only for the bacteria that have now retreated to the marshes and the hot springs, or perhaps, as on Venus, was unsuitable even for them.

This is not gratuitously sensationalist. I am at least as tired of gloom and doom as everybody else. It is, however, a modest plea that we should take the lessons of history seriously. Our planet, like all other planets that we know about, is perfectly capable of being an extremely nasty place, and it is a huge mistake to take our present good fortune as a given.

Suppose, though, that our species does survive the next thousand years. Suppose we do indeed embark upon our allotted 1 million years, our kilometer of time. The human genus has changed radically over the past million years—from *Homo erectus* into us. Would the next million years bring comparable change?

THE FURTHER EVOLUTION OF HUMANKIND

Of course, human beings will continue to change culturally; and this might be considered to be evolution of a kind. After all, if we would entertain different ideas, and behave differently, and have a different attitude to the world—if for example we practiced the politics of reverence—then we would be ecologically quite different. Our impact on our fellow creatures would be altered, and from their point of view at least we would effectively be a different kind of creature. We can argue, too (as others have argued), that the infusion of new ideas to some extent occurs by natural selection. The ideas that produce worldly success do indeed tend to spread, just as science and capitalism have spread these past few centuries (and the present growth of Islam is certainly interesting). But we can also point out, as others have done, that the growth of new ideas proceeds according to Lamarck's fourth law: that is, ideas acquired by one generation are passed on to the next. Darwin himself would have had no problem in accepting that Darwinian and Lamarckian mechanisms might proceed in tandem and indeed, as we have seen, more than toyed with the notion that Lamarckian systems of inheritance obtain generally.

The issue here, however, is that of bona fide Neo-Darwinian evolution: whether it is possible to change the overall gene pool of human beings to such a significant degree that our descendants can properly be considered a different species.

The short answer, on purely theoretical grounds, is yes. Remarkable creatures though we are, we are manifestations of our collective gene pool just like any other animal, and it is possible to subtract alleles from that pool, and add them, and go on doing this until we have a new creature; just as it was possible, by these means, to turn a gomphothere into an elephant or *Australopithecus* into *Homo*. But it is not as easy to envisage the circumstances, or at least the natural circumstances, in which such a change could be brought about.

It is obvious, after all, that the human gene pool is changing. To our shame, rare tribes of aboriginal people worldwide continue to disappear, and they must take at least a few recondite alleles with them. Mutation continues, too, and since the human population is now so large, the gene pool as a whole must be accumulating mutations more rapidly than ever before. Some groups, too, are now breeding much faster than others—Kenyans faster than Germans, for example—so some alleles are becoming relatively more common than others. Even so, we cannot argue that such fluctuations truly represent evolution. The loss of alleles, sad though it is, is genetically marginal. At the same time, few of the new mutations contribute significantly to the life of our species. The shift in frequency of alleles within the pool is only the usual "noise" in the system. No one seriously supposes, after all, that northern Europeans are going to disappear altogether.

In short, the fluctuations of the human gene pool are providing raw material for evolution—the genetic variation. But the key ingredient is missing. Natural selection is simply not acting forcefully or consistently upon that variation. The loss of alleles in aboriginal peoples is random; at least, we cannot argue that particular alleles are being lost because they are disadvantageous. At the same time, the people who seem to be successful are not necessarily breeding particularly quickly. Neither can we argue that the success of particular present-day groups is correlated with their genes. The materially successful people of California belong to all conceivable human genotypes. Finally, and I think crucially, the human population is just too big to change significantly by Neo-Darwinian mechanisms. Natural selection works best on middle-sized, isolated or semi-isolated populations, like those of the australopithecines on the margins of Africa's retreating Pliocene forest. It is hard to see how it could operate on a population of 5 billion individuals who, by courtesy of the world's airlines, are all in close genetic contact with each other.

There are, however, two feasible sets of circumstances in which we might envisage Neo-Darwinian change. The first is if there is after all some kind of world catastrophe: an ecological crash. In such circumstances we might envisage that the human species would be reduced to patchy populations, separated by various kinds of badlands. Excess ultraviolet radiation or the still-present specter of nuclear war (or even peacetime nuclear catastrophe) might provide the conditions. Then we could envisage the isolated groups evolving afresh, just as hominid groups of comparable size, with comparable degrees of stress, evolved in the past.

It is interesting to speculate how those isolated groups might evolve. There is absolutely no good reason to assume that the trend of the past few million years would be continued: that intelligent human beings would produce a race of superintelligents. That is actually an unlikely option, since it is hard to see why natural selection should favor geniuses in the straitened circumstances of a devastated future, any more than it would favor the emergence of a literary codfish beneath the Arctic ice. Besides, it seems likely (though nobody knows) that genius is a genetically difficult trick to pull. Overall, indeed, isolated human populations of the future seem far more likely to generate a significantly less intelligent lineage.

The brain, physiologically, is an expensive item. One way and another, it is said to commandeer 20 percent of total metabolic effort. Unless this prodigal brain can be put to good use, it is a luxury that future hominids in straitened circumstances may prefer to do without. In the same way, many island birds from many different groups have abandoned the apparently self-evident advantage of flight. How glorious to fly! But if you do not need to fly because there are no predators to escape from, then you are advised to stay on the ground because it is cheaper. In the same way, *Australopithecus afarensis* apparently gave rise to the robust paranthropines, as well as to the hominines. Of course, we cannot tell at this range whether the paranthropines were less bright than *afarensis*. The point is, though, that the hominine emphasis on the brain was only one of the options open to the hominids. Muscularity was another. Neither can we assert that the paranthropines were a failure. They lasted a million years and might be with us still. Neither were the hominines bound to succeed. I believe that we can identify their clear advantages, as outlined in chapter 6. But advantage does not come with guarantees. Every lineage needs good luck. "Time and chance," as Ecclesiastes reminds us, "happeneth to all men."

We can, however, envisage a quite different set of circumstances in which human beings might continue to evolve: not by natural selection but by artificial selection. In other words, our descendants could in theory breed a new kind of hominine, probably (to speed things up) with the aid of genetic engineering. However, I mention this only in the interests of logical completeness. Discussion of its ramifications belongs in other books.

But the theoretical possibility of future evolutionary change does raise an issue that is of universal significance. What, in theory, might human beings evolve into? More broadly, of what is life capable?

PROGRESS, POTENTIAL, AND DESTINY: WHAT MIGHT LIVING THINGS DO?

I argued in chapter 6 that the concept of progress in evolution is worth entertaining. Some creatures really are demonstrably better than others at carrying out definable tasks: better both in engineering terms (for example, extracting more useful calories from a given quantity of food) and logistically (according to the analyses of game theory). On the whole, too, the better creatures evolve from the less good, even though there seem at times to have been some interesting reversals.

But one reason why these somewhat obvious and, I believe, useful ideas have become unfashionable is because people seem to confuse progress with destiny. In fact, progress in general is likely to happen. Better creatures are bound to appear through the chance processes of mutation, and, having appeared, they are liable to succeed. Furthermore, as discussed in chapter 4, some lines of development are more likely than others. In short, animals must obey the rules of engineering, which are founded in the laws of physics; and in practice only a limited range of body forms, or ecomorphs, is feasible. Put the two ideas together and it looks as if the gene pools of ancestral creatures are bound to be pushed in particular directions. Indeed, as de Duve comments in *Vital Dust*, "Should things start all over again, here or elsewhere, the final outcome could not be the same. But how different would it be?" (p. 287).

In other words, we can predict to a large extent that any suite of creatures given the freedom to evolve will eventually fill at least some niches and that they will do so in at least some of the ways that we have already observed. But we absolutely cannot predict which particular creatures are liable to evolve. It was never possible to predict that life on Earth would, within the lifetime of the Earth, have produced *Homo sapiens*, or indeed the hominids as a whole. I have presented the story of human evolution as one of opportunities taken, of the right gene pool in the right place at the right time. What was always absolutely unpredictable, however, was that such a set of circumstances would ever arise. If India had not crunched into Asia, there would be no Himalayas. If there were no Himalayas, then the world might still be covered in tropical forest. Given that there were plenty of primates around in the Eocene, we can reasonably predict that there would be plenty around now. But the predominant forms surely would be monkeys, supremely adapted to the trees. There is absolutely no reason to assume that any of them would have developed along hominine lines. More generally, there is no reason whatever to suppose that intelligence of human proportions would ever have appeared among any lineage. Indeed, there is no reason to assume that such a quality would have appeared before the next meteor collision, which if Raup and Sepkoski are right, is due in another 13 million years. In fact the world might have begun its final cooling, with the death of all life, before circumstances ever arose that could produce creatures like us. In short, there is progress, but there is no destiny. Time and chance always play their part.

Yet there is another notion that seems to me intriguing and relevant. It is obvious that nothing can happen, except the things that are possible. To put the point more simply: everything that does happen must be possible. Perhaps this seems too self-evident to be worth stating. Yet it is interesting. For if it is possible for phenomenon B to result from phenomenon A, then we have to conclude that phenomenon A had the potential to produce phenomenon B. Indeed, we might argue, as I think Aristotle essentially did, that the potential to produce B must be numbered among the innate properties of A.

With such notions we might produce an argument that seems to me very close to the thesis of Teilhard de Chardin, who contrived to be a paleontologist, a Darwinian, and also a Jesuit priest. Thus, it seems to me entirely unexceptionable to point out that intelligence of the kind we recognize in human beings is a product of living flesh. Intelligence can be seen as emergent property, something that results when some of the molecules contained in flesh are suitably arranged. By the same token, we can argue that the molecules of that flesh must have had the potential to produce intelligence. Look at those molecules more closely and you find they are compounded from carbon, oxygen, hydrogen, nitrogen, phosphorus, sulfur, and a few metals—a pretty commonplace array. Yet, in Chardinesque fashion, we have to concede that this chemistry-set collection, suitably arranged, has the potential to produce thought. We could go even further back and observe that these elements are themselves compounded of fundamental particles. These particles, then, had the potential, once suitably arranged, to produce the highest flights of genius.

Why is this relevant? Well, earlier in this book I mentioned the children's TV character Dr. Who, the extraordinarily accomplished scientist who is able to travel through time. Suppose he traveled back to the time of the Big Bang, when the Universe began, and all the fundamental particles were still separate. Could he have predicted just by looking at those particles (and leaving aside for a moment his knowledge of the future) that they had the potential to form themselves into the elements of the periodic table? I am not at all sure how. I feel he would have to have waited a few years, or a few million years, to see how things turned out. Once the elements had formed, in the depths of successive generations of stars, could he have predicted that they would when suitably combined produce living things? Again, I think not. If he visited Earth 2 billion years ago or so, and saw the first organisms with nu-

cleated cells, could be have predicted that from these humble creatures would evolve animals that can think? Again, one asks, "How could
he?" Where are the clues in those primitive creatures to indicate future
possibilities? So I would like now to make precisely the same point in
the context of present-day creatures, and indeed of ourselves. That is:
we simply cannot tell, by looking at existing life, exactly what life is capable of. We have seen that Robert Broom and other eminent biologists of the twentieth century felt that human beings, bright creatures
that we are, represent the ultimate in biological evolution. The same
point, stated only slightly differently, is made in the Old Testament: we
are supposed to have been created "in God's image," and God, by definition, is ultimate. But what reason do we have to assume that this is
the case? What do we really know of life? I submit that we are no more
able to predict the possibilities—the potential—of life, than Dr. Who
would have been in trying to predict the potential of newly emerging
elements.

Specifically, we think that our brains are wonderful, and so they are.
Yet we can identify areas of extreme feebleness. We are extremely poor
at math and feel that those who gain academic honors in that subject
must be very bright indeed. But why, then, do those "geniuses" use cal-
culators? Because a five-dollar pocket calculator with a tiny microchip
can carry out the mechanics of math a hundred times quicker than
they can. Of course, the calculator lacks imagination, but the point
stands nonetheless: here is a comparatively simple trick of math that
even the greatest mathematicians simply cannot do. We assume, because in general we are smarter than most other animals, that we must
be brighter than them in all respects. Yet this, too, is nonsense. Memory is generally conceded to be a component of intelligence. But animals that hide food for the winter, such as nuthatches, sometimes have
memories of topography that are quite out of our league. In short, it is
very easy to envisage a dozen ways in which we could in theory improve
the components of our present-day intelligence by leaps and bounds.
There is no a priori reason why a human being should not combine
the qualities, say, of Einstein, Shakespeare, Mozart, Darwin, J. M. W
Turner, a nuthatch, and a pocket calculator. Indeed, there is no a priori reason why such a paragon should not be considered ordinary.

But the qualities of known geniuses could well prove mundane, compared to what *might* be possible. For example, homing pigeons and
many other creatures have been shown to possess some magnetic sense
by which they can orient themselves relative to the Earth's field. Could
not future lineages—in theory—develop such a sense to a high de-

gree? On the other hand, no creature present or past is known to respond to, or transmit, long-wave electromagnetic radiation, i.e., radio waves. Yet there seems to be nothing in those possibilities to offend any known laws of physics. In theory then, we could envisage future creatures that carry their own two-way radios in their heads. Combine this with a more advanced magnetic sense, and we can envisage creatures that might be able to pinpoint their positions anywhere on the Earth's surface. It is very difficult to envisage the natural circumstances that would favor an emergence of such creatures, of course, but that is not my point. I am simply asking the more fundamental question, "Of what is life really capable?"

So we can see the future evolution of living things on Earth as a kind of obstacle race. Living things as a whole might have the potential to do a whole range of things that at present are unknowable and unpredictable. The question is, which and how many of those things will in practice be realized? It would be nice if there were creatures in the future that were intelligent enough to take note. Indeed, it would be nice, speaking chauvinistically, if those creatures included future members of our own lineage.

However, the chances that any untapped potential will be realized, by whatever lineage, depend very much on our actions over the next few hundred years. But those actions must not be geared only to the needs of the next few hundred years. The events of this planet can take far longer than that to unfold. We surely have the potential to survive as a species for at least a million years; there is no reason to doubt that. But if we want our descendants to claim that million years and do so in the company of other creatures, then we must think from the beginning in such terms. In short, as I commented in the Prologue, we cannot claim to be taking our species and our planet seriously until we acknowledge that a million years is a proper unit of political time.

INDEX